Copolymers, Polyblends, and Composites

Copolymers, Polyblends, and Composites

Norbert A. J. Platzer, *Editor*

A symposium sponsored by the Division
of Industrial and Engineering Chemistry
and co-sponsored by the Division of Polymer
Chemistry, the Division of Organic
Coatings and Plastics Chemistry,
and the Division of Cellulose,
Wood, and Fiber Chemistry at the
167th meeting of the American Chemical
Society, Los Angeles, Calif., April 1–5,
1974.

ADVANCES IN CHEMISTRY SERIES **142**

AMERICAN CHEMICAL SOCIETY

WASHINGTON, D. C.   **1975**

Library of Congress CIP Data

Copolymers, polyblends, and composites.
  (Advances in chemistry series; 142 ISSN 0065-2393)

  Includes bibliographical references and index.

  1. Block copolymers—Congresses. 2. Graft copolymers—Congresses. 3. Fibrous composites—Congresses.
  I. Platzer, Norbert A. J., 1911- II. American Chemical Society. III. Title. IV. Series: Advances in Chemistry series; 142.

QD1.A355 no. 142 [QD382.B5] 540'.8s [547'.84]
75-17726
ISBN 0-8412-0214-1   ADCSAJ 142 1-482 (1975)

Second Printing 1977

Advances in Chemistry Series

Robert F. Gould, *Editor*

Advisory Board

FOREWORD

ADVANCES IN CHEMISTRY SERIES was founded in 1949 by the American Chemical Society as an outlet for symposia and collections of data in special areas of topical interest that could not be accommodated in the Society's journals. It provides a medium for symposia that would otherwise be fragmented, their papers distributed among several journals or not published at all. Papers are refereed critically according to ACS editorial standards and receive the careful attention and processing characteristic of ACS publications. Papers published in ADVANCES IN CHEMISTRY SERIES are original contributions not published elsewhere in whole or major part and include reports of research as well as reviews since symposia may embrace both types of presentation.

CONTENTS

PREFACE

This year's U.S. production of thermoplastics, thermosets, and synthetic rubber is expected to be 29 billion pounds. About 80% of this is based on only a few common monomers. To improve performance, the polymer industry rarely changes to a new, probably more expensive polymer, but instead it shifts from mere homopolymers to copolymers, polyblends, or composites. These three types of multicomponent polymer systems are closely inter-related. They are intended to toughen brittle polymers with elastomers, to reinforce rubbers with active fillers, or to strengthen or stiffen plastics with fibers or minerals.

ADVANCES IN CHEMISTRY SERIES volume 99 contains 37 papers presented at our first symposium on multicomponent polymer systems in 1970. During the past four years, significant research and development work has been devoted to this area, and this volume of 39 papers may be considered a continuation of the previous volume.

Multiphase Polymer Systems

From the viewpoint of macrostructure, homo- and copolymers are generally defined as single-phase systems. Polyblends and composites are always classified as multiphase systems consisting of a polymeric continuous phase or matrix and of a dispersed phase. The latter can be either another polymer or any other foreign material such as glass fibers, fillers, or minerals.

From the viewpoint of microstructure, most polymers have to be considered as multiphase systems because we find micro-inhomogeneity with domains of folded chains, order and disorder, frozen-in stresses, and different crosslinking densities. This micro-inhomogeneity is pronounced in oriented polymers such as fibers. The latter are built of microfibrils which are composed of amorphous layers and crystalline regions and which A. Peterlin describes in the first chapter of this volume. Furthermore, polymeric materials consist of macromolecules of different chain length. Their processing behavior and end-use properties are influenced not only by their average molecular weight but also by their molecular weight distribution. G. V. Schulz discusses the latest refinement in the determination of molecular weight distribution.

Separation into two phases may occur not only in solids but also in solution. This has been observed in copolymers of maleic acid with alkyl vinyl ethers by U. P. Strauss and P. Lane. The microstructure of two-phase poly(methyl methacrylate) is the subject of D. T. Turner and R. P. Kusy.

Homogeneous single-phase polyblends are very rare. Liquid–liquid phase separation of optically homogeneous polyblends of a styrene/acrylonitrile copolymer with poly(methyl methacrylate) has been studied by L. P. McMaster. A quantitative test method of the dynamic mechanical properties of multiphase polymer systems was developed by L. Bohn. He was able to demonstrate the correlation between shear modulus and gel volume of brittle polymers toughened with elastomers. Making two phases more compatible either through copolymerization or through a coupling agent is the subject of N. G. Gaylord.

Copolymers

The most common technique for improving the properties of a polymerized monomer is to allow it to copolymerize with another monomer. We distinguish between random, graft, block, and alternating copolymers as well as network structures in Figure 1. These copolymers differ in their performance characteristics. The effect of their structure on properties is discussed by S. Russo and B. M. Gallo using copolymers of styrene and methyl methacrylate as examples.

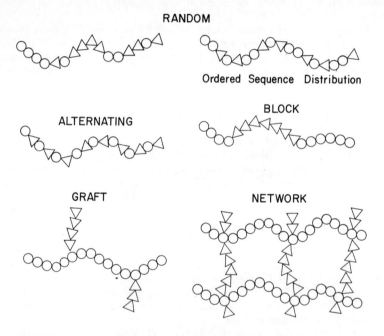

Figure 1. Types of copolymers

Random Copolymers. Random copolymers are produced in bulk, solution, aqueous suspension, or emulsion using free-radical initiators of the peroxide type or redox systems. Initiation through irradiation is also possible as C. Schneider and M. Krebs report on copolymers of vinylene carbonate with halogen-substituted olefins. In emulsion polymerization it is frequently desirable to produce highly concentrated latices. To maintain fluidity, small particles have to coalesce into larger ones. H. Schlueter describes a technique for controlled agglomeration of SBR latex using polyethylene oxide interacting with potassium oleate.

During batch copolymerization, composition may drift with conversion because of differences in comonomer reactivity and can result in less valuable heterogeneous products. By controlled feeding of the monomers to the polymerization reactor, it is possible to regulate the sequence distribution in random copolymers and to obtain homogeneous copolymers. B. N. Hendy shows in his chapter that acrylonitrile/styrene copolymers of high AN content can be made in emulsion through calorimetrically controlled monomer feeding. These copolymers are used as barrier resins in blown bottles for carbonated beverages. Using the different reactivity ratios of aminimides and of vinylpyridine or vinylpyrrolidone, new types of reactive copolymers were prepared by H. J. Langer and B. M. Culbertson. They are being used as adhesive for steel cord in rubber tires.

Alternating Copolymers. The reactivity of polar monomers, *e.g.* acrylonitrile, methyl methacrylate or maleic anhydride can be enhanced by complexing them with a metal halide (*e.g.*, zinc or vanadium chloride) or an organo-aluminum halide (ethyl aluminum sesquichloride). These complexed monomers participate in a one-electron transfer reaction with either an uncomplexed monomer or another electron-donor monomer, *e.g.* olefin, diene, or styrene, forming alternating copolymers with free-radical initiators.

Graft Copolymers. In graft copolymerization, a preformed polymer with residual double bonds or active hydrogens is either dispersed or dissolved in the monomer in the absence or presence of a solvent. On this backbone, the monomer is grafted in free-radical reaction. Impact polystyrene is made commercially in three steps: first, solid polybutadiene rubber is cut and dispersed as small particles in styrene monomer. Secondly, bulk prepolymerization and thirdly, completion of the polymerization in either bulk or aqueous suspension is made. During the prepolymerization step, styrene starts to polymerize by itself forming droplets of polystyrene with phase separation. When equal phase volumes are attained, phase inversion occurs. The droplets of polystyrene become the continuous phase in which the rubber particles are dispersed. R. L. Kruse has determined the solubility parameter for the phase equilibrium.

Type and configuration of the dispersed rubber have also a significant influence on the grafting reaction. For example, the following three configurations of polybutadiene exist:

cis-1,4   trans-1,4   1,2-(vinyl)

Currently a polybutadiene of medium cis-1,4 configuration of about 36%, made with butyllithium catalyst in solution, is the rubber most widely used in impact polystyrene. The cis-1,4 configuration is characterized by a lower glass transition temperature ($-108°C$) than the trans-1,4 configuration ($-14°C$) resulting in satisfactory impact strength at low temperatures. This rubber contains also about 12% 1,2-vinyl configuration which is more reactive than 1,4 configurations and is predominantly responsible for the grafting as well as for crosslinking reactions which D. J. Stein, G. Fahrbach, and H. Adler discuss in their chapter. In a very thorough study they investigated the effect of the vinyl configuration on grafting and crosslinking using poly-

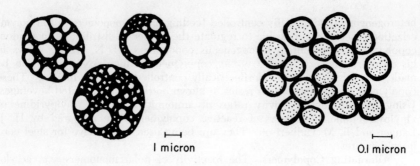

├─┤   ├─┤

I micron   0.I micron

Figure 2.   Rubber particles

Left: in impact polystyrene made by mass (solution) polymerization. Right: with grafted shell in ABS made by emulsion polymerization.

butadiene of 1–90% vinyl configuration. To avoid highly crosslinked rubber, it is imperative to use a polybutadiene of low vinyl content and to maintain low polymerization and processing temperatures. A polybutadiene of high cis-1,4 configuration of about 98%, made with a cobalt or Ziegler-type catalyst, is being used in specific impact polystyrene grades. It contains only 1% vinyl content which makes it more difficult for grafting. A polybutadiene of high vinyl configuration of about 70% has been evaluated for photodegradable impact polystyrene for disposable containers and is made by anionic polymerization with a chromium–aluminum catalyst.

A polybutadiene with 60% trans-1,4, 20% cis-1,4, and 20% vinyl configurations is made in free-radical emulsion polymerization and used as substrate in the manufacture of ABS. The latex particles have less than one-tenth the size of the rubber particles dispersed in impact polystyrene as illustrated in Figure 2. At the large porous rubber particles, grafting takes place first inside the voids and later as warts on the outside. At the small solid particles, the graft copolymer forms a shell around the solid rubber core. This shell grows with increasing grafting level and provides a more uniform distribution of the particles in the SAN matrix. Both impact polystyrene and ABS represent polyblends consisting of a graft copolymer embedded in an ungrafted matrix. An analytical method of separating the two phases by means of an ultracentrifuge was developed by B. Chauvel and J. C. Daniel. Interaction between the two phases and their effect on the rheological behavior of ABS was investigated by A. Casale, A. Moroni, and G. Spreafico. By separating the two phases in a reversible gel technique and by selectively degrading the polybutadiene backbone, G. Riess and J. L. Locatelli were able to compare the grafted with the ungrafted SAN. They observed that the grafted portion was characterized by a higher molecular weight than the ungrafted one because of preferential solvation of polybutadiene by styrene. Rate of polymerization could be correlated with monomers concentration, initiator, and again with the vinyl content of polybutadiene. M. Catoni, G. Pizzigoni, and I. Ronzoni analyzed composition and physical properties of 48 commercial ABS grades and developed mathematical models for them. The values measured agreed well with those calculated.

Instead of polybutadiene, polyolefins may be used as substrate upon which styrene and acrylonitrile can be grafted. This, however, is difficult by conventional techniques because no double bonds are available for the grafting reaction. For this reason, EPDM elastomers (ethylene/propylene/diene monomer) have been proposed because the number of free double bonds may be adjusted at will. Most of them contain 10% of a diene monomer such as ethylidene norbornene or 1,4-hexadiene. F. Severini, A. Pagliari, C. Tavazzani, and G. Vittadini discuss the use of EPDM as substrate upon which styrene with acrylonitrile or vinyl chloride have been grafted. Use of polyethylene or EVA results in an inhomogeneous mixture of grafted and ungrafted portions partially crosslinked. H. Alberts, H. Bartl, and R. Kuhn discovered that grafting can be achieved in the presence of mono-olefins as regulators yielding a uniform graft copolymer.

Acrylate rubbers such as poly(butyl acrylate) or poly(ethyl hexylacrylate) are characterized by better aging characteristics than polydienes. Ethyl hexylacrylate and acrylonitrile were grafted onto PVC in solution by R. G. Bauer and M. S. Guillord. They observed that this graft copolymer was transparent in contrast to a mere polyblend of PVC and an AN/acrylate copolymer.

Block Copolymers. Block copolymers are generally made ionically using either butyllithium or a Ziegler-type catalyst. When styrene and a diene monomer such as butadiene are charged together with the catalyst solution into a batch reactor, the reactivity ratios are such that the butadiene molecules polymerize first and with almost total exclusion of any styrene present. Only when all the butadiene monomer is consumed does the bulk of the styrene enter the polymer chain. Thus, a block copolymer is formed. Because the polystyrene block behaves like styrene homopolymer, it is more soluble than random BS-rubber or polybutadiene in styrene monomer. Therefore, the block copolymer becomes more compatible and lends itself well as backbone for making graft impact polystyrene or ABS. G. Riess and Y. Jolivet demonstrated that diene–styrene block copolymers may be used as dispersing agents of rubber particles in a polystyrene matrix, acting like an emulsifier at the interface.

Since there is no termination step in ionic polymerization, tri-block copolymers can be made. For example, SBS-block copolymers are manufactured by charging the monomers in succession: styrene is polymerized first; butadiene is added to the live polystyrene, and finally, styrene again. In these triblock copolymers, a microphase separation may occur as illustrated in Figure 3. The rigid polystyrene-end segments are joined by the elastomeric polybutadiene center sections. The polystyrene segments associate with each other giving large aggregates or domains. At room temperature, these block copolymers behave almost like crosslinked rubber. When heated above the glass transition temperature of polystyrene (above 100°C), the domains soften and may be disrupted by applied stress allowing the block copolymer to flow. This process is reversible, and these thermoplastic elastomers may be injection molded into a final product without a cure step. L. E. Nielsen tested five commercial thermoplastic elastomers and found their creep characteristics dependent on their crystallinity and found them to be higher than that of crosslinked rubber. Microphase separation and the formation of crystalline domains were observed by A. Takahashi and Y. Yamashita in styrene–tetrahydrofurane block copolymers, by T. Hashimoto, E. Hirata, T. Ijitsu, T. Soen, and H. Kawai in block copolymers of ethylene oxide and isoprene, and by A. H. Ward, T. C. Kendrik, and J. C. Saam in α-methylstyrene–dimethylsiloxane block copolymers.

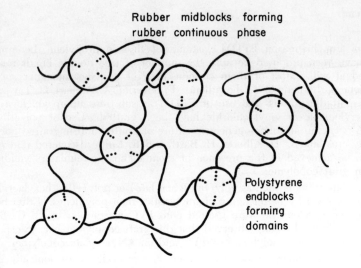

Rubber midblocks forming
rubber continuous phase

Polystyrene
endblocks
forming
domains

Figure 3.   Block copolymer: styrene / butadiene / styrene

It is possible to prepare block copolymers by free-radical initiation, as R. B. Seymour, G. A. Stahl, D. R. Owent, and H. Wood discuss in their chapter. Methyl methacrylate macroradicals were made with peroxide and azo initiators in diluents, and different vinyl monomers were polymerized onto them. Block copolymers of two ethylene imines, one having a long (lauroyl) side chain and one with a short (propionyl) side chain were synthesized by M. H. Litt and T. Matsuda in a two-step cationic polymerization process. Block and random copolymers of episulfides were prepared by E. Cernia, A. Roggero, A. Mazzei, and M. Bruzzone using anionic catalysts of metalated sulfoxides and sulfones.

Multiblock copolymers consisting of an ordered sequence of flexible and rigid units are made in condensation reactions. Segmented polyurethanes, as used in Spandex fibers, are made in three steps. First, a flexible linear polyester or polyether of molecular weight 500–4000 and with reactive hydroxyl group at both ends is prepared. Secondly, it is allowed to react with a diisocyanate resulting in the formation of the soft-segment prepolymer. In the third step, the hard segments are formed by reaction of the prepolymer with low molecular weight glycols or diamines. This results in a polyurethane or polyurea with hydrogen bonding sites which provide the tie points responsible for the long-range elasticity. Aromatic sulfon ether diamines with an ordered sequence of flexible and rigid units have been prepared by G. L. Brode and co-workers from *p*-aminophenol and dichlorodiphenyl sulfone.

Crosslinked Copolymers. Crosslinking one linear polymer with a second polymer can be achieved by generating free-radical sites on a preformed polymer in the presence of a monomer. Grafting occurs on this site and combination termination results in crosslinking. C. H. Bamford and G. C. Eastmond have prepared such crosslinked copolymers of polycarbonate and polychloroprene using molybdenum and manganese carbonyls as initiators.

Polyblends

In contrast to copolymers where the components are linked by strong covalent bonds, the components in polyblends adhere together only through van der Walls forces, dipole interaction, or hydrogen bonding. They generally consist of a polymeric matrix in which another polymer is imbedded, and they are made by compounding on mill rolls, in extruders, or in Banbury mixers. We distinguish between four types of polyblends as given in Table I.

Table I.   Polyblends

Matrix	Dispersed Phase	Improvement	Examples
rigid	soft	tougher	impact polystyrene, ABS rigid PVC
rigid	rigid	faster melt flow higher impact strength reduced shrinkage	PPO + impact PS PVC + ABS or MBS thermoplast in polyester
soft	soft	longer wear	SBR + natural rubber SBR + cis-polybutadiene
soft	rigid	increased strength reduced abrasion, cut growth and flex cracking	SBR + PS

Polyblends with Rigid Matrix. The combination of a rigid matrix with a soft dispersed phase is selected to toughen brittle polymers with elastomers. The matrix gives hardness and stiffness while the dispersed phase provides toughness. Compatibility between the two phases is important and effected by using graft or block copolymers as the dispersed phase. Impact polystyrene and ABS belong into this group.

Polyblends in which both phases are rigid are frequently called polyalloys. Poly(phenyl oxide) is blended with impact polystyrene to improve melt flow. Complete compatibility between the two phases is rare and was observed between poly(methyl methacrylate) and poly(vinylidene fluoride) by D. R. Paul and J. O. Altamirano. Thermoplastics are added to polyesters to reduce mold shrinkage.

Polyblends with Soft Matrix. Polyblends in which both phases are soft are mixtures of different rubbers. Treads of automobile tires are made of polyblends of SBR with either natural rubber or cis-polybutadiene. Covulcanization of EPDM with various rubbers is discussed in the chapter of M. E. Woods and T. R. Mass. Relaxation behavior of blends of EVA rubber with styrene/ethylene–butylene/styrene block copolymer and of poly(ethylene oxide) with ethylene oxide/propylene oxide/ethylene oxide block copolymer were studied by M. Shen, U. Mehra, L. Toy, and K. Biliyar.

The addition of a rigid polymer to a soft matrix results in an increase of the modulus of elasticity of the polyblend and is known as rubber reinforcement. The rigid dispersed particles act like active fillers against abrasion, tear, cut growth, flex cracking, and tensile failure. M. Morton, R. J. Murphy, and T. C. Cheng investigated the behavior of vulcanized SBR reinforced with polystyrene, PMMA, and polyacenaphthylene particles. They found that a higher modulus of the dispersed polymer raises both tensile and tear strength.

In addition, they measured the effect of particle size and surface energy on the interfacial adhesion.

Composites

In contrast to polyblends, composites consist of a polymeric matrix in which a foreign material is dispersed. They also may be classified into four groups as shown in Table II.

Table II. Composites

Matrix	Dispersed Phase Filler or Reinforcing Material
1. Thermoplastics	short glass fibers, beads, hollow spheres minerals
2. Thermosets	wood flour, paper, asbestos, clay
3. Vulcanized rubber	carbon black, calcium silicate
4. Polyesters, polyepoxides, structural resins	long glass fibers, mat or cloth; graphite fibers boron whiskers

Reinforced Thermoplastics. Thermoplastics are reinforced with either mineral particles such as ground wollastonite or feldspar or with short glass fibers, beads, or hollow spheres. They can be injection molded into strong structural parts replacing cast metals. The polymeric matrix itself must have a higher elongation than the reinforcing material to utilize the strengthening effect completely. The dispersed reinforcing material changes the melt flow of the viscoelastic polymeric matrix. L. R. Schmidt studied injection molding of polypropylene reinforced with glass beads and short fibers. He determined the mold filling characteristics that influence the physical properties of the molded piece.

As thermoplastics are reinforced with minerals, concrete can be reinforced with thermoplastics tripling its strength, reducing water absorption by 99%, and improving freeze–thaw resistance enormously. Polymer concrete is being made by either mixing monomer with cement or by impregnating precast concrete with monomer, followed by polymerization. According to M. Steinberg, 8% methyl methacrylate, styrene, or acrylonitrile are sufficient to improve the properties. This technique is already being applied to tunnel linings, pipe, highway, and bridge constructions. It offers interesting possibilities for solid waste utilization in sewer pipes, wallboards, and radioactive waste containers.

Fillers in Thermosets and Rubber. Thermosets such as phenolformaldehyde, melamineformaldehyde, and ureaformaldehyde resins are filled with wood flour, α-cellulose, or paper to add bulk, prevent cracking, and reduce cost. Asbestos imparts heat resistance, mica gives excellent electrical properties, aluminum powder improves heat transfer, and powdered silica or china clay reduces water absorption.

Rubber is filled with carbon black or calcium silicate which act also as reinforcing agents. For example, the tensile strength of vulcanized SBR can be raised tenfold through compounding with 50% carbon black. Elastomers of polar nature such as polychloroprene or nitrile rubber interact even stronger with active fillers.

Reinforced Structural Resins. Polyesters and polyepoxides are reinforced with E- or S-glass strands, mats, or cloth raising their tensile strength 10–20 times and their modulus of elasticity 10–100 times, approaching the properties of steel. Reinforced with graphite, boron, or the new poly-p-benzamide fiber or whiskers, novel composites surpass steel.

High temperature polyimides from an aromatic diamine and an aromatic dianhydride containing a hexafluoroisopropylidene group have been evaluated as melt fusible laminating resins in composites with graphite fibers and quartz fabric. These polyimides with glass transition temperatures ranging from 285° to 385°C have been introduced under the code name NR-150 as discussed by H. H. Gibbs and C. V. Breder. Another heat resistant material that withstands constant exposure at 260°C has been prepared by J. R. Griffith, J. G. O'Rear, and T. R. Walton from aromatic ortho-dinitrile coupled with copper forming a phthalocyanine network.

Conclusion

Starting with the different types of copolymers and continuing with polyblends and composites, this volume spans a range of polymer chemistry. I hope it contains valuable information for each of the readers and that it will inspire them to further progress in polymer science and technology.

NORBERT A. J. PLATZER

Monsanto Co.
Springfield, Mass.
September 1974

The Composite Structure of Fibrous Material

A. PETERLIN[1]

Camille Dreyfus Laboratory, Rt. 1, Research Triangle Park, N. C. 27709

The basic element of fibrous material, microfibril, is composed of alternating amorphous layers and crystalline blocks. The stacked lamellae of the spherulitic or cylindritic starting material are transformed into densely packed and aligned microfibrillar bundles (fibrils) of the highly oriented fibrous material. During the necking process, the few tie molecules connecting the stacked lamellae yield the interfibrillar tie molecules that connect the microfibrils laterally. Their number is small compared with that of intrafibrillar tie molecules, but it increases substantially during plastic deformation of the fibrous structure when the fibrils are sheared and axially displaced. Fibril shearing displaces the microfibrils in the fiber direction and enormously extends the interfibrillar tie molecules by chain unfolding without any substantial change in microfibrillar structure.

The crystalline polymer solid, oriented or not, is a two-component system with the crystalline blocks or lamellae separated by thin amorphous layers containing chain folds, free chain ends, and tie molecules (Figure 1). The microfibrils of highly drawn polymer consist of crystalline blocks alternating with amorphous layers. Stacks of parallel lamellae of the unoriented solid may be included in larger morphological units, *i.e.*, spherulites and cylindrites. The amorphous material is concentrated not only in the amorphous layers between blocks or lamellae but also in the boundaries between adjacent microfibrils and stacks of parallel lamellae and between adjacent fibrils, spherulites, or cylindrites. There is also a finite contribution to the density defect, which determines the volume crystallinity α_v of the sample, from true crystal defects of blocks and lamellae; lattice distortion as a consequence of stresses originating in chain folds on the surface, point and line vacancies at the end of a macromolecule, interstitials, kinks, and boundaries between adjacent mosaic blocks. Therefore, α_v must not be taken too literally as the volume fraction of the crystalline component nor $1 - \alpha_v$ as that of the amorphous component. Both components deviate from ideality, the crystal density being less than expected for an ideal crystal and that of the amorphous component being greater than expected for a relaxed, supercooled melt.

[1] Present address: Polymers Division, National Bureau of Standards, Washington, D.C. 20234.

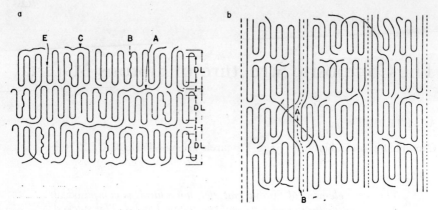

Figure 1.   Schematic model of crystalline and amorphous components in (a) spherulitic and (b) fibrous materials

a. Stack of densely packed, parallel lamellae of the microspherulitic structure (model of the starting material): A, interlamellae tie molecule; B, boundary layer between two mosaic blocks; C, chain end in the amorphous surface layer (cilium); and E, linear vacancy caused by the chain end in the crystal lattice
b. Microfibrillar model of fiber structure: A, interfibrillar, and B, intrafibrillar tie molecule

Moreover, the amorphous and crystalline components are not independent of each other. The amorphous surface layers between adjacent lamellae are formed by free chain ends, chain folds, and tie molecules. The free chain ends are fixed at the other end in the crystal lattice. Chain folds and tie molecules are fixed at both ends, the former on the same lamellae, the latter on different ones. These constraints modify deeply the properties of the amorphous layer. Even above T_g, the mobility of the amorphous chains is substantially less than in a rubber with the same average number of mers in the chain segments between subsequent crosslinks.

The shearing motion and the normal deformation that eventually lead to lamellar separation are relatively easy deformational modes of the amorphous layers that are impeded mainly by tie molecules. The latter, if stretched beyond their contour length, must either rupture or pull out of the crystal lattice. In the first stage of normal deformation, the inability of the amorphous layer to contract laterally reduces its density; this is indicated by a substantial increase in small-angle x-ray scattering (1).

The heterogeneity of the crystalline polymer solid is accentuated still more in the case of mechanical properties by the enormous mechanical anisotropy of the crystals and the large difference in the elastic moduli of the crystalline and amorphous components. With polyethylene, the elastic modulus of the crystals is 345^2 or $240^3 \times 10^{10}$ dynes/cm^2 in the chain direction ($E_{||}$) and 4×10^{10} dynes/cm^2 in the lateral direction ($E_\perp$) ($2, 3$). The elastic modulus of the amorphous component (E_a) of polyethylene is 10^9–10^{10} dynes/cm^2 (4). This is significantly less than $E_{||}$ and $E_\perp$ but at least 10 times the elastic modulus of a rubber that has about five monomers in the chain segments between the crosslinks. This is quite surprising since room temperature is far above the glass transition temperature of polyethylene (T_g is either $-20°C$ or $-120°C$), and therefore one would expect a fully developed rubbery behavior. The most probable explanation of this effect are the above-mentioned strong lateral constraints on the chain sections in the amorphous layers which are fixed in the crystal lattice of the lamellae at both ends (chain folds and tie molecules) or at least at one end (free chain ends).

Since the extremely thin crystal lamellae do not form coherent links through the whole sample, the effect of the amorphous component on the mechanical properties of the crystalline polymer solid is even more important. At the boundaries of the lamellae, the forces have to be transmitted through the much softer amorphous layers. The larger surface-to-volume ratio of lamellae with a thickness of 100–200 A compensates to a substantial degree for the much smaller mechanical rigidity of the amorphous material. This is because the resistance of the lamella is proportional to its thickness, and the force transmitted through the amorphous layer is proportional to its surface. Under normal conditions, the transmitted forces are great enough to deform lamellae by chain slip and tilt; these finally lead to lamellar fracture. This effect is the main process when spherulitic material is drawn; the lamellae are transformed into the bundles of microfibrils of the resultant fibrous structure.

The situation is still more extreme when the amorphous component is plasticized by a swelling agent (5). Since the swelling agent does not enter the crystals, their mechanical properties are hardly affected. However, the reduction in the elastic modulus of the amorphous component and the increased mobility of the chains that are partially separated by molecules of swelling agent make the deformation of amorphous layers so much easier that, as a rule, the forces transmitted by them to the crystal lamellae are too small to cause large scale plastic deformation (*e.g.*, lamellar fracture and pulling out of microfibrils). The bulk plastic strain in such a case is mainly the consequence of lamellar sliding, rotation, and even physical separation. All these modes require only plastic deformation of amorphous layers without much effect on the crystal lamellae. In such a case, one obtains a highly porous and almost rubbery elastic material which requires high strain and small stress to break.

Plastic deformation in the presence of a liquid is very different from deformation at higher temperature. Higher temperature makes the amorphous material softer in very much the same manner as the swelling liquid, but it also softens the crystal lattice which is unaffected by the liquid. Therefore, the mode of deformation and the basic properties of the deformed material are not conspicuously altered by temperature increase although some values (*e.g.*, draw ratio, long period, and degree of orientation) can be greatly affected.

The complicated morphology of crystalline polymer solids and the coexistence of crystalline and amorphous phases make the stress and strain fields extremely nonhomogeneous and anisotropic. The actual local strain in the amorphous component is usually greater and that in the crystalline component is smaller than the macroscopic strain. In the composite structure, the crystal lamellae and taut tie molecules act as force transmitters, and the amorphous layers are the main contributors to the strain. Hence in a very rough approximation, the Lennard-Jones or Morse type force field between adjacent macromolecular chain sections (6, 7) describes fairly well the initial reversible stress–strain relation of a spherulitic polymer solid almost up to the yield point, *i.e.* up to a true strain of about 10%.

Fibrous Materials

Electron microscopy reveals that the microfibril is the basic structural element of the fibrous structure (Figure 1) and not the lamella. Indeed, the latter is primarily an optical artifact that does not contribute significantly to the mechanical properties of the fiber. Electron micrographs of surface replicas of slightly etched drawn fibers and films with fibrous structure reveal only highly aligned, very long microfibrils with lateral dimensions d of 100–200 A. The length l_{mf} is in the tens of microns. The microfibrils can be separated from each other rather easily. After prolonged etching, one can also observe the lamellae oriented more or less perpendicular to the draw direction (8). But they may also be oriented at an oblique angle, thus yielding the characteristic four point, small-angle scattering pattern. The appearance of lamellae after prolonged etching is the consequence of the almost complete etching away of the amorphous layers between the crystalline blocks of the same microfibril so that the axial connection is nearly completely lost. But the lateral connection of each block with crystalline blocks of adjacent microfibrils is impaired much less because the boundary contains hardly any fully amorphous material which can be etched away. Consequently, one observes stacked lamellae with a rather wavy surface, and there is little or no trace of microfibrillar structure.

The orientation of the crystal lattice $f_c = (3 \langle \cos^2 \theta \rangle - 1)/2$ is not a good parameter for describing mechanical properties although it is often cited in characterizing drawn material. It is useless when annealing is done at higher temperature close to the melting point. Under these conditions, the crystal orientation is reduced only marginally, particularly if the sample is clamped so that it cannot shrink; however, the values of the mechanical properties are drastically reduced down to those of unoriented material (9). Concurrently, infrared dichroism (10, 11, 12, 13) shows a nearly complete loss of amorphous orientation; this would indicate either that the taut tie molecules have almost completely relaxed and assumed a nearly equilibrium random conformation or that they have been reduced in number. Electron micrographs of surface replicas do reveal a rotation of lamellae around an axis perpendicular to the fiber axis and a lateral growth beyond the boundary of the fibril (14). Although no rotation is observed on lamellae in the interior of the sample, one has the impression that the lamellae have attained a physical individuality not present in the material as drawn as well as a substantial independence from each other which is possible only if the axial connection by tie molecules has been lost to a large extent and the microfibrils have ceased to exist.

One must conclude that in crystalline polymers the orientation of the crystalline component is not the cause of the high strength and elastic modulus in the direction of orientation. The best example is an oriented single crystal mat which may have a high orientation of crystal lattice similar to that in a moderately drawn film without any trace of high strength in the chain direction. The enhanced mechanical properties of the drawn sample result from the fibrous structure, *i.e.* from presence of strong microfibrils that derive their axial strength from the numerous taut tie molecules that connect the folded chain crystal blocks across the amorphous layers alternating with the blocks in the axial direction of the microfibrils. The presence of taut tie molecules and hence of full strength microfibrils is quantitatively detectable by the orientation of the amorphous component; this therefore seems to be the best criterion for describing the mechanical properties of fibrous material.

A detrimental effect on strength similar to that caused by annealing at high temperature may occur even during drawing if it is done at too high a temperature close to the melting point so that the material can relax substantially in spite of the drawing stress which prevents shrinkage. The tie molecules are less taut and, very likely, also less frequent. This reduces the axial strength of the microfibril and hence also that of the fibrous material. The effect is to some extent compensated for by the better packing of the aligned microfibrils, which makes the sample more homogeneous so that load bearing and distribution are more uniform than in a sample drawn at a lower temperature. The annealing effects are indeed a limiting factor in the choice of a higher temperature for drawing which would make the material softer and hence easier to draw. It is fortunate that the higher rate of drawing hampers the annealing effects to some extent so much that high speed, high temperature drawing is in many cases an extremely advantageous process technically for fabricating high strength fibrous material.

In a stress field, the microfibril reacts almost as rigid structure that is much stronger than any other element in the polymer solid. This is a consequence of (1) the large fraction of taut tie molecules connecting consecutive crystal blocks like a quasicrystalline bridge, and (2) the special dense packing of aligned microfibrils; it is not a consequence of some higher crystal perfection of the blocks. Indeed one knows that drawing increases the crystal defect concentration and thus produces a lower-than-ideal crystal density and consequently also lower strength and elastic modulus of the crystals (*15*). Removal of crystal defects and restoration of crystal perfection during annealing make the crystals stronger, but they cannot prevent the drastic reduction in elastic modulus and strength of the fibrous material caused by the disappearance of tie molecules.

The dense packing of microfibrils with the enormous surface $(4l_{mf}d)$ to cross section (d^2) ratio $(4l_{mf}/d \simeq 4000)$ seems to overcompensate for the crystal defects. The longitudinal displacement of a single microfibril is opposed by the sum of the van der Waals forces along the whole microfibril which is so great that such a displacement is less easy than with stacked lamellae. Moreover, it blocks individual crystal deformation because such deformation always requires some longitudinal displacement of the whole or a part of the microfibril. Consequently, deformation of fibrillar structure is much more a cooperative than a highly localized effect of a single crystal block. That not only makes the microfibril more resistant to deformation, but it also homogenizes the stress and strain field. The combination of all these effects results in higher tensile strength and elastic modulus of the fibrous material.

As long as one can avoid longitudinal void formation, the elastic modulus in the transverse direction is also larger than in a spherulitic sample (*16*). In spite of the fact that the crystal lattices of adjacent microfibrils are not in crystallographic register, the van der Waals forces between them are much stronger than in the thicker amorphous layer between two stacked lamellae of the spherulitic material and only slightly weaker than the forces in the amorphous layer between crystal blocks of the same microfibril which is bridged by the large fraction of taut tie molecules.

Each stack of parallel lamellae yields a bundle of parallel microfibrils (Figure 2) that differ a little in average draw ratio from the adjacent bundle (fibril) as a consequence of slightly different micronecking conditions caused by the local variation in lamellar orientation and orientation of the microneck-

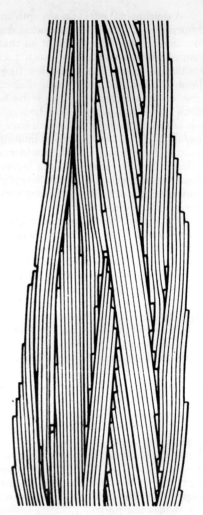

Figure 2. Schematic model of fibril and microfibril packing in fibrous material

The long and narrow microfibrils are bundled into fibrils; their ends, located mainly on the outer surface of fibrils, act as point defects (point vacancies) of the microfibrillar lattice

ing zone (*14, 17*). Within each fibril, the crystal blocks and amorphous layers of adjacent microfibrils tend to fit laterally so that the surface free energy at the boundary is minimized. The interfaces are primarily crystal–crystal and amorphous–amorphous. The crystal–crystal boundary is distorted just enough so that the lattice coherence for wide-angle x-ray scattering is interrupted. As a consequence of such lateral fit, the blocks are arranged in lamellae roughly perpendicular to the fiber axis. Variation in local stress and deformation flow field in the neck may make the lamellae fluctuate around an oblique angle (four-point, small-angle x-ray diffraction pattern). A wavy surface produces the line-shaped diffraction pattern. But the lamellae are more an optical artifact than a true building element of the fibrous structure. They can be made visible by electron microscopy by prolonged fumic nitric acid etching (*8*); this procedure removes the amorphous component without significantly attacking the crystalline regions and the slightly less ordered boundaries between adjacent microfibrils. But they do not play any substantial role in determining the mechanical properties of fibrous material.

The finite dimensions of the stacked lamellae of the original sample impart a finite length to the microfibrils and a sigmoidal shape to the fibrils (18). The packing of microfibrils within each fibril is more perfect than the lateral packing of fibrils because fibrils differ more in lattice orientation and draw ratio than do microfibrils. Moreover, most microfibrils have their ends at the outer boundary of the fibril, thus producing a high local concentration of structural defects which is important in crack propagation. The point vacancy of the microfibrillar lattice at the end of the microfibril is the primary candidate for microcrack formation under applied stress. The coalescence of such microcracks along the boundary of the fibril, which is favored by their proximity and the relative weakness of the boundary, tends to produce long, axially oriented voids that separate the fibrils and finally lead to the fibrous fracture so characteristic of highly drawn fibrous material.

The strength of fibrils and microfibrils is determined by different structural elements. With microfibrils, the intrafibrillar tie molecules are the main source of strength. With fibrils, however, the few interfibrillar tie molecules are of minor importance; but the enormous surface of the long (10μ), thin $(100-200 \text{ A})$ microfibrils produces autoadhesive forces so great that the deformation of fibrils by axial displacement of microfibrils is very difficult to achieve.

Deformation of Fibrous Material

The same reasoning can be applied to the elastic and plastic deformation of fibrous material. Fibrils with the more perfect lateral packing of microfibrils are expected to be more resistant to deformation than the lattice of fibrils. It is easier to displace the fibrils relative to each other in the stress direction than to deform them since that would require shear displacement of the microfibrils. Such sliding motion of fibrils is opposed by the friction resistance in the boundary. Despite the weakness of the van der Waals forces, the friction resistance over the length $(\simeq 10\mu)$ of the microfibril which is about 1000 times larger than the thickness of a crystal lamella $(\simeq 100 \text{ A})$, will be much larger than the forces opposing plastic deformation in spherulitic material. Hence the strain hardening effect in crystalline polymers observed during transformation of a spherulitic specimen into fibrous material is a direct consequence of the large surface-to-volume ratio of the strong fibrils and microfibrils which efficiently hampers shearing displacement of microfibrils and longitudinal displacement of fibrils. It also explains the increase in elastic modulus and stress to yield by a factor nearly equal to, or even higher than, the draw ratio (*see* the following example).

In the special case of branched polyethylene, the elastic modulus at 45° to the fiber axis is exceptionally small; this means that shear compliance along the fiber axis is very high (16). Such deformation involves reversible shearing displacement of adjacent fibrils. Since a similarly high shear compliance does not occur with linear polyethylene and isotactic polypropylene, the difference may be attributable to the substantial difference in draw ratio (4.5 in branched, 20 in linear polyethylene and propylene) which results in proportionately shorter microfibrils and fibrils in the former. The shorter the microfibril, the shorter the fibril and the smaller the surface-to-cross section ratio and hence the smaller the resistance to shear displacement.

The plastic deformation of fibrous material deserves more detailed analysis. When nylon 6 is drawn, two stages are apparent (Figure 3). First a neck propagates through the sample and changes the more or less spherulitic into a fibrous structure. The draw ratio λ^* is 2–3. During this morphologic transformation the load for drawing is practically constant which corresponds to the lower yield stress of spherulitic material. High chain orientation and a change in long period, depending on the temperature of drawing, are established (18).

After this transformation is completed, the load F increases very rapidly up to the break which occurs at a draw ratio λ_b of about 5. The stress, $\sigma = \sigma^*\lambda/\lambda^*$, increases still more rapidly as a consequence of the decreasing cross section of the material. In this second stage between λ^* and λ_b, the drawing is uniform in contrast with the first stage where a neck propagates through the sample. The chain orientation improves but the long period remains unchanged.

Since no drastic morphological change occurs during the homogeneous drawing of fibrous structure, one can conclude that the basic elements, *i.e.* the microfibrils and fibrils, remain basically unaffected. Plastic deformation can proceed only by pulling the adjacent elements in the draw direction in a manner described by the affine transformation (Figure 4).

$$\vec{r} = (x,y,z) = (\lambda_f^{-1/2}x, \lambda_f^{-1/2}y, \lambda_f z) \tag{1}$$

where $\lambda_f = \lambda/\lambda^*$. The displacement of the centers of mass of the fibrils according to Equation 1 is opposed by frictional forces on the outer boundary of the fibrils. The same applies to microfibrils.

The forces are proportional to the lateral surface area of fibrils and microfibrils. Since they increase with better, more perfect contact of adjacent elements, one would expect a higher resistance to the displacemnt of microfibrils than to that of fibrils. Microfibrils do not vary much in their draw ratio; they are packed more closely, and to some extent they are connected by interfibrillar tie molecules. Fibrils vary much more in their draw ratio and in their outer boundary there are a great many structural defects caused by the irregular ends of microfibrils.

Since the deformational forces are proportional to the cross section ($A = d^2$) and the resistance to the lateral surface area ($4dl$), one would expect a larger displacement effect on fibrils than on microfibrils. The average length of both is practically the same, $l_f = l_{mf}$, but fibril thickness is about ten times that of microfibrils, $d_f \simeq 10d_{mf}$. If the mentioned difference in friction coefficient is also considered, one can conclude that most plastic deformation of fibrous structure is the consequence of the nearly affine displacement of fibrils and that only a small fraction is attributable to that of microfibrils.

A closer analysis (19) of the stress field inside the fibril caused by the affine displacement reveals that the main forces on the microfibrils produce a shear displacement (Figure 5). It is relatively small, *e.g.* about 2000 A, for adjacent microfibrils of nylon 6 which may be 30,000–60,000 A long and 100–200 A thick. But it has a dramatic effect on interfibrillar tie molecules unfolded over the same length of 2000 A which corresponds to a molecular weight of about 20,000. In nylon 6 with a long period $L = 100$ A, such an unfolded chain crosses about 20 amorphous regions; it is therefore counted 20 times as a taut tie molecule. Even a small fraction of interfibrillar tie molecules,

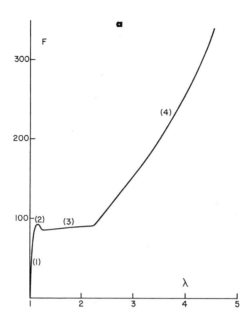

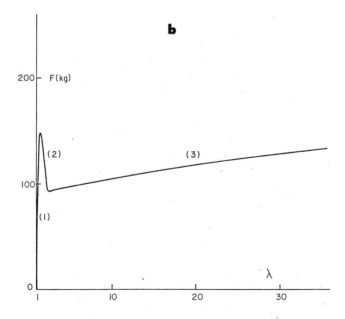

Figure 3.   Load-elongation curves of (a) *nylon 6 and* (b) *linear polyethylene*

*1, elastic range; 2, upper yield point; and drawing ranges of (3) spherulitic and
(4) fibrous materials*

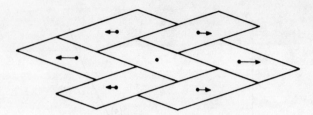

Figure 4. Schematic model of displacement of centers of mass of short fibrils according to affine transformation (Equation 1)

0.5–1%, in the completed fibrous structure yields a large fraction, 10–20%, after such a deformation.

The longitudinal displacement of fibrils is much larger than that of micro-fibrils. It may be as large as the length of the fibril, 3–6 μ. Such displacement not only extends completely any interfibrillar tie molecules connecting adjacent fibrils, but it also smooths, by chain unfolding, the nonhomogeneities at the ends of microfibrils. Besides the increase in the fraction of taut tie molecules in the boundary, there is also better, more perfect contact of adjacent fibrils because of the removal of the most conspicuous defects caused by the ends of microfibrils located in the fibril surface. One may speculate that such an effect would produce the rapid increase in resistance to drawing which appears so drastically in the second stage of drawing nylon 6. This concept is corrobo-

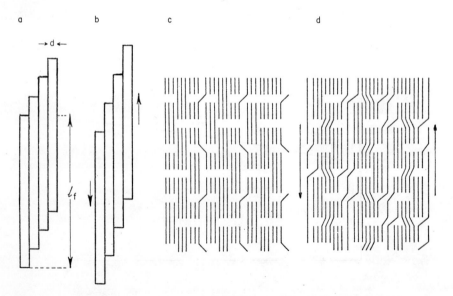

Figure 5. Schematic model of shear deformation of fibrils (a and b) and the concurrent unfolding of interfibrillar tie molecules (c and d)

The initial stages (a and c) show the skewed shape of fibril as formed and the presence of a modest fraction of interfibrillar tie molecules (0.5 per amorphous layer of one microfibril). Shear displacement by three long periods increases this number to 1.5.

rated by the observation that the diffuse, small angle x-ray scattering of a highly strained fibrous nylon sample, which is caused by opening of the point defects at the ends of microfibrils, is substantially less with high strength fibers. During fabrication, these fibers are drawn to a higher draw ratio than low strength fibers, and hence removal of such nonhomogeneities by chain unfolding is more complete (20).

The clear separation of transformation from spherulitic to fibrous structure and the drawing of fibrous structure observed with nylon 6 are not found with polyethylene or polypropylene. This could lead to the conclusion that in these polymers the deformational process is basically different from that in nylon 6. But, closer inspection of material drawn to gradually increasing draw ratios reveals a closer similarity than one can derive from load–elongation curves and direct observation of necking.

In linear polyethylene, the dependence of transport properties of gases and vapors on draw ratio (21) demonstrates very convincingly that at λ^* between 8 and 9 the transformation from spherulitic to fibrous structure is completed. Below λ^* the sorption, the diffusion coefficient, and particularly its concentration dependence correspond to data of spherulitic material. Above λ^*, all properties are completely different and they correspond to those of the fibrous morphology. Hence, observed draw ratios up to $\lambda^* = 25$ (at 60°C) include substantial drawing of fibrous structure with $\lambda_f = \lambda/\lambda^*$ of up to about 3. Most of this drawing occurs in the mature neck immediately after, or maybe even simultaneously with, the structural transformation. A minor contribution comes from the plastic deformation of the already necked material (*see* the strain–time curves of single volume element of the drawn sample in Figure 6) (22). The ratio of drawing in and after the neck is smallest in the virgin, very soft neck. It increases rapidly with the mature, very sharp neck. It seems that the postneck drawing disappears gradually since fracture of the drawn sample in constant rate tensile experiments occurs almost instantaneously as soon as the necking has transformed the whole sample into fibrous structure.

Absence of the rapid increase in load with mature neck indicates that the increase in frictional resistance to affine displacement of fibrils is smaller in polyethylene than in nylon 6. According to the model of fibrous structure, this means that there is less of a smoothing effect at the microfibril ends by unfolding of chains, less improvement in lateral contact between adjacent fibrils, and a smaller increase in the fraction of taut interfibrillar tie molecules. Such an effect is compatible with the model which predicts fibril length proportional to λ^* and lateral dimensions proportional to $\lambda^{*-1/2}$. In polyethylene with $\lambda^* \simeq 9$, the fibrils are more than three times longer than in nylon 6 with $\lambda^* = 2$-3. Hence, with an equal number of microfibrils per fibril, the surface density of microfibrillar ends in polyethylene is at most one-third that in nylon 6. One would expect almost the same reduction in the effect of smoothing at such ends on the contact quality and frictional resistance of fibrils.

The reduction in formation of new taut tie molecules per amorphous layer manifests itself in the almost constant sorption and diffusion coefficient above λ^*. The constancy of sorption means that there is a constant fraction of almost unperturbed locations in the amorphous component where the molecules of the sorbate can be accommodated. The constancy of diffusion coefficient means that there is a constant fraction of taut tie molecules which, by their better packing, reduce the fractional free volume and which, by their tautness, increase the energy needed for hole formation, *i.e.* the activation energy of diffusion.

2

Refinement of the Methods for the Determination of Molecular Weight Distributions in Homopolymers

G. V. SCHULZ

Institut für Physikalische Chemie, Universität Mainz, Mainz, West Germany

Precipitation chromatography, as proposed by Baker and Williams and improved in our laboratory, has a systematic error caused by the nonuniformity of the fractions. This error can be eliminated if the fraction nonuniformity is determined by a special kind of mixing experiment. Another interfering effect which is produced in anionic polymerizates by traces of impurities can be eliminated by testing the BW fractions by GPC. Exactly determined molecular weight distribution functions of samples which are synthesized under well defined kinetic conditions provide much information about elementary reactions of the polymerization process. This is demonstrated by some examples of the anionic polymerization of styrene and methyl methacrylate.

Exact knowledge of the molecular weight distribution (MWD) functions is useful and/or necessary for two very different purposes: control of the properties of given polymer samples and fabrics, and investigation of the elementary reaction in polymer formation processes. This kind of use is possible because the MWD function of a given polymer sample is an approximately complete record of its formation process.

However, since the potential content of information in the distribution function is very high, systematic errors in the determination of this function can lead to serious misinterpretations of the kinetics. Therefore it is necessary to eliminate as completely as possible all systematic errors. Statistical errors are less dangerous because they can be detected by the scattering of measured values or by the reproducibility of the distribution curves.

This paper describes the elimination of a systematic error in the Baker–Williams (BW) method, a combination of the BW and the gel permeation chromatography (GPC) methods for elimination of the effects of side reactions on the molecular weight distributions, and, finally, the application of these methods to some problems in the kinetics of anionic polymerization of styrene and methyl methacrylate.

GPC is known to be more suitable for the investigation of broader distributions, as for instance in radical polymerizations. For the very narrow distributions of anionic polymers, only the BW method can be used at present.

Therefore the investigation and control of this method is the main subject of this paper. The elimination of systematic errors in the BW method by investigation of the broadness and the modality of the fraction is described. Finally, it is shown by some selected examples that a method for determining molecular weight distributions can be tested by kinetic controls, and on the other hand, that determination of the MWD can give information about kinetic details which cannot be obtained by kinetic measurements alone.

Elimination of Systematic Errors in the Baker–Williams Fractionation

The Quantities σ and U. Samples prepared by an unperturbed anionic polymerization process have a distribution function, which, as derived by Böhm (1), corresponds to a Gaussian error function. The same in general is the case for fractions produced by precipitation and/or solution. The distribution function then is characterized by two parameters, the variance σ and the number average degree of polymerization $\overline{P}_n$. On the other hand, a given distribution can be characterized by the nonuniformity (Uneinheitlichkeit in German) U, defined by the equation

$$U \equiv (\overline{P}_w/\overline{P}_n) - 1 \qquad (1)$$

where $\overline{P}_w$ is the weight average degree of polymerization (2). It is easy to show that for Gaussian distributions

$$U = (\sigma/\overline{P}_n)^2 \qquad (2)$$

The quantity U has different advantages in calculating the kinetic parameters of a polymerization process. For instance, in an anionic polymerization the living end can exist in different forms which add the monomer with extremely different rate constants. Each of these reactions alone would result in a Poisson distribution. However, if more than one type of living end is in action, we have a multiple state mechanism of polymerization. Then different influences are working in broadening the distribution, each of them with a characteristic variance σ_i. The variance of the resulting sample is the sum of the corresponding squares

$$\sigma^2 = \Sigma \sigma_i^2 \qquad (3)$$

If all influences result in the same degree of polymerization, $\overline{P}_n$, then

$$\sigma^2/\overline{P}_n^2 = (1/\overline{P}_n^2)\Sigma\sigma_i^2 \qquad (4)$$

Hence

$$U = \Sigma \ U_i \qquad (5)$$

It is not necessary that all influences be kinetic in nature. If, for instance, polymerization is performed in a flow tube, a hydrodynamic broadening of the distribution takes place which can be calculated from easily determinable quantities (3). To this broadening there corresponds an additional variance and consequently an additional nonuniformity U_{hydr}. Equation 5 then becomes

$$U = U_{kin} + U_{hydr} \qquad (6)$$

Since the value of U_{hydr} is accessible in all cases, the kinetic part of U can be calculated directly from Equation 6.

In the kinetic equations U and not σ is the relevant quantity. In Figure 1 the quantities σ and U are compared. If σ is constant, the reduced values over

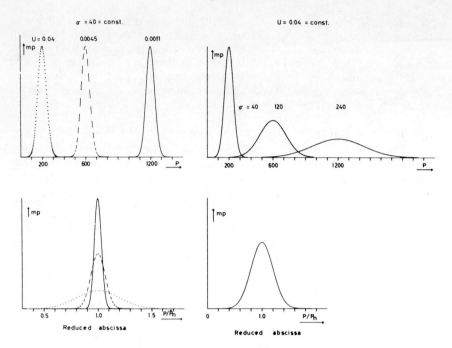

Figure 1.   Gaussian distributions for different $\overline{P}_n$ with constant σ or constant U

the abscissa $P/\overline{P}_n$ are different for different degrees of polymerization (DP). However, when U is constant, the reduced distribution curve is independent of $\overline{P}_n$. Consequently, the three curves for a constant σ corresponds to different kinetics whereas the curves for a constant U correspond to identical kinetic constants (*see* below).

Determination and Correction of a Systematic Experimental Error in the BW Method. For the characterization of narrow distributions, the direct measurement of $\overline{P}_n$ and $\overline{P}_w$ by osmosis and light scattering is not applicable. Even if the experimental error did not exceed 2%, then $U = 0.05$ and $U = 0.03$ could not be distinguished. The GPC method also has its limitations in this range. Therefore, at this time the best method for determining narrow distributions is precipitation chromatography as proposed by Baker and Williams (4) and elaborated on recently in our laboratory (5, 6, 7).

By this method, one cuts the sample to be analyzed into many narrow fractions by using a column in which a solvent–nonsolvent gradient and a temperature gradient are working against one another. From the fractions are measured the mass and the DP. Therefore the elution curve is not related directly to a calibration curve, because all fractions are characterized separately. One can use a GPC column for this characterization thereby economizing the work considerably.

From the weight fractions (m_k) and the DPs (P_k) of the fractions, the averages $\overline{P}_w$ and $\overline{P}_n$ can only be approximated by the usual equations.

We write

$$\overline{P}_{w,app} = \sum_{k} m_k P_k \qquad (7)$$

and

$$\overline{P}_{n,app} = \frac{1}{\sum\limits_{k} m_k / P_k} \tag{8}$$

and we obtain, in analogy to Equation 1, an apparent nonuniformity

$$U_{app} = \sum_{k}(m_k P_k) \cdot \sum_{k}(m_k / P_k) - 1 \tag{9}$$

which is smaller than the real nonuniformity, U_{real}, because of the finite non-uniformity, $\overline{U}_{fr}$, of the fractions. If $\overline{U}_{fr}$ is known, the real nonuniformity can be corrected by the formula of Hosemann and Schramek (8):

$$U_{real} = U_{app} + \overline{U}_{fr}(1 + U_{app}) \tag{10}$$

The error which arises by neglecting this correction is systematic. It must be avoided if a correct evaluation of the measured nonuniformity for kinetic problems is to be achieved.

An attempt to determine $\overline{U}_{fr}$ can be made as shown below, but first the principle of the method should be explained. Figure 2 shows part of a differential weight distribution curve which is divided into fractions. If we could obtain ideal fractions with an approximately rectangular shape, U_{fr} would be on the order of 10^{-4} and could be neglected. However the real fractions can have a nonuniformity comparable with that of the sample. Therefore it is necessary to determine, or at least to estimate, U_{fr}. This is possible by measuring the overlapping effect as shown in Figure 2.

The overlapping of two neighboring fractions, of course, provides no information about the broadness of the fractions. But, if non-neighboring fractions are mixed and the extent of the overlapping is determined, then U_{fr} can be calculated if one assumes that the MWD of the fractions is of Gaussian type. This information can be obtained by measuring the elution curve of the mixture of non-neighboring fractions (Figure 2c). Consequently, it must be considered that two elutions have been made: the first to produce the fractions, the second to obtain the elution curve of the mixture of the fractions. Therefore the measured overlapping effect is larger than the original overlapping.

Unfortunately, the realization of this concept in its simple form is difficult because the mass of a single fraction is very small. Therefore we modified this experiment (*see* Figures 3, 4, 5, and 6) (9). Instead of fractions, we used the middle part of anionically synthesized polystyrenes, cutting the tails on both ends of the distribution by BW fractionation. These two components were mixed and then subjected to two kinds of experiments.

Figure 3 depicts the results of experiments with DPs of 200–500. The average DPs of the components on the left were 244 and 357. That means a ratio of 1.5 between the two maxima. However, the gap between the last fraction of the first component and the first fraction of the second component extends only from 270 to 330. Refractionation of this mixture (lower left) demonstrates clearly that this technique can reveal a bimodal distribution, even if the two maxima are as close to one another as they are in this example. On the other hand, the refractionation yields a fraction lying exactly in the gap at P = 300. This, however, does not prove that there is material at this DP; it is possible that the cutting of the second fractionation takes away parts of the two mixing components, yielding a bimodal fraction. If there is a larger

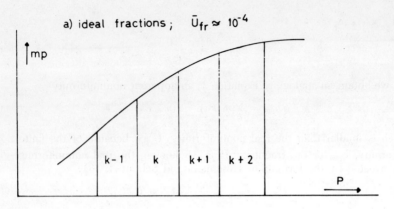

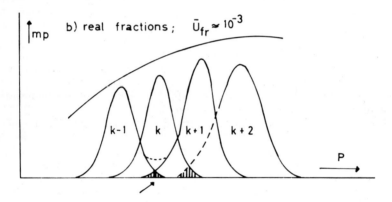

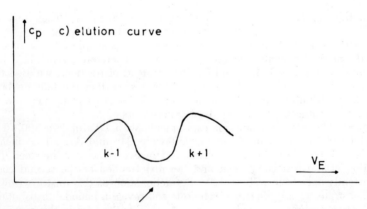

Figure 2. Distributions of ideal (a) and real (b) fractions and the over-lapping of non-neighboring real fractions; c, elution curve of two non-neighboring fractions

gap between the two components (right side of Figure 3), complete separation is observed.

For Figure 4 the same experiment was repeated for a higher range of DPs (around 10^3). Separation efficiency is the same as in the lower range. A bimodal distribution with DP ratio of 1.5 is evidenced.

Now, to return to the question of overlapping. Figures 5 and 6 represent elution curves of the mixtures. They were made by collecting numerous very small fractions from the column, each of which was not sufficient for determining the molecular weight, although each was sufficient for measuring the concentration. There is material at the point of the elution curve that corresponds to $P = 300$ (in the midst of the gap). That means that there is a real overlapping. Since in principle all the original fractions of the mixture contribute material to some extent to the overlapping zone, and since all fractions are of Gaussian type with an average nonuniformity, $\overline{U}_{fr}$, one can calculate this value from the elution curve.

The calculation can be made for each point of the elution curve around the minimum. For a given value of V_e, the elution curve (Figures 5 and 6) yields a value of c, and the curve $P = f(V_e)$ yields a corresponding value of

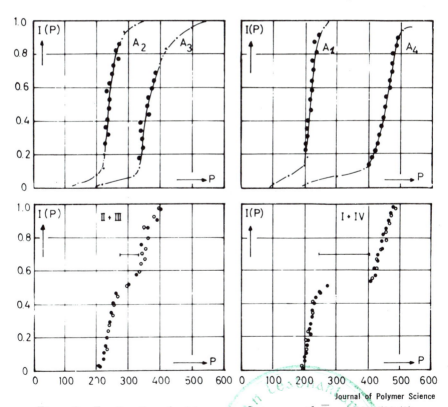

Figure 3.   Fractionation of mixtures of polystyrene with $\overline{P}_w$ of 200–500 (9)

Top, molecular weight distributions of the initial samples (A 1–4); ●, fractions used to prepare mixtures and ●, fractions not used. Bottom, molecular weight distributions of mixtures of II and III and of I and IV; the bar indicates the DP difference between the original fractions lying closest together; the points denote two fractionation experiments.

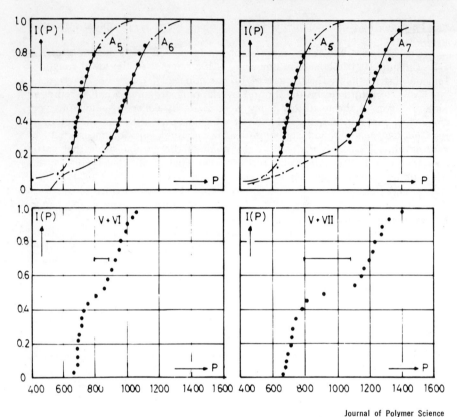

Figure 4. Fractionation of mixtures of polystyrene with $\overline{P}_w$ of 700–1200 (samples A 5–7); see Figure 3 for key to symbols (9)

P and dP/V_e. Moreover, we know the weight fractions (m_k) and the $DPs(P_k)$ of all k fractions of which the mixture is composed (Figures 3 and 4). From these values, one calculates $\overline{U}_{fr}$ by adapting it to the left side of Equation 11, assuming that all fractions have a nonuniformity near an average value, $\overline{U}_{fr}$ (9).

$$\frac{c}{\int_0^\infty c \, dV_e} \cdot \frac{dV_e}{dP} = \sum_k \left[\frac{m_k}{P_k \sqrt{2\pi \overline{U}_{fr}}} \exp\left(-\frac{1}{2 \overline{U}_{fr}} \left[\frac{P_k - P}{P_k} \right]^2 \right) \right] \qquad (11)$$

Applying this procedure to the elution curves in Figures 5 and 6, one obtains an average $\overline{U}_{fr} \leqslant 2 \times 10^{-3}$. This value is small, but neglecting it would give a systematic error. Therefore we corrected the values of U_{app}, as calculated by Equation 9, using Equation 10 with $\overline{U}_{fr} = 1 \times 10^{-3}$. The examples given below indicate that this correction gives reasonable values.

The described experiments and their interpretation present a problem which is, perhaps, a little philosophical: it is the question of whether in principle it is possible to correct a method by that method itself. This is not impossible, but surely such an operation has its limitations, and these are not known exactly.

One way out of this dilemma is to combine two different methods, for instance GPC and BW. This correction is also limited since the errors of both methods are not known exactly. However in special cases this combination can lead to considerable improvement (*see* below).

Another solution might be the use of completely uniform samples for the calibration. Unfortunately, such samples are not yet available. A third possibility is to use samples with an *a priori* known MWD. That is possible, to some extent, if one prepares polymer samples under well-defined kinetic conditions. For example, one can produce polystyrenes under conditions which give a Poisson distribution. This is discussed below.

A Combination of the BW and the GPC Methods for Controlling the Fractions of an Anionic Polymerizate

Each anionic kinetic experiment is disturbed to some extent by two side reactions brought about by traces of impurities. Traces of water in the monomer solution cause a rapid killing of some living ends simultaneously with the

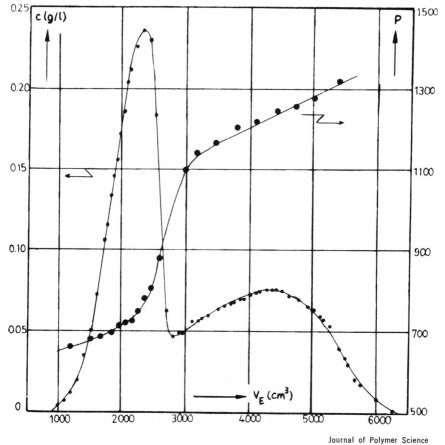

Figure 5.   Concentration, c, and DP as functions of the elution volume: mixture II + III; $q_o = 45$, $q_\infty = 63$ vol % benzene (9)

initiation process. On the other hand, traces of oxygen entering the system at the termination reaction can cause the combination of two growing polymer chains. If the propagation is bivalent (as *e.g.*, if sodium naphthalene is used as initiator), the main peak of the differential distribution curve is accompanied by two smaller peaks at half and double the DP of the main peak (*see* Figure 7).

Detection of this kind of distribution is not difficult, and, moreover, it is helpful for checking the experimental conditions. However, for exact determination of the nonuniformity of the main peak, which alone is significant for the kinetics of the process, these side peaks can cause some trouble. This comes from an uncertainty in the cutting of the fractions. The fractions at the beginning and at the end of the main peak can contain some material of the side peaks, or, if the side peaks are very small, all their material. If this effect is not eliminated, the found nonuniformity of the main peak is enhanced. This effect

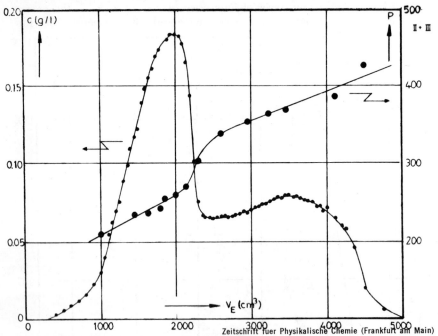

Figure 6.   Concentration and DP as functions of the elution volume: mixture V + VII; $q_o = 55$, $q_\infty = 70$ vol % (1)

can be detected and eliminated by using a GPC column (*10*). The GPC method cannot determine exactly the distribution of the main peak, but it can detect without difficulty a bimodal distribution in which the two maxima lie at DPs differing by a factor of two. Moreover, this method makes it possible to estimate the portions belonging to the different components of the sample.

Figure 8 exemplifies this method. The two end fractions are bimodal and permit the separation of the material belonging to the side peaks. The middle fractions are unimodal and symmetrical. From the corrected end fractions and the middle fractions, the apparent nonuniformity can be calculated by Equation

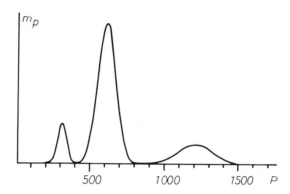

Figure 7. Distribution of an anionic polymerizate with two side maxima

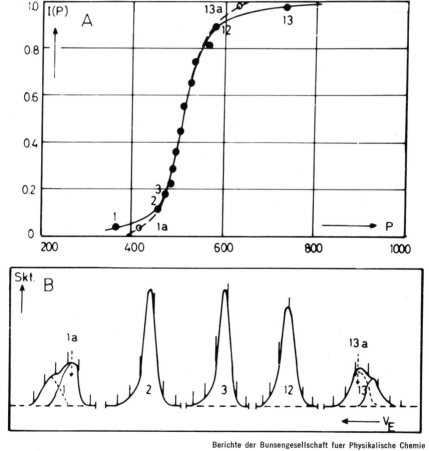

Berichte der Bunsengesellschaft fuer Physikalische Chemie

Figure 8. Correction of the BW fractions of an anionic polymerizate of polystyrene by gel permeation chromatography (10)

9, thereby allowing calculation of the real nonuniformity by Equation 10 with $\overline{U}_{fr} = 1 \times 10^{-3}$ as explained above.

Some Examples Illustrating the Value of MWD Determinations for Kinetic Problems

Rate Constants for the Transitions between the Different Ionic States of the Active Chain End in the Anionic Polymerization of Styrene. The cases treated in this section demonstrate the high information capacity of MWDs which makes this determination extremely useful for solving kinetic problems. In addition, these examples reveal how it is possible to test experimentally determined MWDs by kinetic measurements.

For instance, by the described method one can determine the MWD of polymers which are produced under kinetic conditions for which a Poisson distribution is expected. That is the case in anionic polymerization when only contact on pairs are present and when the dissociation is suppressed by addition of counter ions. Moreover, the reaction time must be equal for all growing molecules. These conditions are realized for the polystyrene samples shown in Figure 9 (10, 11, 12). The curves are calculated for the corresponding Poisson distributions, and the points are the experimental values. For a Poisson distribution, the nonuniformity is given by the equation

$$U_{pois} = 1/\overline{P}_n \tag{12}$$

Of course, in these extreme cases the limitations of the method are easily recognized. That can be seen in Table I. Since the determination of $\overline{P}_n$ involves no difficulties, the calculated nonuniformity is given exactly. The deviations between the values in the two last columns of Table I are smaller than $\pm 1 \times 10^{-3}$ which is less than would be expected from the experimental error.

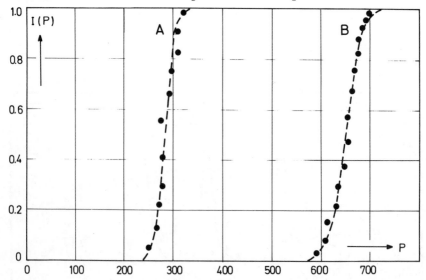

Berichte der Bunsengesellschaft fuer Physikalische Chemie

Figure 9.   Integral distribution functions of polystyrenes synthesized under Poisson conditions in two solvents (10)

– – –, *calculated Poisson distribution;* ●, *experimentally determined value*

The Poisson distribution is only a limiting case in anionic polymerization, realizable only if not more than one kind of active endgroup is in the system. Generally, two or three kinds of active endgroups exist: the contact ion pair, the solvent separated ion pair, and the free anion with propagation constants which differ by some orders of magnitude. The polymerization then proceeds *via* a multistate mechanism with an MWD which is characteristically broader than the Poisson distribution.

Table I.   Experimentally Observed Poisson Distributions

Anionic Polymerization of Styrene (Counter Ion, Na⁺)

Reaction Conditions	$\bar{P}_n$	$U_{app}, \times 10^3$	$U_{real}, \times 10^3$	$U_{pois} = \bar{P}_n{}^{-1}, \times 10^3$
Dioxane, 25°C, no addition of sodium ion	285	2.0	3.0	3.5
Tetrahydropyran, 35°C, addition of $NaB(C_6H_5)_4$	650	1.3	2.3	1.6

When only two kinds of ion pairs are in the system (dissociation being suppressed by addition of counter ions), the polymerization proceeds *via* a two-state mechanism. In this case the nonuniformity is composed of two terms:

$$U_{kin} = U_{pois} + U_{cs} \tag{13}$$

The term U_{cs} was first calculated by Figini (13). His equation can be written in the form (10):

$$U_{cs} = \frac{\bar{k}_{(\pm)}}{k_{cs}} \left(1 - \frac{k_{(\pm)c}}{\bar{k}_{(\pm)}} \right)^2 c^* \frac{2 - X_p}{2 X_p} \tag{14}$$

where $\bar{k}_{(\pm)}$ is the measured overall propagation constant of the ion pairs $k_{(\pm)c}$ is the propagation constant of the contact ion pair, c^* is the overall concentration of living ends, and X_p is the degree of conversion.

The rate constant k_{cs} of the transition from the contact to the solvent separated ion pair is obtainable from Equation 14 since all other quantities on the right side of the equation can be determined by kinetic experiments and U_{cs} can be determined by the described fractionation technique. The kinetic measurements also provide the equilibrium constant K_{cs} between the two ion pairs (14), and so the reverse constant, k_{sc}, may be derived from:

$$k_{sc} = k_{cs}/K_{cs} \tag{15}$$

From Table II one can see that U_{real} increases with decreasing temperature which means that the equilibrium between the two kinds of ion pairs

Table II.   Rate Constants k_{cs} and k_{sc} for the Transitions between the Contact and the Solvent Separated Ion Pairs for Polystyrylsodium in Tetrahydropyran

Sodium Ions Added for Suppressing the Dissociation (12)

T, °C	$U_{app}, \times 10^3$ (Eq. 9)	$U_{real}, \times 10^3$ (Eq. 10)	$U_{pois}, \times 10^3$ (Eq. 12)	U_{cs} (Eq. 13)	k_{cs}, sec^{-1} (Eq. 14)	K_{cs}	$k_{sc} \times 10^{-4}, sec^{-1}$ (Eq. 15)
23	1.3	2.3	1.5	(0.8)	(2.2)	0.3	(7)
1	2.1	3.1	1.9	1.2	1.4	0.8	1.2
−23	5.2	6.2	2.2	4.0	0.8	5	0.26
−43	8.0	9.0	2.2	5.8	0.5	12	0.10

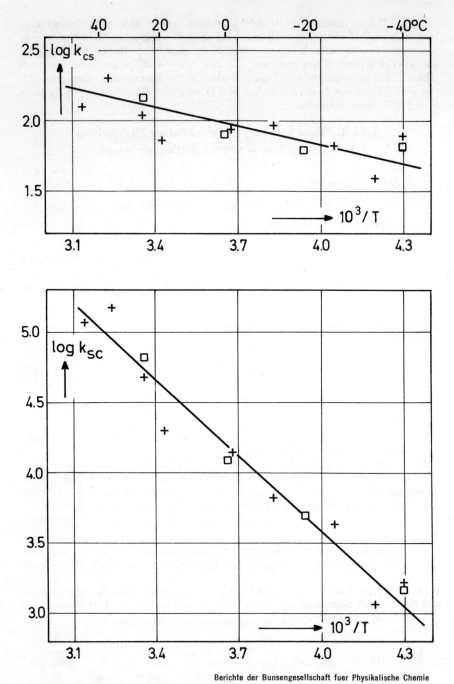

Figure 10. Arrhenius plot of the rate constants for the transition from the contact to the solvent separated ion pair (k_{cs}) and for the reverse reaction (k_{sc}) (polystyrene sodium in tetrahydrofuran with addition of sodium ions) (10)

shifts in the direction of the solvent separated pair (*see* the K_{cs} column). Whereas the equilibrium constant increases with lower temperatures ($\Delta H_{cs} <$ 0), the two rate constants decrease because of the positive activation energy. When experiments of this kind are performed over a greater temperature range, one obtains good Arrhenius lines for the transition rates k_{cs} and k_{sc}. This is shown in Figure 10 for the system polystyrene sodium in tetrahydrofuran (*12*).

By the same method one can determine the dissociation and reassociation rate constants, k_d and k_a (*15*). It may be mentioned that the rate constant for the association of the free ions to the solvent separated ion pair, as determined by the method described here, yields exactly the values which are expected for a diffusion-controlled reaction (*10*).

A Side Reaction in the Anionic Polymerization of Methyl Methacrylate (MMA). As was reported in many papers, the anionic polymerization of MMA is more complicated than that of styrene because different side reactions are involved in the process. To separate these reactions, G. Löhr in our laboratory has tried to determine the polymerization conditions (*e.g.*, solvent, counter ion, temperature) under which the reaction would proceed in the same, relatively simple way as does styrene. These are: low temperature in tetrahydrofuran with Cs⁺ as counter ion (*16*). Under these conditions, one finds a relatively narrow distribution curve of Gaussian type. However, at increasing temperatures a tail towards lower DPs develops which results in complete disappearance of the usual distribution at $-5°C$ (*see* Figure 11).

This effect can be explained by a side reaction consisting of monomer termination by reaction of a living end with the carbonyl group:

$$Mt = metal$$

The strange form of the distribution is the result of the overlapping of two distributions of macromolecules: those which are terminated by the monomer, and those which remain living until the end of the process. The two corresponding distributions are given by Equations 16 (*17*):

$$m(P) = \begin{cases} m_1(P) + m_2(P) & \text{for } P < P_{max} \\ m_2(P) & \text{for } P \geq P_{max} \end{cases} \qquad (16a)$$

$$m_1(P) = \frac{\kappa^2 P}{\alpha X_p} \qquad (16b)$$

$$m_2(P) = \left(\frac{\kappa P}{\alpha X_p}\right) \frac{(1 - \alpha X_p)}{P_{max}(2\pi U_{int})^{1/2}} \exp\left[-\frac{(P - P_{max})^2}{2 U_{int} P^2_{max}}\right] \qquad (16c)$$

where $\kappa = k_{t,m}/k_p$; $\alpha = [M]_o/c_o{}^*$; and $P_{max} = (1/\kappa)\ln(1 - \alpha\kappa X_p)$.

The total distribution and the two partial distributions are plotted in Figure 12. It is not difficult to separate the partial distributions so that the Arrhenius parameters of the side reaction can be obtained. On the other hand, the nonuniformity of the main peak clearly indicates that the non-disturbed propa-

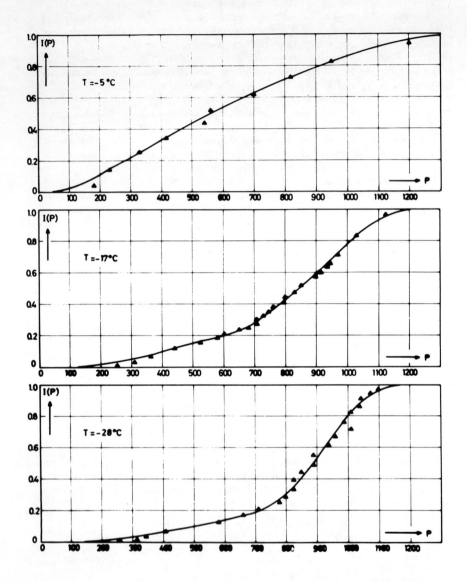

Makromolekulare Chemie

Figure 11.   Integral distribution functions obtained by BW fractionation of anionic poly(methyl methacrylate)s prepared at different reaction temperatures under otherwise identical reaction conditions; initiator, cumylcesium in the presence of about 10^{-3} mole/l cesium triphenylcyanoborate (19)

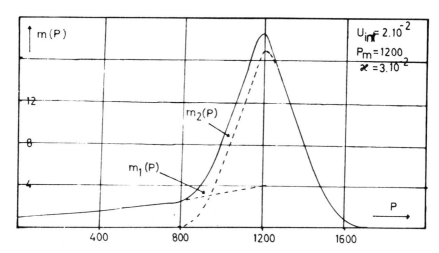

*Figure 12.   Schematic representation of the distribution of a polymer produced in
an anionic polymerization with simultaneous termination by the monomer (see
Equations 16 a, b, and c)*

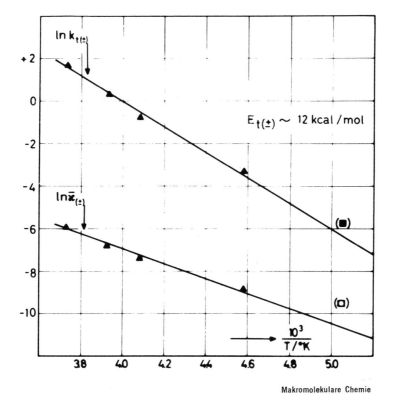

Makromolekulare Chemie

*Figure 13.   Arrhenius plots of the rate constant for the termination re-
action by the monomer (methyl methacryl cesium in THF) (19)*

gation does not proceed *via* a one-state mechanism since its distribution is considerably broader than a Poisson distribution. Until now we are not sure whether this is caused by a two- or a multistate mechanism since one can imagine different positions of the counter ion at the active end group.

From the values of the constant κ derived from the distributions in Figure 11 by the Equations 16a, 16b, and 16c, the Arrhenius lines and activation parameters for the constants $k_{t,m}$ and k_p can be obtained. The Arrhenius lines are plotted in Figure 13. Since in these experiments the dissociation of the active end groups was completely suppressed by additional cesium ions, the constant $k_{t,m}$ can be regarded as the termination rate constant of the ion pairs.

Literature Cited

1. Böhm, L. L., Z. *Phys. Chem. Frankfurt am Main,* in press.
2. Schulz, G. V., Z. *Phys. Chem.* (1939) **B 43,** 25.
3. Löhr, G., Schulz, G. V., Z. *Phys. Chem. Frankfurt am Main* (1969) **65,** 170.
4. Baker, C. A., Williams, R. J. P., *J. Chem. Soc.* (1956) **1956,** 2352.
5. Meyerhoff, G., Romatowski, J., *Makromol. Chem.* (1964) **74,** 222.
6. Schulz, G. V., Deussen, P., Scholz, A. G. R., *Makromol. Chem.* (1963) **69,** 47.
7. Schulz, G. V., Berger, K. C., Scholz, A. G. R., *Ber. Bunsenges. Phys. Chem.* (1965) **69,** 856.
8. Hosemann, R., Schramek, E., *J. Polym. Sci.* (1962) **59,** 13, 29, 51.
9. Böhm, L. L., Casper, R. H., Schulz, G. V., *J. Polym. Sci. Part A-2* (1974) **12,** 239.
10. Böhm, L. L., Löhr, G., Schulz, G. V., *Ber. Bunsenges. Phys. Chem.,* in press.
11. Böhm, L. L., Schulz, G. V., *Makromol. Chem.* (1972) **153,** 5.
12. Komiyama, J., Böhm, L. L., Schulz, G. V., *Makromol. Chem.* (1971) **148,** 297.
13. Figini, R. V., *Makromol. Chem.* (1967) **107,** 170.
14. Barnikol, W. K. R., Schulz, G. V., Z. *Phys. Chem. Frankfurt am Main* (1965) **47,** 89.
15. Schulz, G. V., ADVAN. CHEM. SER. (1973) **128,** 1.
16. Löhr, G., Schulz, G. V., *Makromol. Chem.* (1973) **172,** 137.
17. Löhr, G., Schulz, G. V., *Eur. Polym. J.* (1974) **10,** 121.
18. Löhr, G., *Makromol. Chem.* (1973) **172,** 151.
19. Löhr, G., Müller, A. H. E., Warzelhan, V., Schulz, G. V., *Makromol. Chem.* (1974) **175,** 497.

RECEIVED July 1, 1974.

3

Phase Separation in Solutions of Maleic Acid–Alkyl Vinyl Ether Copolymers

PHILIP LANE and ULRICH P. STRAUSS

Rutgers University, New Brunswick, N.J. 08903

Although the water affinity of weak polyacids usually increases with increasing neutralization, deviations from this behavior were observed with hydrolyzed 1–1 maleic anhydride copolymers of hexyl and octyl vinyl ethers. Solutions of these polyacids in aqueous NaCl were homogeneous at intermediate degrees of dissociation, α, but they became turbid on titration with NaOH provided the NaCl concentration was sufficiently high. The value of α at incipient phase separation decreased with increasing NaCl concentration. Similar results were obtained in LiCl solutions, but substantially higher concentrations of LiCl than of NaCl were needed to produce corresponding effects. No turbidity was obtained with tetramethylammonium chloride (TMACl) or with mixtures of TMACl and NaCl. The results may be interpreted in terms of metal ion–polyanion complexes having specific negative solvent affinities.

In studies on hydrophobic interactions and conformational transitions of selected hydrolyzed maleic anhydride–alkyl vinyl ether copolymers (*1, 2*), we found that under certain conditions neutralization by base caused phase separation in aqueous solutions of the hexyl and octyl copolymers. Both the hexyl and octyl copolymers are insoluble at low pH and require partial neutralization by a base for complete dissolution. Under certain conditions, these copolymers will again precipitate as their degree of neutralization is raised with more base. Insolubility at low pH has been reported for other polyacids (*3, 4*). However, the phase separation at high pH contradicts the wealth of data indicating that neutralization increases the water affinity of polyacids.

Experimental

The synthesis and characterization of the hexyl copolymer (sample C) has been reported (*2*). The octyl copolymer (sample D) was also synthesized and characterized (*5*). The intrinsic viscosities of the two samples in tetrahydrofuran at 25°C were 1.6 and 2.3 dl/g, respectively, which correspond to degrees of polymerization of approximately 1700 and 2700 (*5*).

Dissolution and hydrolysis of these copolymers were effected at room temperature by prolonged shaking with deionized and doubly distilled water

31

to which base had been added. The resulting polymer stock solutions were then mixed with an appropriate electrolyte to produce the starting solutions for the phase separation titrations and for the supporting viscosity studies.

Electrolytes used were NaCl, LiCl, and TMACl. α values for the carboxyl groups are defined as:

$$\alpha = [(MOH) + (H^+) - (OH^-)]/C_p \tag{1}$$

where (MOH), (H^+), and (OH^-) are, respectively, the molarity of added base having in common the cation M^+ of the supporting electrolyte, that of free hydrogen ion, and that of free hydroxyl ion, and where C_p is the concentration of the polyacid in monomoles per liter. (One monomole contains one maleic acid and one alkyl vinyl ether residue.) The values of (H^+) and (OH^-) were obtained by titrating with standard base or acid the electrolyte solutions used as solvents for the polyacids. With this definition, $\alpha = 2$ at complete neutralization of the diprotic acid units.

The phase separation experiments were conducted at 30.00°C by titrating the starting polymer solutions with base until turbid. The value of α at incipient turbidity, α_T, was sharp and reproducible. Back-titration of a turbid solution with hydrochloric acid caused the solution to become clear exactly at α_T, indicating the reversibility of the phase separation effect.

Viscosities were measured with a Cannon-Ubbelhode viscometer at 30.0°C in a bath regulated to within 0.01°C.

Results and Discussion

The results of the phase separation experiments are presented in Table I. The data show that: (1) a minimum concentration of electrolyte was needed for the addition of base to produce turbidity; (2) as the electrolyte concentration was increased beyond this minimum, the value of α_T decreased; (3) higher concentrations of LiCl than of NaCl were needed to obtain turbidity at corresponding values of α; (4) lower concentrations of a given salt were needed to produce turbidity with the octyl than with the hexyl copolymer at corresponding values of α; and (5) at sufficiently high NaCl concentrations, it appeared to be impossible to obtain clear solutions of the hexyl copolymer at any value of α. This may be true for the other systems also, but they have not been investigated.

When base was added to TMACl solutions, no turbidity was observed. Neither was turbidity attainable in a solution of the hexyl copolymer which was 0.2M in both NaCl and TMACl, even though turbidity was observed in solutions 0.2M in NaCl alone. When turbidity was induced with NaOH in the presence of NaCl, the addition of TMACl would clarify the solution.

This phenomenon is highly specific to the nature of the counterion. The general theory of the excluded volume of polyelectrolytes can be summarized by the expression (6):

$$\left(\frac{1}{2} - \chi\right) = \left(\frac{1}{2} - \chi\right)^* + A \frac{i^2}{C_s} \tag{2}$$

where the left-hand side of the equation is a quantity proportional to the excluded volume of the polyelectrolyte, the asterisked quantity is the non-electrostatic contribution, and the last term is the electrostatic contribution to $(\frac{1}{2} - \chi)$ with C_s being the ionic strength, i the effective thermodynamic degree of ionization, and A a positive constant depending on the solvent and the solute. At incipient phase separation, the total excluded volume $(\frac{1}{2} - \chi)$ should have a slightly negative value which, for a given polymer sample, should be

Table I. Turbidity Threshold Conditions at 30°C

Polyacid	Electrolyte	Electrolyte Conc., M	α_τ	pH_τ
hexyl copolymer,	NaCl	0.140	a	a
C_p = 3.15 monomole/l		0.200	1.875	10.22
		0.280	1.762	9.68
		0.400	1.670	9.09
		0.600	1.582	8.40
		0.700	1.575	8.19
		0.800	1.569	8.01
		1.000	1.501	7.59
		1.360	b	b
		1.500	b	b
	LiCl	0.400	a	a
		1.050	1.939	9.34
		1.100	1.889	8.70
		1.550	1.620	7.72
		1.640	1.528	7.41
		2.090	1.437	7.20
octyl copolymer,	NaCl	0.140	a	a
C_p = 3.00 monomole/l		0.200	1.626	9.59
		0.300	1.579	9.24
		0.381	1.446	9.05
		0.428	1.400	8.20
	LiCl	0.400	a	a
		0.850	1.816	9.01
		1.050	1.629	8.08
		1.550	1.349	7.27
		1.640	1.280	7.14
		2.090	1.234	6.74

[a] No turbidity.
[b] Turbid at all values of α.

constant with varying pairs of values of α_τ and its corresponding electrolyte concentration. If we consider any of the systems in Table I, we note that with increasing α_τ the last term in Equation 2 must increase because C_s decreases and i would be expected to increase slightly or remain constant over the range of α_τ values covered. For the left-hand side of Equation 2 to remain constant, the non-electrostatic contribution on the right-hand side must, therefore, become increasingly negative with increasing α_τ.

A way in which this could occur is through specific solvent incompatibility of complexes of alkali metal ions with polyanion groups. The existence of such complexes has been demonstrated by dilatometry (7). Furthermore, since the effective metal ion concentration near the polyion changes only slightly with the total metal ion concentration while the number of available binding sites increases linearly with increasing α_τ in the range under consideration, the number of metal ion complexes would also be expected to increase as α_τ increases. It is noteworthy that potentiometric titration experiments revealed that lithium ion was bound more strongly than sodium ion whereas our data indicate that sodium ion was more effective in producing phase separation. These findings imply that the solvent incompatibility of the sodium complex must be considerably greater than that of the lithium complex. A similar effect was observed with polyphosphates (8, 9). The difference between sodium and lithium ions in their ability to achieve phase separation arises from the larger size of the sodium ion which facilitates its participation in the cross-linking of nonadjacent chain segments.

The solubilizing effect of the TMA$^+$ ion may be explained as follows: The TMA$^+$ ions displaced some metal ions from the ionic atmosphere of the macroion, and, as a result, the number of bound metal ions was reduced. Any condensed (10) TMA$^+$ ions were bound more weakly than the metal ions they displaced, and their effect on the solvent incompatibility of the macroion was much smaller.

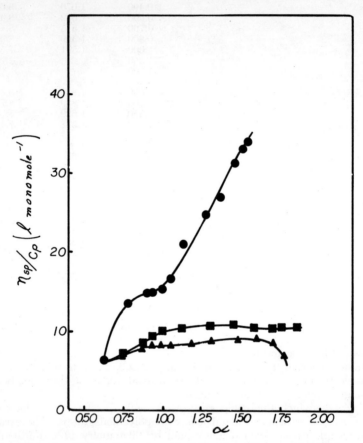

Figure 1. Reduced viscosity of hydrolyzed copolymer of hexyl vinyl ether and maleic anhydride as a function of α in 0.2M electrolyte solutions at 30°C

$C_p = 3.2 \times 10^{-3}$ *monomole/l;* ●, *TMACl;* ▲, *NaCl; and* ■, *LiCl*

These conclusions are supported by the viscosity data plotted in Figure 1. The reduced viscosity of the hexyl copolymer is presented as a function of α in 0.2M solutions of TMACl, NaCl, and LiCl. The polymer concentration of 3.2×10^{-3} monomole/l was low enough to allow interpretation of the results in terms of the molecular dimensions of the polymer molecules. The findings demonstrate strikingly the differences in the effects of the TMA$^+$ ion and the alkali metal ions. Whereas the polyacid showed an enormous expansion with increasing α in the presence of TMA$^+$ ion, this expansion was suppressed almost completely by the alkali metal ions. The difference between the effects of sodium and lithium ion was much smaller; however, it was sufficient to produce phase separation with the former but not with the latter at the electrolyte concentration used in this viscosity study.

What is the effect of potassium ion? On the basis of a limited number of experiments, we can report that with the hexyl and octyl copolymers no phase separation was observed at high α in $0.2M$ KCl solution, and that therefore K$^+$ appears to be less effective than Na$^+$ in promoting phase separation at high α. This finding is not inconsistent with our interpretation of the data for Li$^+$ and Na$^+$. The crosslinking of nonadjacent chain segments depends on both the size of the cation and the strength with which it is bound. Potentiometric titration studies indicate that potassium ion was more weakly bound than sodium ion in the range of α studied. In this range, the binding strength of Li$^+$, Na$^+$, and K$^+$ was inversely related to their size; this appeared to impart a maximum insolubilizing efficiency on sodium ion.

Literature Cited

1. Dubin, P., Strauss, U. P., *J. Phys. Chem.* (1970) **74**, 2842.
2. Dubin, P., Strauss, U. P., *J. Phys. Chem.* (1973) **77**, 1427.
3. Ikegami, A., Imai, N., *J. Polym. Sci.* (1962) **56**, 133.
4. Michaeli, I., *J. Polym. Sci.* (1960) **48**, 291.
5. Dubin, P., Ph.D. Dissertation, Rutgers University, 1970.
6. Flory, P. J., *J. Chem. Phys.* (1953) **21**, 162.
7. Begala, A. J., Strauss, U. P., *J. Phys. Chem.* (1972) **76**, 254.
8. Strauss, U. P., Woodside, D., Wineman, P., *J. Phys. Chem.* (1957) **61**, 1353.
9. Strauss, U. P., Ander, P., *J. Phys. Chem.* (1962) **66**, 2235.
10. Manning, G., *J. Chem. Phys.* (1969) **51**, 924.

RECEIVED July 22, 1974.

4

Microstructure of Two-Phase Poly(Methyl Methacrylate)s

R. P. KUSY and D. T. TURNER

Dental Research Center, University of North Carolina,
Chapel Hill, N. C. 27514

Glassy polymers prepared from proprietary mixtures of a polymeric powder and a liquid monomer are used in dental and other clinical applications where replacements are needed for hard tissue. Subsequent polymerization yields a two-phase structure in which the dispersed phase is carried over from the powder. Structure was revealed by optical microscopy of a plane surface formed by fracture or polishing. Quantitative microscopy provided information about volume fraction of the dispersed phase and average grain diameter. Trans- and intergranular fractures were examined by fractography and crack propagation studies. Material with intergranular fracture had lower tensile strength than materials with exclusively transgranular fracture. Poor bonding between the grains and the matrix probably resulted from rapid polymerization with insufficient time for diffusion of monomer into the powder.

For clinical applications in dentistry, it is convenient to work with mixtures made from a polymeric powder and a liquid monomer. The mixture attains a doughy consistency, and it can then be formed into intricate shapes. Subsequently, as a result of the combination of reactants in the powder and liquid, the monomer polymerizes to provide a rigid load-bearing material. This is usually a two-phase material in which the dispersed phase is attributable to particles originally present in the powder. Many proprietary materials comprise two phases of poly(methyl methacrylate), although other ingredients may be included to modify appearance and properties.

The microstructure of this type of material was studied as early as 1952 by Fischer and Isenbarth (1). These authors demonstrated by thin section transmission microscopy that a two-phase microstructure was present in a number of materials available commercially for restoring tooth structure. In 1955 Helmcke reported similar findings with electron microscopic studies of fracture replicas (2, 3). Then in 1958, Smith examined at low magnification fracture surfaces in materials made from denture base polymers; attention centered on a system of ridges concentric with a mirror region (4). In retrospect, this phenomenon was similar to that observed in one-phase samples of poly(methyl methacrylate) (5). Subsequently, in 1961, Smith showed that the microstructure of these materials could be revealed by reflected light after the surface was treated with nitric acid (6).

Recently, polyphase acrylic polymers have become increasingly important. They are the preferred materials for restoring anterior teeth where esthetic considerations are paramount. However, they are not satisfactory for the restoration of posterior teeth, especially in load-bearing locations where mechanical properties, such as abrasion resistance, are important (7). These materials are also used to make hip prostheses, and they are finding increasing application in orthopedics (8).

The immediate objective of this work was to characterize quantitatively the microstructure of polyphase acrylic polymers. The long-term aim was to relate microstructure to selection of materials, processing conditions, and service properties. This report is confined to preliminary observations on crack propagation and tensile strength. More detailed accounts are being published elsewhere (9).

Experimental

The following materials which are commercially available for making or repairing denture bases were studied: I — flash acrylic (Yates Manufacturing Co., Chicago, Ill.), II — repair resin, and III — Lucitone (II and III both from L. D. Caulk Co., Milford, Del.). The powder (generally 19 g) was added to a liquid monomer (10 ml) and mixed according to the manufacturer's instructions. The resulting doughs were pressed into specimen plates (65 × 62 × 5 mm) in accordance with American Dental Association specification no. 12. Material III was maintained under pressure for 16 hrs at 70°C. Materials I and II were treated similarly, but without heat. Dumbell samples of ½" gauge length were cut by band saw and polished with emory paper. The samples were placed in distilled water for one week at 37°C. They were then allowed to equilibrate in air for several months before testing. A few fracture surfaces were also examined with an ETEC scanning electron microscope.

Tensile strength was measured on an Instron machine at a strain rate of 10^{-2}/sec. Fracture surfaces were examined directly by reflected light using a Zeiss orthoplan microscope.

Surfaces were polished by standard metallographic techniques to a $1/4\mu$ finish. Microstructure was revealed by the following techniques: 1) heating at 100°C (5 min), 2) contact with concentrated nitric acid (15 sec), 3) contact with fumes from concentrated nitric acid (1 min), 4) dropwise addition of ethanol (5 min), or 5) dropwise addition of acetone–benzene–toluene–xylene (40:20:20:20 v:v).

Cracks were propagated in a sample with a prepolished surface by making a saw cut and then driving in a wedge, either by turning a screw or by impact. After fracture, the microstructure of the polished surface was revealed by exposure to nitric acid fumes.

Results and Discussion

Quantitative Characterization of Microstructure. Material I had a featureless surface when it was carefully polished, apart from occasional features resulting from porosity. When it was heated near the glass transition temperature (100°C), the surface became dull and was pimply when viewed at room temperature (Figure 1). This presumably corresponds to the orange-peel appearance described by Smith and attributed to the relief of localized stress concentrations around each granule (6). Comparison of the number, size, and shape of the grains in Figure 1 with the pearls originally present in the powder indicates that the annealing method provides a good qualitative guide to micro-

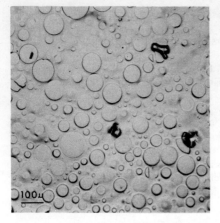

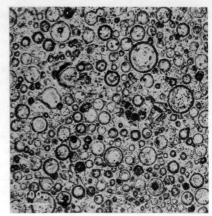

Figure 1.   Photomicrograph of material I
after heat treatment at 100°C

Figure 2.   Photomicrograph of material I
after immersion in ethanol

structure. However, the findings were not suitable for quantitative analyses which relate features of a planar surface to a three-dimensional distribution.

Immersion of a polished surface of material I in ethanol for 5 min resulted in a surface covered with a powdery, white debris; this gave a darker photograph with many artifacts (Figure 2). A cleaner surface essentially free of artifacts was obtained by the dropwise addition of ethanol which was allowed to run off the sample, thereby preventing the accumulation of debris (Figure 3). Similar results were obtained with material II using either alcohol (Figure 4) or the mixture of organic liquids. Contact with liquid nitric acid caused over-etching of the surface, but exposure to the fumes gave excellent results (see Figures 8 and 9).

There was an interesting difference between the grains in Figures 3 and 4. In Figure 3, some of the grains of material I were in various stages of disintegration. The grains marked A exemplify dissolution leading to a feature that

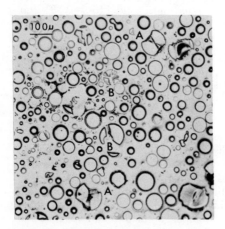

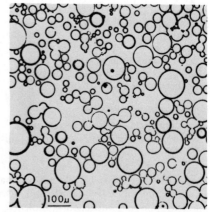

Figure 3.   Photomicrograph of material I
after dropwise addition of ethanol

Figure 4.   Photomicrograph of material
II after dropwise addition of ethanol

resembled an apple core; other grains appeared to fracture (B). These features were not observed in Figure 4, presumably because polymerization of material II proceeded so rapidly that little or no dissolution of the grains could occur. In fact, the mixture did harden within a few minutes after preparation.

Figure 4 was analyzed by quantitative microscopy. The volume fraction of spherical grains (V_f) was estimated from the fraction of grid intersections which superpose on the circles of Figure 4. A number average particle diameter, $\bar{d}$, was estimated from the equation $\bar{d} = 3V_f/N_i$ where N_i is the average number of interfaces between the two phases per unit length. N_i was estimated by reference to a $15°$ grid. A maximum value for the sphere diameters (d_{max}) was taken as the largest diameter of the circles in Figure 4, but this was probably an underestimate. In Table I, these quantities are compared with values determined directly from the powder $(\bar{d}$ and $d_{max})$ and the made-up volume fraction of powder in the initial mixture. This latter volume was calculated from the equation $V_f = W_p/(W_p + W_L)$, where W_p and W_L are the weights of powder and liquid in the initial mixture, respectively. This equation entailed two main assumptions: 1) that the powder was comprised solely of spheres, and 2) that conversion of monomer to polymer was complete. Microscopic examination indicated that the first assumption is approximately correct. Furthermore, it is safe to assume $>90\%$ conversion (10). Therefore, this approximation is not critical in making the point that a relatively small portion of the powder dissolves in the monomer to provide the matrix phase (about $1/3$). Moreover, as the average particle size nearly doubled in the final product, it may be concluded that the smaller spheres were preferentially dissolved. Consistent with this, the values of d_{max} indicate that the largest spheres did not decrease greatly in diameter.

Table I.   Comparison of Spheres in the Initial Mixture and in the Final Product

Material	V_f	$\bar{d}, \mu$	d_{max}, μ
Initial mixture	0.60	<20	160
Final product	0.44	42	140

Fractography. Fracture surfaces provide information about microstructure and the mechanism of failure. Generally, however, the commercial, 2-phase acrylic polymers used for dental applications form surfaces which are too un-

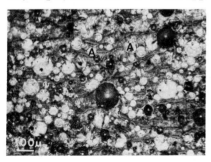 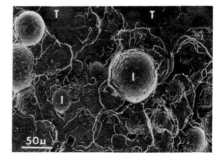

Figure 5.   Intergranular and transgranular defects in material II
a. *Fractograph (black circles, intergranular and white circles, transgranular failures)*
b. *Scanning electron micrograph (T, transgranular and I, intergranular fractures)*

even to be examined successfully by reflected light. This is so with material I; material II, however, gives a suitable fracture surface (*see* Figure 5a). The white circles are spheres which were fractured transgranularly and hence reflect a larger fraction of incident light, whereas the dark circles are spherical protrusions or depressions caused by intergranular fracture between the spherical particles and the matrix. This latter identification was carefully checked by matching the two fracture surfaces (9). Another feature of Figure 5a is the long features (*see* A-A, for example) which extend from left to right. It was argued, in the case of poly(methyl methacrylate) alone, that these are formed by a stick–slip fracture mechanism (11). We observed that this mode was largely confined to the matrix. From this, together with evidence of intergranular fracture, it was concluded that the bonding between the matrix and the spheres was weak.

Fracture surfaces of material II were also examined by scanning electron microscopy, and the features noted were consistent with the occurrence of trans- and intergranular fracture (Figure 5b).

Material III was particularly interesting because only transgranular fracture was observed. Near the flaw where fracture initiated was a smooth, mirror region (Figure 6), and farther away was a rougher, mist region (Figure 7). The parabolic markings were common on the fracture surface of 1-phase poly(methyl methacrylate). In material III, the bonding between the spheres and the matrix was sufficiently strong to prevent intergranular fracture.

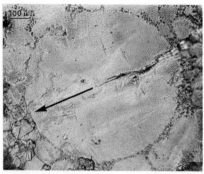

Figure 6. Fractograph of material III showing the fine features of a large particle in the mirror region

Note the fibrous appearance. Arrow indicates the direction of fracture propagation.

Figure 7. Fractograph of material III showing the fine features of a large particle in the mist region

Note the parabolic markings. Arrow indicates the direction of fracture propagation.

Crack Propagation and Tensile Strength. The incidence of trans- and intergranular fracture was demonstrated by crack propagation experiments. For example, material I, which did not provide a suitable surface for fractographic studies, did fracture transgranularly (Figure 8). Incidentally, the irregular surface created by the crack might account for the lack of resolution by optical microscopy. Figure 9 confirms that material II fractured both trans- and intergranularly. The microstructure in Figures 8 and 9 was revealed by exposure to nitric acid fumes.

Finally, the tensile strengths of materials I, II, and III were determined (Table II). All the two-phase materials were weaker than one-phase poly-(methyl methacrylate). However, only the weakest of the two-phase materials

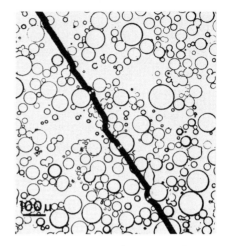

Figure 8. Transgranular crack propagation in material I
Surface etched with nitric acid fumes

Figure 9. Trans- and intergranular nature of crack propagation in material II
Surface etched with nitric acid fumes

(material II) exhibited intergranular fracture, and this was because of weaker bonding between the matrix and the spherical particles. One factor leading to weak bonding is probably the rapid rate of polymerization which allows insufficient time for monomer penetration into the spheres. This effect was predicted by Fischer and Isenbarth (1).

Table II.   Tensile Strength[a]

Sample	Tensile Strength, dyne/cm² × 10⁸
Poly(methyl methacrylate)[b]	7.0 ± 0.3
Material I	4.5 ± 0.4
Material II	3.5 ± 0.3
Material III	4.9 ± 0.6

[a] Average of 6 specimens.
[b] Plexiglas from Rohm and Haas.

Conclusions

The microstructure of commercially available, two-phase acrylic polymers can be revealed by various treatments of polished surfaces. These phases can be characterized quantitatively.

Fractography and crack propagation studies can differentiate between trans- and intergranular fracture in polyphase acrylics. A two-phase acrylic which exhibited intergranular fracture had a lower tensile strength than ones which fractured only transgranularly.

Future Work

A deficiency of this study is that the composition and structure of the proprietary materials studied was not known in sufficient detail. It would be desirable to use carefully characterized starting materials.

Literature Cited

1. Fischer, C. H., Isenbarth, R., *Deut. Zahn. Mund. Kieferheilk.* (1952) **17**, 112.
2. Helmcke, J. G., *Deut. Zahnaerztl. Z.* (1955) **10**(3), 1269.
3. *Ibid.* (1955) **10**(4), 1885.
4. Smith, D. C., *Nature London* (1958) **182**, 132.
5. Wolock, I., Kies, J. A., Newman, S. B., "Fracture," B. L. Averbach *et al.*, Eds., pp. 250-264, Wiley, New York, 1959.
6. Smith, D. C., *Brit. Dent. J.* (1961) **111**, 9.
7. Leinfelder, K. F., private communication.
8. Charnley, J., "Acrylic Cement in Orthopaedic Surgery," Churchill Livingstone, London, 1972.
9. Kusy, R. P., Turner, D. T., *J. Dent. Res.* (1974) **53**, 520.
10. "Guide to Dental Materials and Devices," 5th ed., p. 88, Amer. Dent. Ass., 1970-71.
11. Andrews, E. H., "Fracture in Polymers," pp. 177-198, American Elsevier, New York, 1968.

RECEIVED March 29, 1974. Work supported by PHS research grant no. DE02668 from the National Institute of Dental Research, in part by general research support grant no. RR5333 from the General Research Support Branch of the National Institutes of Health, and in part by the Materials Research Center, University of North Carolina, under grant no. GH33632 from the National Science Foundation.

Aspects of Liquid–Liquid Phase Transition Phenomena in Multicomponent Polymeric Systems

LEE P. McMASTER

Union Carbide Corp., Bound Brook, N. J. 08805

Mechanisms of liquid–liquid phase separation were studied in the binary styrene–acrylonitrile copolymer/poly(methyl methacrylate) system. Evidence is presented which suggests that spinodal decomposition occurs in this system. The Cahn theory provides an interpretation of key experimental results. Both a dispersed, two-phase structure and a highly interconnected, two-phase structure can be formed. These two structures coarsen at significantly different rates. The dispersed-phase structure coarsens by Ostwald ripening, an extremely slow process in the polymer–polymer system. The interconnected structure coarsens more rapidly. Data suggest that the mechanism of coarsening is viscous flow driven by interfacial tension.

Recognizing that polymer–polymer systems exhibit phase behavior similar to that of other condensed phase systems, it should be possible to prepare multiphase polymeric materials of unique structure by phase transition. The purpose of this work is to show that much of the theoretical and experimental analysis of phase transitions in metallurgical and ceramic systems can provide an interpretation of the behavior of polymeric systems. In addition, there are several features of polymeric systems which distinguish them from low molecular weight systems.

Two alternate mechanisms of liquid–liquid phase transition are possible in multicomponent condensed phase systems (1). These two mechanisms are nucleation and growth and spinodal decomposition. The former mechanism is associated with a meta-stable equilibrium of the homogeneous system. The system is stable to continuous composition changes, but it is unstable to sufficiently large composition fluctuations. A sufficiently large composition fluctuation is called a nucleus which grows by a normal diffusion-controlled growth process. The limit of metastability is the spinodal. Within the spinodal, the system is unstable with respect to continuous composition changes. It was shown by Cahn (1, 2) that such continuous composition changes will be preferred in this composition region and that phase separation will proceed spontaneously by an alternate mechanism called spinodal decomposition. One

For the monodisperse polymer–polymer case, the binodal and spinodal curves define three regions of thermodynamic stability. Regions outside the binodal are thermodynamically stable to both small and large amplitude composition fluctuations. The lowest free energy state is that of a homogeneous system. The region between the binodal and spinodal is a thermodynamically metastable region. Although the system is stable to small amplitude composition fluctuations, phase transition can occur *via* a sufficiently large amplitude composition fluctuation. In this region, an initially homogeneous sample phase separates by a nucleation and growth mechanism (*1*). The unstable region is located entirely within the spinodal. In this region, phase separation can proceed *via* continuous growth of small amplitude composition fluctuations. In fact, observation of continuous growth of certain small amplitude composition waves constitutes a proof that the spinodal has been crossed. Spinodal decomposition is the dominant mechanism of phase separation. The boundary between the metastable and unstable region is given by:

$$\frac{\partial^2 f}{\partial \phi^2} = 0 \qquad (2)$$

where f = free energy per unit volume, and ϕ = component volume fraction. In the metastable region, $\frac{\partial^2 f}{\partial \phi^2} > 0$; in the unstable region, $\frac{\partial^2 f}{\partial \phi^2} < 0$.

Figure 1 illustrates the competitive nature of nucleation and growth and spinodal decomposition as mechanisms of phase transition. Consider a homogeneous sample of composition far from the critical composition. The sample can be phase separated by raising its temperature to thrust it into the two-phase region. At compositions far from the critical point, the sample has to pass through a large metastable region before it can pass into the unstable region. Because of the nature of nucleation kinetics, it may not be feasible to heat the sample rapidly enough to suppress phase separation by nucleation and growth before the sample reaches the unstable region. Hence, far from the critical composition, nucleation and growth is the expected dominant mechanism of phase separation.

On the basis of the classical theory of nucleation and growth (*11*), phase decomposition within the metastable region is expected to proceed by formation of a critical nucleus and subsequent growth of the nucleus into a spherical droplet whose composition remains invariant during the growth process. In the absence of significant coalescence of disperse phase droplets, further growth of these droplets after achievement of bulk thermodynamic equilibrium is governed by the rate of Ostwald ripening (*12*). Hence, a highly dispersed two phase structure is anticipated.

On the other hand, near the critical composition, the binodal and spinodal merge so that the metastable region reduces considerably in size. In this case, it is easier to pass directly into the unstable region by a thermal excursion before phase separation begins. Once in the unstable region, spinodal decomposition is the mechanism of phase separation.

The Kinetics of Spinodal Decomposition. Cahn's kinetic theory of spinodal decomposition (*2*) was based on the diffuse interface theory of Cahn and Hilliard (*13*). By considering the local free energy a function of both composition and composition gradients, Cahn arrived at the following modified linearized diffusion equation (Equation 3) to describe the early stages of phase separation within the unstable region. In this equation, Ω is an Onsager-type

$$\frac{\partial \phi}{\partial t} = \Omega \left(\frac{\partial^2 f}{\partial \phi^2} \nabla^2 \phi - 2K \nabla^4 \phi \right) \tag{3}$$

phenomenological coefficient representing proportionality between a flux and a gradient in chemical potential. The coefficient, K, is the gradient energy coefficient arising from the effects of localized composition gradients (13). This equation resembles the normal diffusion equation. The second term in the parentheses is significant only when diffusion distances are relatively short. By comparing the first term with the conventional diffusion equation, the diffusion coefficient is seen to be equivalent to:

$$D \sim \left| \Omega \frac{\partial^2 f}{\partial \phi^2} \right| \tag{4}$$

The absolute value is necessary for the diffusion coefficient to be positive since $\partial^2 f / \partial \phi^2 < 0$ in the unstable region. Solution of the Cahn equation has the form (2),

$$\phi - \phi_o = \exp[R(\beta)t] \cos (\beta \cdot X), \quad \lambda_i = \frac{2\pi}{\beta_i} \tag{5}$$

where the growth rate $R(\beta)$ is given by,

$$R(\beta) = - \Omega \beta^2 \left\{ \frac{\partial^2 f}{\partial \phi^2} + 2K \beta^2 \right\} \tag{6}$$

Because K is restricted to non-negative values (2, 14), the sign of $R(\beta)$ is governed by $\partial^2 f / \partial \phi^2$. Negative values yield positive values of $R(\beta)$ for some wave numbers. Hence, phase separation should proceed spontaneously within the spinodal at a scale governed by those values of β that yield a positive growth rate. In fact, because of the exponential nature of the growth rate, the scale should be dominated by the most rapidly growing wave length,

$$\lambda_m = \frac{2\pi}{\beta_m} = 2 \sqrt{2} \pi \left[- \frac{1}{2K} \frac{\partial^2 f}{\partial \phi^2} \right]^{-1/2} \tag{7}$$

The terms within the brackets of Equation 7 can be used to estimate the scale of phase separation (6). Flory-Huggins theory (15) gives:

$$\frac{\partial^2 f}{\partial \phi^2} = RT \left[\frac{1}{V_1 \phi_1} + \frac{1}{V_2 \phi_2} - \frac{2\chi_{12}}{V_1} \right] \tag{8}$$

With $\chi_{12} = \chi_{12,s} + \chi_{12,h}/T$, Equation 8 becomes,

$$\frac{\partial^2 f}{\partial \phi^2} = \frac{2R \, \chi_{12,h}}{V_1} \left(\frac{T - T_s}{T_s} \right) \tag{9}$$

where T_s = temperature at the spinodal.

By a derivation of the free energy of an inhomogeneous system somewhat analogous to that of Cahn and Hilliard, Debye (16) has shown:

$$K = \left(\frac{RT \, \chi_{12}}{V_1} \right) \frac{l^2}{6} \tag{10}$$

In this equation, l is a range of molecular interaction. For low molecular weight liquid–liquid and vitreous glass systems (16, 17), l is 5–20 A (10 A is a typical value). For polymer solutions, Debye has shown that $l = \bar{R}/\sqrt{6}$ where $\bar{R}$ is the root mean square end-to-end distance of the polymer chain. Similarly, for polymer–polymer systems, $l = \bar{R}/\sqrt{3}$ on the basis of a derivation analogous to that given by Debye for polymer–solvent systems. In the latter case,

The scale of phase separation can also be estimated using the Flory equation of state thermodynamics (*see* Ref. *10* for details of the analytic expressions for $\partial^2 f / \partial \phi^2$ and χ_{12}). With appropriate expressions for $\partial^2 f / \partial \phi^2$ and χ_{12}, Equation 10 can be substituted into Equation 7 and evaluated numerically. Some typical values for sets of parameters originally used by McMaster are shown in Figure 2. For all cases considered, the scale of phase separation is somewhat larger than the Flory-Huggins value (with $\chi_{12,s} = 0$). In no case was the scale of phase separation more than *ca.* 2.5 times the Flory-Huggins scale. Similar results are obtained if non-zero values of $\chi_{12,s}$ are used in Equation 11. Hence, Equation 12 provides a reasonable lower bound on the scale of phase separation.

An interesting morphological characteristic of spinodally decomposed systems is the unique periodicity and high level of phase interconnectivity observed experimentally and predicted by the general solution to Equation 3. By taking the full infinite series solution to Equation 3 and using random numbers to generate an initial three dimensional distribution of small amplitude composition fluctuations, Cahn (*2*) showed that a highly interconnected three dimensional network of characteristic size, λ_m, was predicted so long as the minor phase constituted at least 15% of the volume. For volume fractions less than 15%, the minor phase transformed to an isolated structure.

The Cahn theory predicts that composition changes gradually in both directions from the mean. Particle size during the earliest stages of phase separation is invariant; rather, the amplitude of the composition difference gradually increases. There is no sharp interface between composition extremes during early stages of phase separation. It is only later in phase separation that nonlinear terms omitted from Equation 3 cause a sharpening of phase boundaries and a leveling off of composition extremes (*21*). At the same time, coarsening of the structure begins. This coarsening starts before composition equilibration has been reached. Coarsening of this type has been observed in both metallurgical solid solutions and in vitreous glass systems by small angle x-ray scattering techniques (*18, 19*).

After phase equilibration has been reached, the high level of phase interconnectivity provides a coarsening mechanism unavailable to the dispersed phase structures. This mechanism is viscous flow driven by interfacial tension. One objective of the experimental program was to investigate particle coarsening characteristics of both dispersed and interconnected structures.

Experimental Technique

The system styrene–acrylonitrile copolymer (SAN) 28% acrylonitrile/poly(methyl methacrylate) exhibits thermodynamic solubility relationships adequate for studying phase transition phenomena. The molecular weight properties of the polymers used in this study (Table III) were measured by gel permeation chromatography. The cloud-point curve for binary mixtures of these two polymers was determined by a technique developed previously (*10*).

Table III. Molecular Weight Distribution Characteristics of Experimental Polymers

Polymer	Commercial Code	M_n	M_w	M_z	a_i	a_1/a_2
Styrene–acrylonitrile	Union Carbide RMD–4511 (28% AN)	88,600	223,200	679,600	3.05	1.93
Poly (methyl-methacrylate)	DuPont Lucite 140	45,600	92,000	145,000	1.58	—

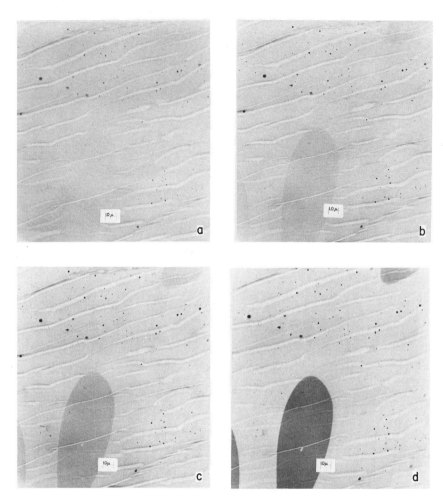

Figure 4.   TEM of phase separated sample as a function of irradiation time
25% SAN/75% PMMA. Irradiation time, sec: a, 15; b, 30; c, 45; and d, 60.

Briefly, thin 1–2 mil films of the polymers were cast onto microscope slides from 1,2-dichloroethane. The films were observed at a low angle of back or forward scattering with a microscope illuminator. Film temperature was raised (lowered) at 0.2°C/min or less until the first (last) faint cloudiness in the film appeared (disappeared). Adjustment of the data to obtain the cloud-point curve are described by McMaster (*10*).

The phase transition characteristics of two polymer concentrations were studied experimentally, 25% SAN near the critical point and 75% SAN far from the critical point. Homogeneous mixtures of the two polymers were prepared by milling at 140°C. Homogeneous, 25-mil films of the two concentrations were prepared by compression molding at 150° and 210°C, respectively; films prepared at these temperatures were homogeneous, single-phase specimens. Samples were then transferred without removing them from the platens to a second press at a higher temperature. Below 220°C, the second press was heated by condensing steam; above 220°C, by electric heaters. Steam

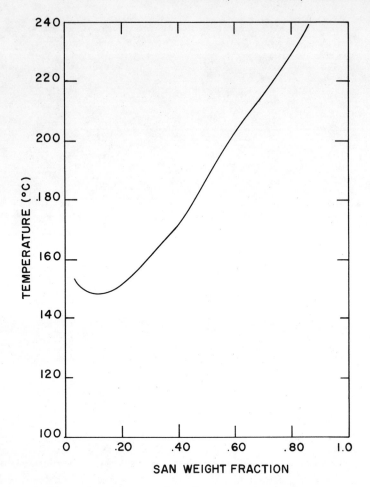

Figure 5.   Experimental cloud point curve (SAN/PMMA)

heating provided more uniform spatial temperature distribution, but adequate results were obtained with electric heating if samples were taken from the same relative position in the film. The time–temperature history of the sample was recorded by a thermocouple embedded within the polymer film. After a specified time interval, the film was quenched with cooling water. Once the sample passed through its glass transition temperature (105°C), the phase structure was frozen. Two typical time–temperature histories are shown in Figure 3. Time at temperature is the total time the sample was within 10°C of the equilibrium temperature. Although an assumption of this sort is arbitrary, the resultant error was inconsequential except for very short total cycles (less than 3 min); even in these cases, the rate of phase separation was so temperature sensitive that phase separation occurring at lower temperatures was usually insignificant.

Visual inspection of films which had been heat-treated above the binodal revealed that phase separation had occurred to varying degrees depending on the nature of the heat treatment and the composition of the film. Some effort was made initially to stain the films with several chemical stains. These efforts were unsuccessful, but it was found that the phase domains could be observed by transmission electron microscopy (TEM) without staining. The following procedure was used.

The films were microtomed to 150 A on a Reichert OMUZ ultramicrotome. Samples were then mounted on a copper grid and placed in an RCA EMU3 transmission electron microscope. Fixed electron beam intensity was used. Phase-separated samples initially appeared homogeneous, but they gradually developed when exposed to a constant beam intensity for increasing time until maximum contrast between the phases was observed. This development process is illustrated in Figure 4(a–d); a phase-separated sample was exposed to the electron beam for 15, 30, 45, and 60 sec. The mechanism of development is not understood, but this does not appear to be an artifact. Note in Figure 4 that the structure does not change during the exposure process. The contrast gradually increases but the structure remains fixed. The process seems to manifest itself as a gradual increase in the scattering cross-section of the SAN-rich phase—perhaps by selective reaction of the SAN-rich zones with the

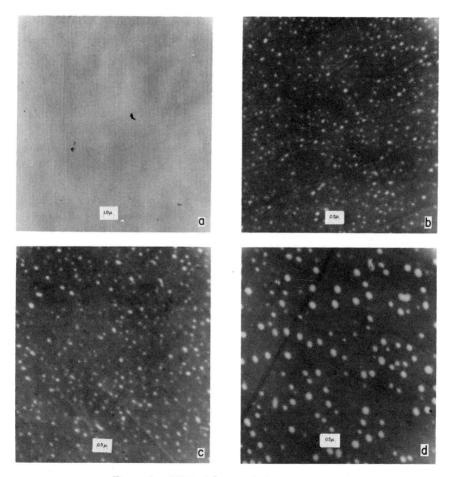

Figure 6.   TEM of dispersed phase structures
75% SAN/25% PMMA; T = 265°C. Time, min: a, control; b, 3.4; c, 12.4; and d, 122.4.

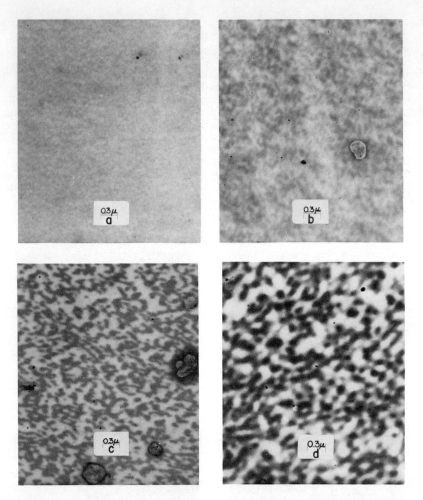

Figure 7.   TEM of stages of phase esparation for interconnected structures
αtβ)
25% SAN/75% PMMA; T = 180°C. Time, min: a, 3.4; b, 4.4; c, 13.4; d, 32.4; and e, 62.4.

vacuum oil under the irradiating conditions. Irrespective of the mechanism, all examinations of the samples by other techniques verified the structure. In particular, the same structures were observed by optical microscopy and visual inspection of coarsened structures possessing phase domains of 50–500μ.

Experimental Results

The molecular weight properties of both polymers are presented in Table III. Figure 5 is the experimental cloud-point curve. The system was particularly sluggish because of reduced mobility of the components in the lower temperature range, so precision is somewhat diminished. According to the molecular weight distribution statistics and computational results like those shown in Figure 1, the critical point should be shifted approximately 10% up the

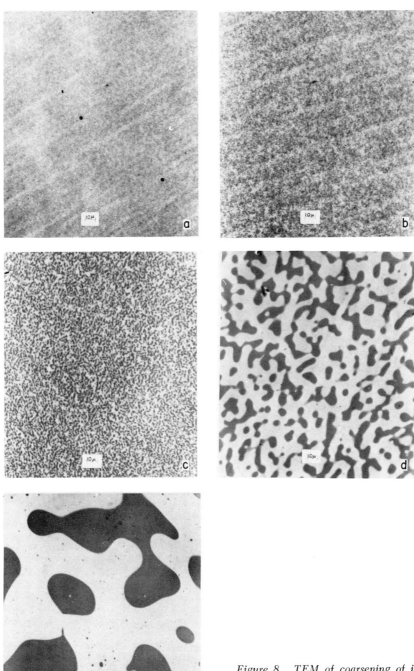

*Figure 8.   TEM of coarsening of in-
terconnected structures*
25% SAN/75% PMMA; T = 210°C. Time,
min: a, 1.9; b, 2.4; c, 6.4; d, 31.4; and e,
241.4.

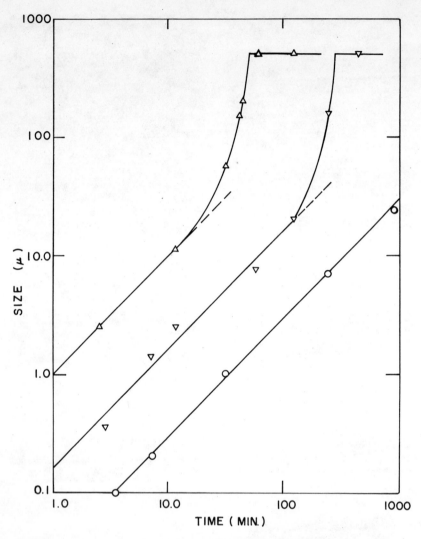

Figure 9. Coarsening of interconnected structures (particle size vs. time; $\xi = \alpha t\beta$)

Symbol	T, °C	α	β
O	210	0.0295	1.0
▽	250	0.150	1.0
△	280	1.00	1.0

right ascending branch of the cloud-point curve. This would place the critical concentration at 20–25% SAN.

Some typical results of TEM examination of the 75% SAN composition are presented in Figure 6(a–d). Figure 6a shows the initially homogeneous sample of 75% SAN composition. Succeeding photographs show the same composition after exposure times of 3.4, 12.4, and 122.4 min at 265°C. The structure is obviously a dispersed phase structure of PMMA-rich droplets in an SAN-rich matrix. Similar structures were obtained with heat treatment over the entire temperature range studies (250°–315°C). It was not practical to observe formation of the phase structure during the phase separation process because of the fairly high mobilities of the polymers in this temperature range.

TEM micrographs of the 25% SAN composition treated at 180°C are presented in Figure 7. Specks of debris should not be confused with the phase separating structures. These heterogeneities are dirt particles which were in the polymers originally or were introduced during milling. The initially homogeneous specimen is not pictured since it resembles Figure 6a. Note that there is no sharp interface in samples given shorter heat treatments. Small amplitude composition modulations are apparent, and their size seems relatively constant for times up to 4.4 min (Figure 7a). For longer times, the interface between the two phases sharpens, composition differences continue to grow, and early stages of particle coarsening can be observed. The scale of phase separation can be estimated from Figure 7a. The size of the domains in these early stages of phase separation is approximately 500–1000 A.

Phase interconnectivity is obvious in Figure 7, but it is shown even more dramatically in Figure 8, a similar TEM sequence at 210°C. Bulk thermodynamic equilibrium was reached much sooner, and after achieving equilibrium, structure coarsening was far greater. After the 2.4-min heat treatment, all samples had reached bulk equilibrium. But structure coarsening continued beyond the stage shown in Figure 8e until breakdown in interconnectivity began when particle size approached the thickness of the film sample. An overall increase in particle size of 10,000 occurred after incipient phase separation up to the point of complete breakdown of phase interconnectivity.

Coarsening characteristics are illustrated at three temperature levels in Figure 9 where the characteristic dimension of the minor phase is plotted as a function of time. Size is proportional to time until a diameter of *ca.* 20μ is reached; that is, the growth rate is constant. For sizes greater than 20μ, the coarsening rate increases dramatically. In fact, when the data are replotted on semi-logarithmic paper (Figure 10), the growth rate is seen to be exponential. The particles finally cease to grow, breakdown in interconnectivity of the structure is complete, and one is left with roughly spherical droplets of minor phase dispersed in the major phase. Droplet size at point of complete breakdown in interconnectivity appears to be related to the minimum dimension of the polymer film. In Figures 9 and 10, film thickness was approximately 600μ and final droplet size was approximately 500μ. When film thickness was doubled, final droplet size also doubled.

Interpretation of Results

The Cahn theory provides a selfconsistent interpretation of the experimental results obtained in this study. It is not the only possible interpretation, but it does provide logical explanations for all observations. Hence, it is invoked in interpreting the results. Complete verification will require further experimentation.

The phase structures in Figure 6 were obtained from a 75% SAN composition. Because this composition is far from the critical composition, one should expect unequal distribution of phase volumes. In fact, the minor phase constitutes approximately 10% of total volume. Hence, it is not possible to prove conclusively that one particular mechanism of phase transition produced the observed structure. This structure could have formed by nucleation and

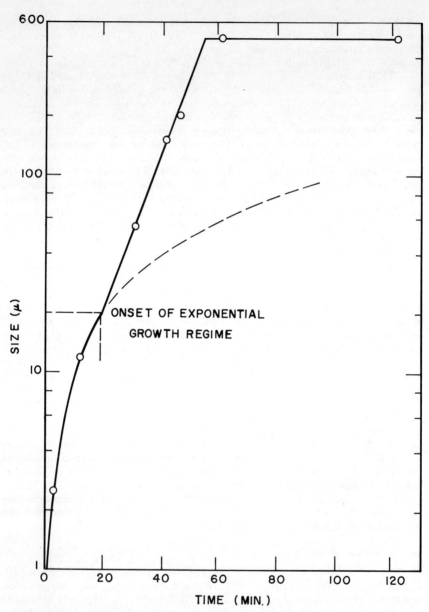

Figure 10. Coarsening of interconnected structures in exponential growth regime
Particle size vs. time; 25% SAN/75% PMMA; T = 280°C

growth, or it could have formed by spinodal decomposition since no phase
interconnectivity is expected at low phase volume ratios. The only way to
prove which mechanism was operative is to follow the kinetics of the early
stages of phase separation; unfortunately, this was not possible for this
composition.

After achieving bulk thermodynamic equilibrium, there is very slow coarsening of the structure to reduce the surface free energy of the system. A possible mechanism for this coarsening is Ostwald ripening, that is, gradual dissolution of the smaller droplets and gradual growth of the larger droplets by diffusion of the dispersed-phase rich component (PMMA) through the matrix phase. A semiquantitative analysis of this coarsening process can be made using the data of Lifshitz and Slyozov (*12*). They demonstrated that the following time dependence of the volume average droplet size should be expected:

$$d^3 = d_o{}^3 + \frac{64\ \sigma\ (X_e V_m)Dt}{9RT} \tag{13}$$

where d_o is droplet diameter after bulk thermodynamic equilibrium, σ is interfacial tension between droplet and matrix phases, X_e is equilibrium mole fraction of droplet rich component in the matrix phase, V_m is molar volume of droplet phase, and D is diffusion coefficient for matrix phase.

The slope of the droplet growth curve can be estimated from Figure 6. By using suitable values ($\sigma = 0.3$ dyne/cm, $X_e = 0.10$, and $V_m = 87{,}400$

$$\frac{64\ \sigma\ (X_e V_m)D}{9RT} = 2.26\ \times\ 10^{-4}\ \mu^3/min \tag{14}$$

cm^3/mole), the diffusion coefficient can be roughly estimated. This interfacial tension value was not measured experimentally, and the assumed value is somewhat lower than the values measured recently for other similar but less compatible polymers (*22, 23*). Wu (*23*) found that the interfacial tension of polystyrene/poly(methyl methacrylate) was 1.2 dyne/cm at 180°C. Given the increased compatibility of the system in this study, the assumed value seems most reasonable. Furthermore, it is consistent with a similar value which was used in analyzing the coarsening of spinodally decomposed structures. The values of X_e and V_m were taken from an extrapolation of Figure 3 and from Table II. With these values, the diffusion coefficient is:

$$D = 9\ \times\ 10^{-12}\ cm^2/sec\ at\ 265°C \tag{15}$$

This value is compared below with similar rough estimates obtained from spinodally decomposed systems.

Figures 7 and 8 were obtained with a 25% SAN composition near the critical composition. At the critical composition, roughly equal phase volumes are expected. The SAN-rich phase constitutes roughly 35% of the phase volume in the micrographs of Figure 8. However, the key to the mechanism lies not in the phase volume ratio; rather, it lies in the observed kinetics of the phase separation process. Figure 7 shows clearly that phase separation proceeds by a gradual change in composition over fairly well-defined regions in space. The scale of the phase separation (500–1000 A) is quite consistent with the data in Table II for spinodal decomposition in polymer–polymer systems. These observations, along with the observed high level of phase interconnectivity, are all consistent with theoretical predictions based on the Cahn theory, and they confirm semiquantitatively that spinodal decomposition is indeed the mechanism of phase separation.

Figure 7 demonstrates that Equation 3 is strictly applicable only in the very earliest stages of phase separation. The micrographs show simultaneous growth of composition differences and characteristic particle size after 4.4 min

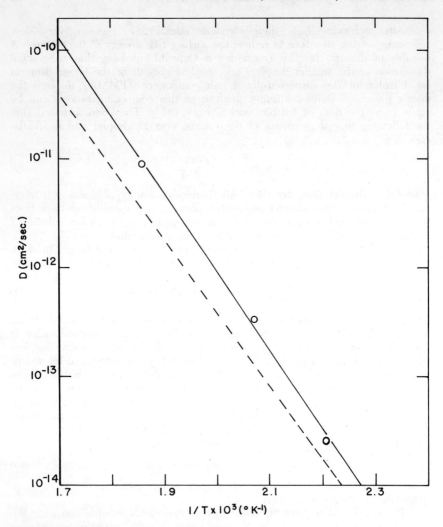

Figure 11. Diffusion coefficient from spinodal decomposition and Ostwald ripening
——, *Experimental;* ----, *Bueche equation*

heat treatment (Figure 7b), long before bulk thermodynamic equilibrium is reached. The full non-linear version of Equation 3 is required to describe the kinetics of these simultaneous processes (21). Such an analysis is beyond the scope of the experimental technique used in this work.

The diffusion coefficient in this polymer–polymer system can be roughly estimated from the rate of spinodal decomposition. To make the estimate, particle size change during spinodal decomposition will be ignored, and average particle size will be used. If the most rapidly growing wavelength from Equation 7 is substituted into Equation 6, then:

$$R(\lambda_m) = \left(- \Omega \frac{\partial^2 f}{\partial \phi^2} \right) \frac{2\pi^2}{\lambda_m^2} \qquad (16)$$

From Equation 4 and by analogy with the normal diffusion equation, $R(\lambda_m) = \pi^2/t$ to reach 99% equilibrium. Equation 16 then reduces to the familiar relationship of the diffusion coefficient to the mean square displacement distance,

$$\lambda_m{}^2 = 2Dt \tag{17}$$

From Figures 7 and 8, $\lambda_m \approx 1000$ A and $t \approx 30$ and 2.5 min for the 180° and 210°C heat treatments, respectively (these times were estimated as the times beyond which no significant increase in phase contrast could be observed). Substitution of these values in Equation 17 yields:

$$D = 2.6 \times 10^{-14} \text{ cm}^2/\text{sec at } 180°C \tag{18a}$$

$$D = 3.3 \times 10^{-13} \text{ cm}^2/\text{sec at } 210°C \tag{18b}$$

As rough as these estimates are, they can still be combined with the data from Ostwald ripening to obtain the temperature dependence of the diffusion coefficient shown in Figure 11. The slope of the line suggests an activation energy for diffusion of $E_D = 33$ kcal/mole. The zero shear rate viscosities of the polymers were also measured in this temperature range, and their temperature dependencies give a comparable activation energy for viscous flow of $E_\eta = 30.5$ kcal/mole. As a final comparison, the Bueche equation (24) can be used to estimate comparable diffusion coefficients.

$$D = \left(\frac{\rho N}{36}\right)\left(\frac{\overline{R}^2}{M}\right)\left(\frac{kT}{\eta}\right) \tag{19}$$

The value of $(\overline{R}^2/M)^{1/2} = 0.7 \times 10^{-8}$ cm was used to derive the diffusion coefficients plotted in Figure 11. Given the approximations used in deriving these diffusion coefficients, the relative agreement of the various estimating procedures is most rewarding.

Once bulk equilibrium has been achieved, a coarsening process begins which is far more rapid than that observed for the dispersed phase structure. Figure 9 indicates that particle size increases linearly with time during early stages of particle coarsening. This functional dependence cannot be explained by Ostwald ripening (*see* Equation 13). This coarsening can occur because the high level of phase interconnectivity allows viscous flow of both major and minor phases by an interfacial tension driven mechanism.

A simplified analysis of the complex viscous flow occurring during this coarsening process can be made by applying Tomotika's findings (25). He analyzed the breakup of an infinite cylindrical thread of one viscous, Newtonian fluid dispersed in a second viscous, Newtonian fluid by a linearized stability analysis. As indicated in Figure 12, breakup occurs by the exponential growth of certain infinitesimal transverse disturbances. Tomotika found that the dominant disturbance assumes a minimum value of,

$$\left.\frac{\lambda_{\max}}{a}\right|_{\min} = 10.7 \tag{20}$$

for a viscosity ratio $\eta'/\eta \approx 0.25$ where $a =$ filament radius. For viscosity ratios greater or less than this value, the dimensionless ratio of Equation 20 increases monotonically to infinity as that ratio $\ln(\eta'/\eta) \to \pm\infty$.

When Tomokita's simple structure is compared with the complex interconnected structure shown in Figure 12, one observation is obvious. The three-dimensional interconnectivity of the structure provides numerous branch points; many are closer together than the dominant wavelength. Because of this short range interconnectivity, breakdown of structure is prevented during

TOMOTIKA STABILITY ANALYSIS

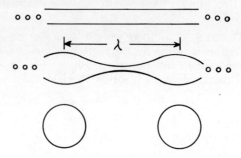

COMPARED WITH REAL STRUCTURES

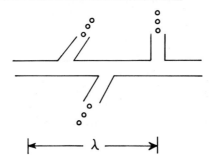

Figure 12. Interpretation of coarsening phenomena for interconnected structures

early stages of the coarsening process. Structure breakdown can occur only when the dominant wavelength (or at least those wavelengths with appreciable growth rates) is shorter than the range of interconnectivity.

The Tomotika analysis can still be used to estimate the coarsening rate during early stages of the coarsening process. Even with interconnectivity, growth rate can be estimated by recognizing that the dominant wavelength is continuously changing as the diameter of a growing filament is fed by other filaments which are decreasing in size. From the Tomotika analysis, the growth rate of a disturbance is (top of next page):

Table IV.   Comparison of Observed and Predicted Growth

T, °C	η, $p \times 10^{-3}$	η', $p \times 10^{-4}$	η'/η	σ, dyne/cm	f
210	138	42.3	3.07	.30	.50
250	8.60	4.00	4.65	.30	.50
280	2.47	1.22	4.93	.30	.50
280	2.47	1.22	4.93	.30	—

$$\xi = \xi_0 \exp(\alpha_{max} \, t) \tag{21}$$

Differentiating this equation, the initial rate of growth of this disturbance is:

$$\left. \frac{d\xi}{dt} \right|_{t=0} = \xi_0 \alpha_{max} \tag{22}$$

From the Tomotika analysis,

$$\alpha_{max} = \frac{\sigma}{2 \, \eta' a} \, G(\eta'/\eta) \tag{23}$$

where $G(\eta'/\eta)$ is a unique function of the viscosity ratio. Substitution of Equation 23 into Equation 22 yields:

$$\left. \frac{d\xi}{dt} \right|_{t=0} = \frac{\sigma f}{2\eta'} \, G(\eta'/\eta) \tag{24}$$

where $f = \xi_0/a$, the ratio of the magnitude of the initial disturbance to the filament radius. From the structural characteristics (Figures 7 and 8), it seems reasonable that this ratio be taken as invariant during the coarsening process. Hence, Equation 24 suggests that the growth rate is constant which agrees with Figure 9.

Equation 24 can be used to estimate the rate of coarsening during the constant growth rate regime if suitable values of σ and f are used. We set $\sigma = 0.3$ dyne/cm (a value consistent with that used in the Ostwald ripening analysis) and $f = 0.5$ to obtain comparisons with the experimental data in Table IV. The computed and observed growth rates in this constant growth rate regime agree quite well. It is believed that these values of σ and f are reasonable, but, even given the arbitrariness of these values, the excellent agreement of the temperature dependence of the growth rates with constant values of σ and f is most satisfying. The results of this analysis show, at the very least, that the growth rate can be predicted as a function of temperature provided the growth rate at one temperature and the temperature dependencies of the viscosities are known.

The nature of the switch from constant growth rate to the exponential growth rate regime is not well understood. It appears to be associated with a gradual breakdown in interconnectivity so that the dominant wavelength and the scale of interconnectivity become comparable. In this regime, if $a = 10\mu$ (diameter $= 20\mu$) is used in Equation 23, a good comparison with the observed growth rate in Figure 10 is obtained (*see* Table IV). Hence, all comparisons are internally consistent in this regard. However, the predicted and measured final drop sizes (assuming no further growth in filament diameter after onset

Rates During Coarsening, 75% PMMA/25% SAN

Constant growth Rate, μ/min			Exponential growth rate, min^{-1}		
Obser.	*Calcd.*	*Ratio (calcd. obser.)*	*Obser.*	*Calcd.*	*Ratio (calcd. obser.)*
.015	.014	0.92	—	—	—
.15	.165	1.10	—	—	—
.50	.65	1.29	—	—	—
—	—	—	.093	.129	1.39

of the exponential growth regime) do not agree; predicted size is 22μ whereas the measured size is roughly equivalent to the film thickness. One other possible explanation of the enhanced growth rate is that the polymers are degrading at the elevated temperatures. This possibility has not been investigated, but it certainly merits further study.

Conclusions

Liquid–liquid phase transition phenomena in polymer–polymer systems were studied. Evidence is presented which suggests that spinodal decomposition occurs in this system. It was not possible to prove that nucleation and growth also occurred for the dispersed-phase structure because of the rapid rate of phase equilibration.

Far from the critical concentration, a dispersed, two-phase structure was formed. Near the critical point, it was possible to thrust the system into the unstable region of the phase diagram before significant phase separation occurred. In this case, a highly interconnected two-phase structure was formed.

The Cahn theory of spinodal decomposition explains the essential features of phase transition in the unstable region. The scale of phase separation was predicted and exceeded by an order of magnitude the scale in low molecular weight systems. A comparison of the diffusion coefficient estimated by the rate of spinodal decomposition agrees with the value predicted by the Bueche equation. The high level of phase interconnectivity predicted by the Cahn theory was also observed experimentally.

The coarsening characteristics of both dispersed and interconnected structures have been studied. Ostwald ripening was the coarsening mechanism for the dispersed phase structures. The diffusion coefficient estimated from this coarsening pattern agrees well with the value from the Bueche equation. The interconnected structures coarsen by an interfacial, tension-driven, viscous flow mechanism. The growth rate was constant during the early stages of coarsening by this mechanism. Quite good agreement of observed and predicted coarsening rates was obtained by using a modified form of the Tomotika stability analysis. In later stages, the coarsening rate was exponential. Again, the Tomotika data could be applied, but in this case the exact mechanism of coarsening is not completely understood.

Acknowledgments

The author thanks Union Carbide Corp. for permission to publish this work. A special thanks to E. Trzaska, W. Niegisch, and S. Crisafulli for performing the experimental work and electron microscopy.

Literature Cited

1. Cahn, J. W., *Trans. Met. Soc. AIME* (1968) **242**, 166.
2. Cahn, J. W., *J. Chem. Phys.* (1965) **42**, 93.
3. Hopper, R. W., Uhlmann, D. R., *Disc. Faraday Soc.* (1971) **50**, 166.
4. Goldstein, M., *J. Cryst. Growth* (1968) **3–4**, 594.
5. Haller, W., *J. Chem. Phys.* (1965) **42**, 686.
6. van Aarsten, J. J., *Eur. Polym. J.* (1970) **6**, 919.
7. Smolders, C. A., van Aarsten, J. J., Steenbergen, A., *Kolloid Z. Z. Polym.* (1971) **243**, 14.
8. van Emmerik, P. T., Smolders, C. A., *J. Polym. Sci. Part C* (1972) **38**, 73.
9. van Emmerik, P. T., Smolders, C. A., Geymayer, W., *Eur. Polym. J.* (1973) **9**, 309.
10. McMaster, L. P., *Macromolecules* (1973) **6**, 760.

11. Doremus, R. H., "Glass Science," John Wiley, New York, 1973.
12. Lifshitz, I. M., Slyozov, V. V., *J. Phys. Chem. Solids Lett. Sect.* (1961) **19**, 35.
13. Cahn, J. W., Hilliard, J. E., *J. Chem. Phys.* (1958) **28**, 258.
14. Morral, J. E., Ph.D. Dissertation, Massachusetts Institute of Technology, 1969.
15. Flory, P. J., "Principles of Polymer Chemistry," Cornell University Press, Ithaca, N. Y., 1953.
16. Debye, P., *J. Chem. Phys.* (1959) **31**, 680.
17. Zarzycki, J., *Disc. Faraday Soc.* (1971) **50**, 122.
18. Rundman, K. B., Hilliard, J. E., *Acta Met.* (1967) **15**, 1025.
19. Andreev, N. S., Porai-Koshits, E. A., *Disc. Faraday Soc.* (1971) **50**, 135.
20. Cahn, J. W., Charles, R. J., *Phys. Chem. Glasses* (1965) **6**, 181.
21. Cahn, J. W., *Acta Met.* (1966) **14**, 1685.
22. Roe, R. J., *J. Colloid Interface Sci.* (1969) **31**, 228.
23. Wu, S., *J. Phys. Chem.* (1970) **74**, 632.
24. Bueche, F., "Physical Properties of Polymers," Interscience, New York, 1962.
25. Tomotika, S., *Proc. Royal Soc. London* (1935) **150**, 322.

RECEIVED April 19, 1974.

6

Quantitative Evaluation of the Dynamic Mechanical Properties of Multiphase Polymer Systems

LOTHAR BOHN

Hoechst AG, Frankfurt/Main-Höchst, West Germany

This work demonstrates that it is possible to obtain quantitative information about phase volumes and phase structure of multiphase polymer systems from the findings from dynamic mechanical tests using existing theories of the elastic moduli of composites. Furthermore, some special features of the loss spectra of two-phase systems are discussed. Variation in the volume of the dispersed rubbery phase affects not only the magnitude of the loss peak but also its temperature position. A crosslinked and well grafted rubber phase within a rigid matrix undergoes appreciable dilatation on cooling; this causes a shift in the rubber glass transition to lower temperatures.

Dynamic mechanical characteristics, mostly in the form of the temperature response of shear or Young's modulus and mechanical loss, have been used with considerable success for the analysis of multiphase polymer systems. In many cases, however, the results were evaluated rather qualitatively. One purpose of this report is to demonstrate that it is possible to get quantitative information on phase volumes and phase structure by using existing theories of elastic moduli of composite materials. Furthermore, some special anomalies of the dynamic mechanical behavior of two-phase systems having a rubbery phase dispersed within a rigid matrix are discussed; these anomalies arise from the energy distribution and from mechanical interactions between the phases.

Analysis Using Young's or Shear Moduli Only

Theory. Basic theories for the prediction of the modulus of a composite from those of the components were derived by, for example, Hashin in 1955 (1), Kerner in 1956 (2), and van der Poel in 1958 (3). Takayanagi (4, 5), and Fujino et al. (6) developed a very promising and instructive model theory which includes calculation of the loss spectra of composites, and it may easily be extended to anisotropic morphologies. Furthermore, Nielsen and coworkers may be cited for fundamental theoretical and experimental contributions (7, 8, 9, 10).

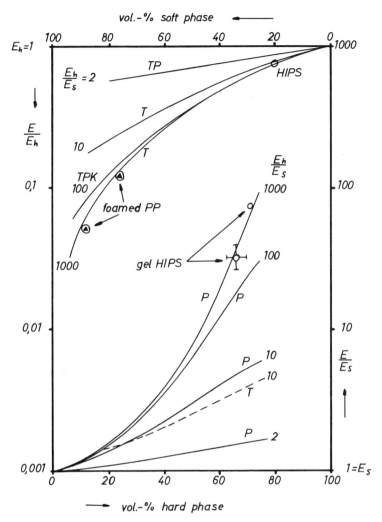

Figure 1.   Modulus E of two-phase systems as a function of the volume fraction of the dispersed phase and the ratio of the hard phase modulus (E_h) to the soft phase modulus (E_s)

According to the theories of K: Kerner (2), P: van der Poel (3), and T: Takayanagi (4, 5)

Figure 1 represents the predictions of the composite modulus from the components according to the different theories. The lower part of the figure depicts the reinforcement of a rubbery matrix by incorporation of isotropic rigid particles. The increase in the composite modulus E is given in relation to the soft matrix phase modulus E_s (at the right side) as a function of the volume content of rigid filler with hard phase modulus E_h. The parameter is the ratio of the moduli of the rigid and rubbery components, E_h/E_s. Takayanagi's approximation (T) as well as the unmodified Kerner equation give values that are much too low for this phase morphology, especially at high

E_h/E_s ratios. On the other hand, calculations based on van der Poel's work (P) agree very well with experimental findings (11). Unfortunately, the van der Poel theory requires the evaluation of a set of eight linear equations, and it can be applied only in tabulated form.

The upper part of Figure 1 depicts the reduction of the modulus of the rigid phase E/E_h by incorporation of a soft rubbery phase, again with E_h/E_s as the parameter. The three theories agree well in this reverse phase position (see the curve for $E_h/E_s = 100$). The composite moduli can therefore be calculated easily by one of the well known equations from Takayanagi's approximative model theory (Equation 1) or from Kerner's theory (Equation 2).

$$\frac{E_h}{E} = 1 + \frac{\phi_s \left(1 - \frac{E_s}{E_h}\right)}{1 - \sqrt{\phi_s} \left(1 - \frac{E_s}{E_h}\right)} \tag{1}$$

When $E_h \gg E_s$:

$$\frac{E_h}{E} = 1 + \frac{\phi_s}{1 - \sqrt{\phi_s}} \tag{1a}$$

$$\frac{E_h}{E} \simeq \frac{G_h}{G} = 1 + \frac{15(1 - \nu_h)\, \phi_s \left(\frac{G_h}{G_s} - 1\right)}{(7 - 5\nu_h)\, \phi_h \left(\frac{G_h}{G_s} - 1\right) + 15(1 - \nu_h)} \tag{2}$$

When $E_h \gg E_s$ or $G_h \gg G_s$:

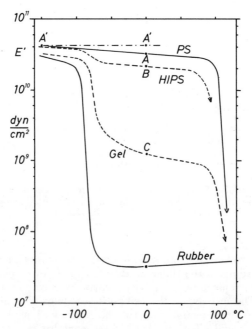

Figure 2.   Dynamic Young's modulus E' of polystyrene (PS), high impact polystyrene (HIPS), separated gel, and crosslinked polybutadiene rubber. Data from Ref. 17.

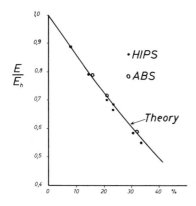

Figure 3. Correlation between modulus reduction and experimentally determined gel content

Gel Volume

$$\frac{G_h}{G} = 1 + \frac{15(1 - v_h)}{(7 - 5v_h)} \frac{\phi_s}{\phi_h} \cong 1 + 1.9 \left(\frac{\phi_s}{1 - \phi_s} \right) \tag{2a}$$

In these equations, the subscripts h and s refer to the hard continuous phase and the soft rubbery phase, respectively, E = Young's modulus, G = shear modulus, v_h = Poisson's ratio (= 0.35 for polystyrene and most rigid polymers), and ϕ_s and ϕ_h = volume fractions of the components.

If the modulus of a dispersed rubbery component is only $1/1000$ that of the surrounding rigid matrix, an air-filled structure is approximated. The two experimental points for foamed polypropylene in Figure 1 fit fairly well into the theoretical curves. It is apparent from Figure 1 that the modulus of a composite of known volume composition reflects clearly the phase distribution, *i.e.*, it indicates which component forms the continuous phase. The moduli may differ up to two decades after reversal of the continuous and the dispersed phases when the ratio of the component moduli is about 1000:1.

It should be pointed out that the relations plotted in Figure 1 apply only for ideal isotropic morphologies. Deviations are an indication of anisotropy, of anomalous structures like interlocking networks, or of an intermediate state of phase reversion. Thus, the statement by some authors (*12, 13*) that the theory did not correlate with their experiments is no argument against the theory, but rather against its application to composites of unknown structure. The morphologies should have been determined by, for example, microscopic or electron microscopic methods. The situation with block copolymers is highly complex. A change in composition, usually by variation of the block length ratio of the two different species, may induce not only a phase reversal but also a fundamental change in the basic morphology to some rodlike or layer structures of high anisotropy. It is not surprising that Kerner's theory was not applicable to such a system (*14*).

Experimental Examples. An example may be given of the applicability of the forementioned theories to a complex but well defined morphology. It is well known that in commercial high impact polystyrene (HIPS) the dispersed rubber phase, the gel, is filled with up to 75% occluded polystyrene (PS) (*15, 16*). In principle, we have a three-phase system! Keskkula and Turley (*17*) succeeded in isolating the gel and determining its dynamic mechanical

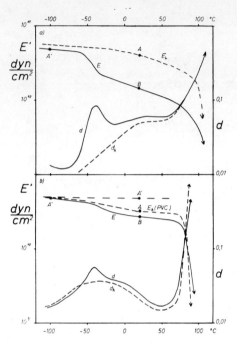

Figure 4.   High impact PVC

a. Dynamic Young's modulus E' and mechanical
loss factor d of an impact modifier containing 47
wt % styrene–butadiene rubber (———) and of the
pure rigid graft phase (– – –)
b. Dynamic characteristics of PVC (– – –) and of
impact PVC containing 15 wt % modifier (———)

properties in addition to those of the original HIPS and the crosslinked rubber (Figure 2). They reported that the gel constituted 19–20 vol % of the HIPS, and that there was about 66 vol % occluded PS within the gel. With $E_h/E_s = 1000$ (modulus A/modulus D), the experimentally determined reinforcement $E/E_s = 32$ (modulus C/modulus D) is in good agreement with Figure 1. The same is true for another investigation of a HIPS gel by Wagner and Robeson (*18*). The experimentally obtained value for the reduction of the modulus of pure PS (at A) to the value of HIPS (at B) by the incorporation of 19–20 vol % gel phase is $E/E_h = 0.75$. This result fits well into the curve with $E_h/E_s = 30$ (moduli at A and C in Figure 2) in Figure 1.

A fairly good correlation between modulus reduction and gel volume is demonstrated in Figure 3 for a number of different HIPS and acrylonitrile–butadiene–styrene (ABS) materials. The gel volume was determined by the usual high speed centrifugation of the methyl ethyl ketone–acetone solution.

Using the modulus drop through the rubber phase transition as an indication of phase volume means approximating the real modulus of the rigid phase at A (Figure 2) by the modulus at A' of the composite in its glassy state (*19, 20, 21*). The greater the temperature dependence of the rigid phase modulus and thus the larger the difference between the moduli at A and A', the greater the deviation from the exact calculation (*see* Figure 4 for an example). Figure 4a is characteristic of the curve for an impact modifier for poly(vinyl chloride) (PVC) containing about 50% styrene–butadiene rubber. Here the use of the modulus at A′ instead of that at A has no appreciable effect. Figure 4b depicts the mechanical properties of a high impact PVC that contains 15 wt % modifier. The modulus of pure, unmodified PVC has a strong temperature dependence just through the transition of the rubber component at −40°C. Whereas the use of the exact modulus of the rigid phase at A in comparison with the composite modulus at B gives, with a reduction $E/E_h = 0.88$, the right rubber concentration of 9–10 vol % (Figure 1), the application of the modulus drop method (*i.e.*, the use of the much higher modulus at A′ for the rigid phase) results in a completely wrong rubber volume which is about three times too high!

Special Effects in the Dynamic Mechanical Characteristics of Systems Having a Rubbery Phase Dispersed within a Rigid Matrix

Effect of the Rubber Content.   On deforming a composite that has a dispersed rubbery phase in a rigid matrix, the magnitude of strain in the rubber will be bound to that in the surrounding rigid matrix. For this reason, the distribution of stored energy between the rubbery and rigid phases depends to a first approximation on the ratio of the moduli of the soft and hard components, E_s/E_h, and of course on the composition by volume. In going through the glass transition range of the rubber, which covers some 20–30°C, this

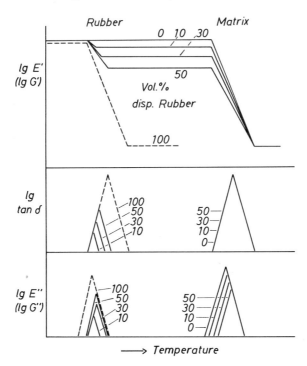

Figure 5.   Schematic: effect of the volume fraction of the dispersed rubbery phase on E′, E″, and tan δ of a two-phase system

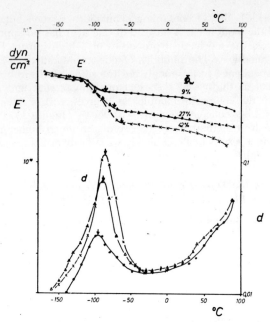

Figure 6. Dynamic characteristics of three ABS samples with different volume fractions of rubbery phase

Arrows indicate the rubber transition in the modulus curves

modulus ratio E_s/E_h ranges from 1 to about 0.001 which indicates that less and less energy is taken up by the softening phase compared with the rigid matrix. At some temperature within the softening range of the rubber, the energy stored becomes negligible, *i.e.*, a rubber particle acts like an empty hole. The composite modulus then remains constant at some level determined by the rubber phase volume (*cf.* Figure 1); the further characteristic of the dispersed phase as well as its absorption capacity are no longer reflected in the composite properties.

The schematic representatition in Figure 5 indicates the consequences for the dynamic mechanical method of analysis. The well known increase in the modulus drop and in the magnitude of tan δ or E'' with increasing rubber content should be accompanied by a shift in the position of tan δ to higher temperatures. The E'' peak temperature at first increases like that of tan δ, then it may remain constant or even decrease again at higher rubber contents. The E'' of rubber has a lower peak temperature than the dispersed rubber in the composite in contrast to the behavior of tan δ. In the matrix transition range, tan δ—something like a reduced absorption of the composite—should remain unaffected both in magnitude and position, whereas the E'' peak should decrease in magnitude and shift slightly toward higher temperatures with increasing rubber content.

In Figure 6, the sharp cutting of the rubber transition in the modulus curves is accompanied by the shift in the tgδ or E'' peak to higher temperatures. Quite similar behavior was reported for block copolymers with up to 40% polybutadiene (*22, 23*). In Figure 7, the shift in the matrix E'' peak to higher temperatures with increasing soft phase volume is apparent. For additional examples, *see* Figures 8 and 10 of Ref. 4. Dynamic mechanical measurements may be represented by modulus and loss modulus (E' and E'' or G' and G'') curves or by modulus and tg$\delta = E''/E'$ or G''/G' curves. By incorporating a dispersed component which exhibits no transition in the temperature range under study, the tgδ curve of the matrix remains practically unaffected. This was observed not only in systems diluted by holes (foamed material) (*24*) and in those reinforced by fillers (*11, 12*), but also in composites with a dispersed rubbery component in the temperature range far enough above the softening of the rubber. For this reason, the tgδ characteristic should be preferred to the loss modulus curves in analyzing a multicomponent system.

Effect of Mechanical Interaction. It was shown both theoretically (*25, 26, 27*) and experimentally (*28*) that, on cooling a composite having a dispersed rubbery phase to temperatures well below the matrix glass transition, the rubber undergoes appreciable dilatation attributable to the expansion mismatch. In compounds with sufficient adhesion between the phases and crosslinked rubber, this dilatation is clearly reflected by a shift in the rubber glass transition to lower temperatures (*29, 30*). Figure 8 exemplifies the ABS system (*29*). Well dispersed but ungrafted rubber particles, which are expected to give insufficient adhesion between the phases, cause a rubber transition 12°C higher than in a well grafted latex rubber exhibiting the dilatation shift to lower temperatures. Additional examples are found in the literature (*31, 32*).

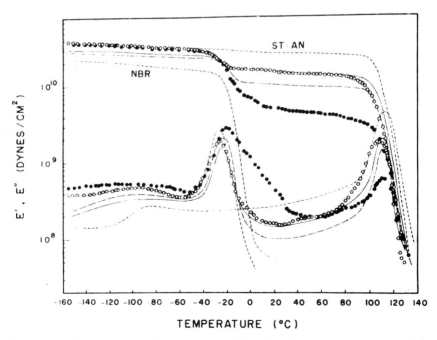

Figure 7. Shear modulus G' and loss modulus G'' of [nitrile–butadiene rubber (NBR)]–[styrene–acrylonitrile (SAN)] copolymer blends. Data from Ref. 4.
NBR (vol %), 36 and 51 respectively

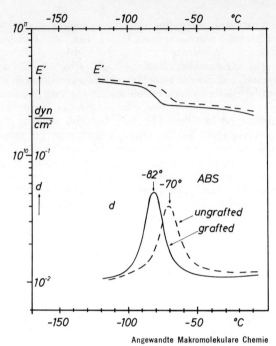

Angewandte Makromolekulare Chemie

Figure 8. Dynamic Young's modulus E' and loss factor d of two ABS materials containing grafted and ungrafted latex rubber respectively (29)

Conclusions

The modulus functions can be quantitatively evaluated to obtain information on the phase structure of composites. Consequently, the complex but well defined three-phase, double-emulsion morphology of high impact polystyrene can be described as agreeing well with theory in all essential features. Evaluation of the modulus drop disregards the temperature dependence of the rigid phase of composite modulus through the transition range. Therefore caution is recommended in drawing quantitative conclusions by this method.

The mechanism of energy distribution between the phases has some consequences for two-phase systems with dispersed rubber within a rigid matrix. Not only the magnitudes but also the positions of the peaks of tan δ and E'' are dependent on the rubber phase volume (Figure 5). A shift to higher temperature with increasing rubber content need not be caused by an increase of T_g resulting from crosslinking (18, 33). The higher E'' peak temperature of the matrix transition of an ABS polymer, compared with the peak position in the unmodified matrix material, does not necessarily indicate a higher matrix T_g in the ABS polymer caused by stresses (33).

The nature (*e.g.*, the softening range) of a hard occlusion in the interior of a rubber particle can never be determined by dynamic mechanical methods. Any dispersed phase can only be characterized in the region of higher modulus. Its low modulus or higher temperature properties are completely lost. Thus, the relaxation spectrum of a composite is generally not a superposition of the component spectra.

Semi-empirical rules, which correlate the static glass transition temperature T_g from differential thermal analysis or dilatometry with the dynamic T_g taken from the tan δ or E'' peak, may be used with caution in analyzing two-phase systems with a dispersed rubbery phase. The dynamic T_g depends on the rubber phase volume, and it may be shifted further toward lower temperature for effectively crosslinked and grafted rubber particles because of dilatation.

Literature Cited

1. Hashin, Z., *Bull. Res. Counc. Israel* (1955) **5C**, 46.
2. Kerner, E. H., *Proc. Phys. Soc. London* (1956) **B 69**, 808.
3. van der Poel, C., *Rheol. Acta* (1958) **1**, 198.
4. Takayanagi, M., *Mem. Fac. Eng. Kyushu Univ.* (1963) **23**, 57.
5. Takayanagi, M., *J. Appl. Polym. Sci.* (1966) **10**, 113.
6. Fujino, K., Ogawa, Y., Kawai, H., *J. Appl. Polym. Sci.* (1964) **8**, 2147.
7. Nielsen, L. E., *J. Appl. Phys.* (1970) **41**, 4626.
8. Nielsen, L. E., Biing-Lin, Lee, *J. Compos. Mater.* (1972) **8**, 136.
9. Nielsen, L. E., Lewis, T. B., *J. Polym. Sci. Part A-2* (1969) **7**, 1705.
10. Lewis, T. B., Nielsen, L. E., *J. Appl. Polym. Sci.* (1970) **14**, 1449.
11. Schwarzl, F. R., Bree, H. W., Nederveen, C. J., Schwippert, G. A., Struik, L. C. E., van der Waal, C. W., *Rheol. Acta* (1966) **5**, 270.
12. Kraus, G., Rollmann, K. W., Gruver, J. T., *Macromolecules* (1970) **3**, 92.
13. Zitek, P., Zehlinger, J., *J. Polym. Sci. Part A-1* (1968) **6**, 467.
14. Angelo, R. J., Ikeda, R. M., Wallach, M. L., *Polymer* (1965) **6**, 141.
15. Molau, G. E., Keskkula, H., *J. Polym. Sci. Part A-1* (1966) **4**, 1595.
16. Bender, B. W., *J. Appl. Polym. Sci.* (1965) **9**, 2887.
17. Keskkula, H., Turley, S. G., *J. Polym. Sci. Part B* (1969) **7**, 697.
18. Wagner, E. R., Robeson, L. M., *Rubber Chem. Technol.* (1970) **43**, 1129.
19. McCrum, N. G., *J. Polym. Sci.* (1958) **27**, 555.
20. Wada, Y., Kasahara, T., *J. Appl. Polym. Sci.* (1967) **11**, 1661.
21. Keskkula, H., Turley, S. G., Boyer, R. F., *J. Appl. Polym. Sci.* (1971) **15**, 351.
22. Matsuo, M., Ueno, T., Horino, H., Chujyo, S., Asai, H., *Polymer* (1968) **9**, 425.
23. Robinson, R. A., White, E. F. T., *Amer. Chem. Soc., Div. Polym. Chem., Prepr.* **10**, 662 (New York, September, 1969).
24. Bohn, L., unpublished data.
25. Matsuo, M., Wang, Tsuey T., Kwei, T. K., *J. Polym. Sci. Part A-2* (1972) **10**, 1085.
26. Beck, Jr., R. H., *J. Polym. Sci. Part B* (1968) **6**, 707.
27. Wang, Tsuey T., Matsuo, M., Kwei, T. K., *J. Appl. Phys.* (1971) **42**, 4188.
28. Schmitt, J. A., IUPAC Symp., Toronto (1968) **2**, A9.
29. Bohn, L., *Angew. Makromol. Chem.* (1971) **20**, 129.
30. Turley, S. G., *J. Polym. Sci. Part C* (1973) **1**, 101.
31. Morbitzer, L., Ott, K. H., Schuster, H., Kranz, D., *Angew. Makromol. Chem.* (1972) **27**, 57.
32. Sternstein, S. S., IUPAC Symp., Leiden (1970) **2**, 973.
33. Matsuo, M., Ueda, A., Kondo, Y., *Polym. Eng. Sci.* (1970) **80**, 253.

RECEIVED April 1, 1974.

7

Compatibilization Concepts in Polymer Applications

NORMAN G. GAYLORD

Gaylord Research Institute, Inc., 273 Ferry St., Newark, N. J. 07105

The general incompatibility of polymers may be overcome by suitable compatibilizing agents, i.e. block or graft copolymers having segments of similar structure or solubility parameter as the polymers being mixed. Presence of a compatibilizing agent influences the properties of polymer blends and alloys as well as polymer melts and solutions. Solution of incompatible polymers in a common solvent as well as dispersion of a polymer in a nonsolvent is promoted by the presence of a compatibilizing agent. Dispersion of additives such as fillers and reinforcing agents (including cellulose, clay, glass fibers, and powdered metals) in a polymer matrix is enhanced by a compatibilizing agent at the additive–polymer interface; interfacial interaction promotes adhesion of solid polymers to polymeric or inorganic substrates.

The general incompatibility of polymers often prevents preparation of useful blends. Introduction of a compatibilizing agent permits the blending of otherwise incompatible polymers to yield compositions with unique properties generally not attainable from either of the components of the polyblend. The presence of a compatibilizing agent influences the properties of polymer melts and solutions as well as polymer blends and alloys. Solution of incompatible polymers in a common solvent as well as dispersion of a polymer in a nonsolvent is promoted by the presence of a compatibilizing agent. Dispersion of additives such as fillers and reinforcing agents in a polymer matrix is enhanced by the presence of a compatibilizing agent at the polymer–additive interface. Adhesion of polymers to surfaces is similarly dependent on compatibilization at the interface. A qualitative overlook of the role of the compatibilizing agent in these polymer applications indicates common denominators and continually expanding potential.

Polyblends and Alloys

For useful polyblends, the term compatibilization refers to the absence of separation or stratification of the components of the polymeric alloy during the expected useful lifetime of the product. Optical clarity of a polyblend is related to the particle size of the dispersed phase and/or the difference in the refractive indexes of the dispersed and matrix phases. Absence of transparency is not indicative of incompatibility in accordance with the broad definition given above and used hereafter.

Block and graft copolymers possessing segments with chemical structures which are the same as those of the polymers to be blended are effective compatibilizing agents. Thus, an AB block or graft copolymer compatibilizes polymers A and B (*see* Diagram 1).

$$
A \underset{B}{\overset{A}{\rule{3cm}{0.4pt}}} B \tag{1}
$$

Since polymer properties are derived from finite length segments of a particular structure, random copolymers do not compatibilize homopolymers and, in many cases, copolymers based on one ratio of comonomers are not compatible with copolymers based on a different ratio of the same comonomers.

Block copolymers are usually superior to graft copolymers as compatibilizing agents since the latter may have restricted accessibility to the backbone segment of the copolymer, particularly when numerous branches are located on a single backbone polymer. Block and graft copolymers are generally prepared by polymerization processes, *e.g.* polymerization of a monomer in the presence of a nonpropagating polymer chain or propagation of a polymer chain in the presence of a monomer differing from that polymerized initially (*1, 2, 3*). Block and/or graft copolymers may be preformed and then added to the mixture undergoing compatibilization, or they may be generated *in situ*. The latter approach permits preparation of compatibilizing agents by the post reaction between two polymers, *e.g.* a compatible polymer blend is prepared from incompatible ethylene–methacrylic acid and methyl methacrylate–methacrylic acid copolymers by milling in the presence of zinc acetate which results in the formation of ionic bonds between the incompatible polymers (*4*) (Diagram 2).

$$
\begin{array}{c}
PE \rule{4cm}{0.4pt} \\
| \\
COOH \\
\end{array}
$$

$$
PE \rule{5cm}{0.4pt} \hspace{1cm} \rule{5cm}{0.4pt} PMMA \tag{2}
$$

$$
\begin{array}{c}
| \qquad\qquad | \\
COO^- \quad {}^+Zn^+ \quad {}^-OOC \\
| \\
COOH \\
| \\
\rule{4cm}{0.4pt} PMMA \\
\end{array}
$$

A compatibilizing agent is also formed *in situ* when incompatible polymers are subjected to shearing forces which rupture polymer chains to generate radicals which undergo coupling.

The molecular weight of the segments in a block or graft copolymer plays an important role in compatibilizing efficiency. A copolymer having a segment with molecular weight greater than 150,000 is generally a poor compatibilizing agent since intra- and intermolecular interactions such as chain entanglements may reduce the accessibility of such a segment to the homopolymer whose compatibilization is desired. The minimum molecular weight of a segment for effective compatibilization will vary with the polymer structure. However, as a qualitative rule-of-thumb, since an oligomer containing 10–15 monomer units has characteristics reasonably similar to those of the higher molecular weight polymer, a segment of a block or graft copolymer containing 10–15 monomer units is an effective compatibilizing agent for the corresponding higher molecular weight homopolymer.

Notwithstanding the general incompatibility of polymers and copolymers, polymers with solubility parameters that differ by 0.5 unit are compatible although their structures may differ. Thus, poly(methyl methacrylate) (PMMA), poly(ethyl acrylate) (PEA), poly(vinyl chloride) (PVC), and poly(butadiene-co-acrylonitrile) (90/10–60/40) form useful compatible blends because their solubility parameters are in the range 9.2–9.4. The difference in solubility parameters resulting in compatibility may be as much as one unit when the polymers are of relatively low molecular weight.

Extrapolation of the concept of the compatibility of polymers with different structures but similar solubility parameters to the construction of effective compatibilizing agents leads to the use of block and graft copolymers having segments with suitable solubility parameters to compatibilize polymers which differ both in structure and solubility parameter. Thus, an AB block or graft copolymer compatibilizes polymers A and C when C has a solubility parameter similar to that of B, polymers D and B when D has a solubility parameter similar to that of A, and polymers C and D when the solubility parameter of C is similar to that of B, and D has a solubility parameter similar to that of A.

Modification of Mechanical Properties. Application of these concepts is illustrated by impact polystyrene: polystyrene (PS) grafted onto polybutadiene (PBD) permits as much as 40% PBD in PS to be incorporated whereas, in the absence of the graft copolymer, incompatibility is detectable by stratification when more than 10% elastomer is blended with PS (Diagram 3). Acrylonitrile–

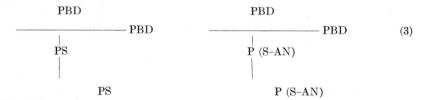

$$(3)$$

butadiene–styrene resin (ABS) consists of the compatibilized blend of poly(styrene-co-acrylonitrile) [P(S–AN)], PBD, and P(S–AN) grafted onto PBD (Diagram 3). The properties may be modified by adding copolymer to the polyblend generated by graft copolymerization.

ABS-type resins are used as impact modifiers for PVC, but the resultant blend has insufficient transparency for application in clear bottles. Transparency can be obtained by grafting PMMA onto crosslinked PBD (5) or poly(butyl acrylate) (6) which has been previously grafted onto PS (Diagram 4). In this case the PMMA branch is compatible with PVC by virtue of its solubility parameter, and optical clarity results from suitable component ratios in the graft copolymer so that the refractive index matches that of PVC. The desired results are not obtained if a copolymer of methyl methacrylate and styrene is

```
——————————————— PBD ———————————————
    |         |                    |
    PS    PS–PMMA              PMMA          (4)
         PVC
```

PVC

PMMA

PMMA   (5)

PEHA

grafted onto the elastomer since the solubility parameter of the grafted co-polymer branch differs from that of PVC.

An impact resistant PVC–acrylic polymer alloy is formed by grafting PMMA onto poly(ethylhexyl acrylate) (PEHA) and then blending the resultant mixture of graft copolymer and homopolymers with PVC (7) (Diagram 5).

Modification of Melt and Solution Properties. In addition to modifying mechanical properties resulting from polymer blending in the presence of a suitable compatibilizing agent, the melt and solution viscosities of the compatibilized blend differ from those of the blend in the absence of the agent because the compatibilized blend is not truly compatible; it is only partially or marginally so.

The solution of a polymer in a good solvent, *e.g.* PS in toluene or PEA in methyl ethyl ketone (MEK), has a high viscosity because the chains are extended and undergo intermolecular entanglement. Solution in a poor solvent, *e.g.* PS in MEK or PEA in toluene, has a lower viscosity because the chains are coiled and undergo intramolecular rather than intermolecular entanglement.

The marginally compatible polyblend is analogous to the poor solvent case in that presence of the discrete dispersed phase results in less interaction between phases and, therefore, lower melt viscosities. Styrene–butadiene block copolymers have lower melt viscosities than random copolymers of the same composition and yield solutions of lower viscosity at the same concentration. This is because of the incompatibility but inseparability of the segments of the chain.

The PVC–acrylic polymer alloy previously described is characterized by high speed injection moldability and, in sheet form, by deep drawing capability, both of which are the result of low melt viscosity. A cellulose–ethyl acrylate graft copolymer compatibilizes cellulose and PEA, cellulose and PVC, as well as starch and PVC (8). These compositions containing PVC obviously have better flow properties than non-thermoplastic cellulose or starch. However, they also have better flow properties than PVC alone.

Polymer Dispersions and Solutions

The effect of the presence of compatibilized incompatible components is apparent in PVC plastisols. Monomeric and polymeric esters are good plasticizers for PVC because they have suitable solubility parameters. A good plasticizer is one which, in sufficient quantity, would almost be a solvent for the polymer. However, a good plasticizer, *i.e.* solvent, in a plastisol results in a high viscosity composition. This may be unsuitable for slush molding or other applications when low viscosity is desirable. The latter is obtained by adding a secondary plasticizer such as a hydrocarbon oil. In reality, the latter is not a plasticizer but actually a non-solvent which converts the good solvent plasticizer to a poor solvent mixture with resultant decrease in plastisol viscosity. A similar result is achieved by using a plasticizer with poor solvent characteristics by virtue of the presence of a long hydrocarbon side chain, *e.g.* dihexadecyl and dioctadecyl phthalate and trimellitate.

Solutions of incompatible polymers in a common solvent cannot be mixed without early demixing on standing which is analogous to the inability to maintain in the solid state the homogeneity of a blend of incompatible polymers undergoing mixing in the molten state. Thus, two layers result when a solution of PS in toluene is mixed with a solution of PEA in toluene. While the PS chains are extended in good-solvent toluene, the acrylic polymer chains are coiled in poor-solvent toluene. Absence of polymer mixing results in demixing. Similarly, when MEK is used as the common solvent, demixing results since MEK is a poor solvent for PS and a good solvent for the acrylic polymer. However, long term solution stability is achieved in the presence of PS–grafted PEA since the latter has segments which are compatible with and mix with the incompatible homopolymers. This effect has been termed oil-in-oil dispersion (*9, 10*).

A PS–grafted poly(ethylene oxide) as emulsifier permits preparation of a water-in-oil emulsion of PS. Solubility of the poly(ethylene oxide) segment in water results in solubilization of the latter in the PS matrix (*11*).

In situ generation of a compatibilizing agent permits preparation of non-aqueous dispersions of polymers in non-solvents. Polymerization of methyl methacrylate in the presence of linseed oil generates a PMMA–grafted linseed oil which promotes the dispersion of PMMA in mineral spirits. The appearance of the high solids dispersion resembles that of an aqueous dispersion or latex, *i.e.* translucent blue-white when polymer particle size is smaller than 0.05μ, white when $0.1–2\mu$, and chalky white when larger than 2μ (*12*).

Stable dispersions of very uniform particle size and solids contents below 80% can be prepared by using a graft copolymer of natural rubber as the compatibilizing or dispersing agent (*13*). Products based on hydroxystearic acid are even better dispersing agents (*14*). High solids dispersions of PMMA, PVC, and poly(vinyl acetate) in hydrocarbon solvents have been prepared by these techniques. Stability of the dispersions apparently results from the steric barrier formed by the solvent-soluble segments in the graft copolymer and, as has been shown, the repulsion forces are 10 times greater than the attraction forces (*15*).

Reinforced and Filled Polymer Composites

Use of wood flour, cotton, and other cellulosic products as fillers in phenolic resin and urea resin compositions is based on the reaction of these methylol-containing resins with the hydroxyl functionality of the cellulose, *i.e.* formation of a cellulose–phenolic resin or cellulose–urea resin compatibilizing agent. A cellulose–ethyl acrylate graft copolymer functions as a compatibilizing agent for incorporating non-thermoplastic, unreacted cellulose as a filler in PEA, PMMA, or PVC (*8*) (Diagram 6). The presence of an inorganic polymer–organic polymer graft copolymer or compatibilizing agent changes the role of

$$\begin{array}{c} \text{cellulose} \\ \text{cellulose} \text{———————} \text{PEA} \\ \text{PVC} \end{array} \qquad (6)$$

the inorganic polymer from a filler to a reinforcing agent, *i.e.* the filler crosslinks the organic polymer.

The interaction of glass fibers containing surface silanol groups with an unsaturated silane couping agent, *e.g.* vinyl triethoxysilane or methacrylato-propyltrialkoxysilane results in the appendage of reactive unsaturation on the fiber surface. When the latter reacts with a styrene-unsaturated (maleate or fumarate) polyester mixture, styrene copolymerizes with the unsaturation in the polyester and on the glass fiber surface. Thus, a glass–*x*–(styrene–polyester) graft copolymer serves to compatibilize the glass and the crosslinked polyester (*see* Reaction 7). When the silane coupling agent contains an amino group,

$$glass—\overset{|}{\underset{|}{Si}}—OH \ + \ (RO)_3Si—X—CH=CH_2 \ \longrightarrow \ glass—\overset{|}{\underset{|}{Si}}—O—\overset{|}{\underset{|}{Si}}—X—CH=CH_2$$

$$\diagdown \quad \diagdown \!\!\!\!\bigcirc \!\!\!\!\diagup \,—CH=CH_2$$

$$\hspace{6cm} H \ \ H \hspace{3cm} (7)$$

$$glass \ \diagup \diagdown \hspace{1cm} polyester—OOC—C=C—COO—$$

glass—*x*—polyester–S

polyester–S

reaction with the glass surface results in formation of surface amine groups which react with and become incorporated into a cured epoxy resin. Thus, a glass–*x*–epoxy resin graft copolymer compatibilizes the glass and the cured epoxy resin (Reaction 8).

$$glass—\overset{|}{\underset{|}{Si}}—OH \ + \ (RO)_3Si—X—NH_2 \ \longrightarrow \ glass—\overset{|}{\underset{|}{Si}}—O—\overset{|}{\underset{|}{Si}}—X—NH_2$$

$$glass \hspace{3cm} \diagup$$

$$\diagdown \!\!\!\!\swarrow \quad epoxy \ resin \hspace{3cm} (8)$$

glass—*x*—epoxy resin

epoxy resin

Silane coupling agents may also be used to generate compatibilizing agents from inorganic fillers and thermoplastic polymers. When a mixture of polyethylene, clay, an alkyl- or allyltrialkoxysilane, and dicumyl peroxide is prepared at 120°C, polymer and peroxide are unchanged. However, the hydroxyl functionality on the clay surface reacts with the silane coupling agent. When the compounded mixture is molded at 165°C, peroxide decomposes to generate free radicals which react with the polyethylene and the alkyl or allyl group on the silane resulting in coupling of the polymer and the silane. The *in situ* generated polyethylene–*x*–clay graft copolymer serves as compatibilizing agent for polymer and filler, the latter actually acting as a crosslinking agent for the polymer (*16*) (Reaction 9).

In situ polymerization of maleic anhydride in the presence of a polyolefin, a free radical catalyst, and a filler such as wood pulp, clay, or glass fibers results

$$
\begin{array}{ccc}
\text{polymer} & \text{polymer} & \left[\begin{array}{l}\text{polymer}\\ \text{XCH}_2\text{Si}\end{array}\right.\\
\text{XCH}_2\text{Si(OR)}_3 & \text{XCH}_2\text{Si}\rceil & \\
\xrightarrow{250°F} & \rceil \xrightarrow{350°C} & \left.\text{filler}\rfloor\right]\\
\text{filler} & \text{filler}\rfloor & \\
\text{peroxide} & \text{peroxide} &
\end{array}
\tag{9}
$$

polymer

polymer–x–filler

filler

in formation of a polyolefin–x–poly(maleic anhydride)–x–filler compatibiliz-ing agent. The poly(maleic anhydride) is grafted onto the polyolefin and interacts with the filler through covalency, *i.e.* reaction of the anhydride group with filler hydroxyl groups, and/or hydrogen bonding between carboxyl and hydroxyl groups (*17*). Fillers, fibrous reinforcing agents, powdered metals, and other materials having surface functionality are readily dispersed in polymers containing carboxyl functionality. The latter may be random copolymers such as poly(ethylene-*co*-methacrylic acid) or graft copolymers such as (acrylic acid-g-polyethylene) and (maleic anhydride-g-polypropylene). As in the pre-vious cases, the polymer–x–filler compatibilizing agent is generated at the polymer–filler interface.

In lieu of covalent or hydrogen bonds, the compatibilizing agent may contain ionic bonds. Thus, interaction of clay with a long chain unsaturated quaternary ammonium compound results in ionic bonding between the clay and the cationic ends of the organic molecule. Reaction of treated clay with a polyunsaturated monomer results in copolymerization of the surface unsatura-tion with the monomer. On blending the resultant encapsulated clay containing residual surface unsaturation by virtue of the presence of unreacted monomer unsaturation with polyethylene, radical sites generated on the latter react with unsaturation on the clay. A clay $^+$N$\sim\!\sim\!\sim$–x–polyethylene compatibilizing agent is thus generated *in situ* (*18, 19*).

Adhesives

Improved interfacial interaction promoted by the presence of a compati-bilizing agent in composites containing fillers or reinforcing agents is used in adhesive formulation. Adhesive for application of PVC floor tile to a concrete surface contains PMMA grafted onto natural rubber (*20*). The graft copolymer undergoes stratification when applied to the back of the tile because of the solubility parameter match between the PVC tile and the grafted PMMA

$$
\begin{array}{c}
\text{PVC}\\
\overline{\text{$\wedge\wedge\wedge$}}\\
\\
\text{PMMA}\\
|\\
\text{NR}\\
\\
\text{NR}\\
\\
\underline{\text{$\wedge\wedge\wedge\wedge\wedge$}}\\
\text{concrete}
\end{array}
\tag{10}
$$

branches. A conventional natural rubber (NR) adhesive is then applied to the floor before placing the tile down (Diagram 10).

The compatibility of PMMA and PVC is also used in adhesives for applying PVC decorative film to wooden or gypsum surfaces; a mixture of PMMA and an epoxy resin makes a useful adhesive. The less than 10% compatibility of PMMA with the epoxy resin limits the level of adhesion. However, by polymerization of methyl methacrylate in the presence of epoxy resin, up to 30% PMMA in the composition is compatibilized because of *in situ* generation of the graft copolymer (Diagram 11).

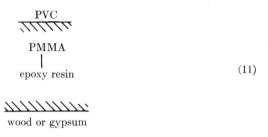

PVC

PMMA
|
epoxy resin

(11)

wood or gypsum

Application of solubility parameter match is also demonstrated in adhesives for polyester tire cord. Whereas resorcinol–formaldehyde resin is used in conjunction with polyvinylpyridine latex as an adhesive for rayon tire cord, this composition is not suitable for poly(ethylene terephthalate) (PET) cord. However, using hexylresorcinol rather than resorcinol results in a match of the solubility parameters of adhesive and fiber with resultant increased adhesion (*21*).

Similarly, a polyurethane containing polyester segments is an effective adhesive for lamination of PET film to wood. The polyester segment of polyurethane is compatible with the polyester film surface while the urethane linkages provide hydrogen bonding interaction between adhesive and substrate (Diagram 12).

PET

NHCOO–polyester–OCONH (12)

wood

Recapitulation

Preparation of polymer blends is limited by the general incompatibility of polymers. Compatibilization can be achieved by suitable compatibilizing agents, *i.e.* block or graft copolymers with segments of structure or solubility parameter similar to those of the polymers being mixed. A compatibilizing agent influences the properties of polymer blends and alloys as well as polymer melts and solutions. Solution of incompatible polymers in a common solvent as well as dispersion of a polymer in a non-solvent is promoted by a compatibilizing agent. Dispersion of additives such as fillers and reinforcing agents (including cellulose, clay, glass fibers, and powdered metals) in a polymer matrix is enhanced by the presence of a compatibilizing agent at the polymer–additive interface. Compatibilization of the interface also promotes the adhesion of solid polymers to polymeric or inorganic surfaces.

Literature Cited

1. Ceresa, R. J., "Block and Graft Copolymers," Butterworths, Washington, 1962.
2. Burlant, W. J., Hoffman, A. S., "Block and Graft Copolymers," Reinhold, New York, 1960.
3. Ceresa, R. J., "Block and Graft Copolymers," in "Encyclopedia of Polymer Science and Technology," Mark, H. F., Gaylord, N. G., Bikales, N. M., Eds., **2**, 485, Interscience, New York, 1965.
4. Rees, R. W., E. I. duPont de Nemours & Co., U.S. Patent **3,437,718** (1969).
5. Himei, S., Takine, M., Akita, K., Kanegafuchi Chemical Industry Co., Japanese Patent **866** (1967); *Chem. Abstr.* (1967) **67**, 22418.
6. Ryan, C. F., Crochewski, R. J., Rohm & Haas Co., U.S. Patent **3,426,101** (1969).
7. Souder, L. C., Larsson, B. E., Rohm and Haas Co., U.S. Patent **3,251,904** (1966).
8. Gaylord, N. G., U. S. Plywood–Champion Papers, Inc., U.S. Patent **3,485,777** (1969).
9. Molau, G. E., *J. Polymer Sci. Part A* (1965) **3**, 1267; *Ibid.*, **3**, 4235.
10. Riess, G., Periard, J., Banderet, A., in "Colloidal and Morphological Behavior of Block and Graft Copolymers," Molau, G. E., Ed., p. 173, Plenum, New York, 1971.
11. Bartl, H., Bonin, W. V., *Makromol. Chem.* (1962) **57**, 74.
12. Schmidle, C. J., Brown, G. L., Rohm & Haas Co., U.S. Patent **3,232,903** (1966).
13. Osmond, D. W. J., Thompson, M. H., Imperial Chemical Industries, Ltd., British Patent **893,429** (1962); U.S. Patent **3,095,388** (1963).
14. Osmond, D. W. J., Waite, F. A., Walbridge, D. J., Imperial Chemical Industries, Ltd., U.S. Patent **3,514,500** (1970).
15. Dowbenko, R., Hart, D. P., *Ind. Eng. Chem. Prod. Res. Develop.* (1973) **12**, 14.
16. Union Carbide Corp., "Silane Adhesion Promoters in Mineral-Filled Composites," *Tech. Bull.* (1973) **F-43598**.
17. Gaylord, N. G., U. S. Plywood–Champion Papers, Inc., U.S. Patent **3,645,939** (1972).
18. Amicon Corp., French Patent **1,539,053** (1968).
19. Hausselein, R. W., Fallick, G. J., *Appl. Polym. Symp.* (1969) **11**, 119.
20. Bevan, A. R., Bloomfield, G. F., *Adhes. Age* (1964) **7** (2), 36; *Ibid.*, **7** (3), 34.
21. Iyengar, Y., Erickson, D. E., *J. Appl. Polym. Sci.* (1967) **11**, 2311.

RECEIVED April 3, 1974.

Structure and Properties of Random, Alternating, and Block Copolymers: The UV Spectra of Styrene–Methyl Methacrylate Copolymers

BIANCA M. GALLO and SAVERIO RUSSO

Istituto di Chimica Industriale, Università, Via Pastore-3, 16132 Genoa, Italy

UV absorption spectra of several styrene copolymers have significant differences and anomalies from those of polystyrene; these are peculiarly evident for styrene–methyl methacrylate random copolymers. Poly(styrene-co-methyl methacrylate) solutions exhibit marked hypochromism at 269.5 nm that is caused by very strong interactions between methyl methacrylate mers and chromophore units, i.e. phenyl rings. Copolymer compositions where hypochromism appears are strongly affected by the solvent. Absorbances (269.5 nm) of copolymer solutions in chloroform, DCE, TCE, THF, and dioxane were compared with optical densities of polystyrene solutions. There is a linear correlation between solvent dielectric constant and copolymer composition corresponding to the maximum hypochromism. Absorbance data on alternating and block copolymers suggest specific interactions between methyl methacrylate and styrene units. Hypochromism at 269.5 nm can severely limit the use of UV spectrometers for evaluating chemical composition of St–MMA random copolymers, with implications for GPC analysis of other styrene copolymers.

The UV absorption in the 260 nm region is frequently used to evaluate styrene content in styrene-based polymers (*1, 2, 3, 4, 5, 6, 7*). Calibration curves for polystyrene solutions are usually based on the assumptions that the UV absorption of the copolymer depends only on the total concentration of phenyl rings, and the same linear relationship between optical density and styrene concentration that is valid for polystyrene holds also for its copolymers. These assumptions are quite often incorrect and have caused sizable errors in the analysis of several statistical copolymers. For example, anomalous patterns of UV spectra are given by random copolymers of styrene and acrylonitrile (*8*), styrene and butadiene (*8*), styrene and maleic anhydride (*8*), and styrene and methyl methacrylate (*9, 10, 11*). Indeed, the co-monomer unit can exert a marked influence on the position of the band maxima and/or the extinction

Table I. Summary of Data

Sample	Ref.	Polymerization Mechanism	Polymerization Temperature, °C	$\overline{M}_n \times 10^{-3}$
Polystyrene	13,14	free-radical polymerization in bulk	50–80	80–140
Polymethyl methacrylate	13,14	free-radical polymerization in bulk	50–80	300–640
Random copolymers	13,14,15	free-radical polymerization in bulk	50–80	80–380
Alternating copolymer	16	monomer complexation	25	250
Triblock copolymer	17	anionic copolymerization	−78–0	300

coefficients related to the absorption of the aromatic ring, even if it does not absorb appreciably in that region of the UV spectrum. Such effects severely limit the use of UV spectrometers for evaluating copolymer composition as in GPC apparatus (5, 6, 7, 12).

The present study accounts for the UV absorption of poly(styrene-co-methyl methacrylate) solutions at 269.5 nm, a peak which is part of the fine vibrational structure of the band centered around 260 nm. Tobolsky et al. (3) recommended this band because of the significant absorption of poly-(methyl methacrylate) at higher frequencies. This band refers to the lowest energy forbidden $\pi \rightarrow \pi^*$ electron transition of the phenyl ring.

Results

Table I summarizes some of the data obtained in this study. Copolymer composition has been evaluated by elemental analysis (15) and confirmed by NMR spectroscopy. Random copolymers show an alternation tendency expressed by the product $r_1 \cdot r_2$ of 0.218–0.252 (15).

Optical density was measured with a single-beam Zeiss spectrophotometer at 20° ± 0.05°C (unless stated otherwise). No effect of molecular weight on absorbance values was detected. Lambert-Beer law was obeyed by every solution, at least up to a concentration of about 2×10^{-2} mole/liter. It was not considered necessary to introduce any correction for scattered light (18).

Optical densities at 269.5 nm for polystyrene solutions at concentrations of $0-1 \times 10^{-2}$ mole/liter and for poly(styrene-co-methyl methacrylate) solutions at a total concentration of 1×10^{-2} mole/liter are presented in Figure 1 as functions of styrene content. The solvents were (from the top): dioxane, chloroform, tetrahydrofuran (THF), tetrachloroethane (TCE), and dichloroethane (DCE). It is evident that the linear relationship between optical density and styrene concentration that is valid for a polystyrene at all concentrations (open circles) does not hold for the statistical copolymers (solid circles). For example, copolymer (25–80 mole % styrene) solutions in chloroform deviate markedly from linearity; the maximum per cent decrease in extinction coefficient (hypochromism) corresponds to a copolymer containing 50 mole % styrene. We define hypochromism as the decrease in absorption intensity at 269.5 nm per chromophore of the statistical copolymer relative to that of the atactic polystyrene. It is also evident from Figure 1 that the alternating copolymer also gives a sharp hypochromism whereas block copolymers and mechanical mixtures of polystyrene and poly(methyl methacrylate) do not deviate from the straight line. Similar results were obtained with the other solvents, but the composition range where hypochromism appears depends on the solvent used.

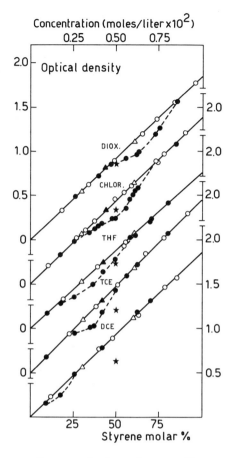

Figure 1.   Optical density at 269.5 nm

●, absorbance data (total copolymer concentration, 1 × 10⁻² mole/liter) vs. styrene content in the copolymer; O, polystyrene solutions; △, mechanical mixtures of polystyrene and polymethyl methacrylate; ▲, block copolymer; ★, alternating copolymer

In all cases, the hypochromic effect is accompanied by a negligible red shift (bathochromism).

In Table II, the maximum per cent hypochromism, that is the decrease in intensity absorption per chromophore referred to the maximum hypochromism, and the corresponding copolymer composition are presented as functions of the solvent used.

Discussion

A first approach to investigating the origin of hypochromism at 269.5 nm in styrene–methyl methacrylate random copolymers is based on its similarity to the hypochromism of isotactic polystyrene as compared with the atactic poly-

Table II.   Hypochromism at 269.5 nm as a Function of Copolymer Composition and Solvent

Solvent	Maximum Hypochromism, %	Styrene Content in Copolymer, mole %
Dioxane	15.5 ± 1.3	~63
Chloroform	19.2 ± 1.3	~50
Tetrahydrofuran	24.0 ± 1.2	~36
Tetrachloroethane	24.8 ± 1.2	~34
Dichloroethane	27.9 ± 1.3	~19

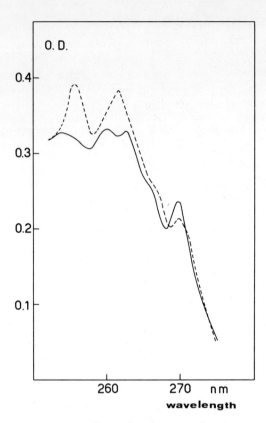

Figure 2. Profile of the absorption band centered around 260 nm (solvent: chloroform)
———, *polystyrene; – – –, styrene–methyl methacrylate random copolymer (50% St)*

mer. As is well known, there are two notable aspects of UV absorption spectra of polystyrene solutions:

(a) The whole band of isotactic polystyrene centered around 260 nm shows a marked hypochromism in comparison with that of the atactic polymer (*19*); and

(b) According to Reiss (*20*), the optical density at 260 nm of isotactic polystyrene in cyclohexane shows a significant, reversible change around 80°C. A similar transition is also supposed to be present in atactic polystyrene. At higher temperatures, the optical densities of isotactic and atactic polystyrene approach that of ethylbenzene. On this basis, some ordered structures (helical sequences) are supposed to be present in polystyrene solutions. Increasing the temperature destroys the preferential interactions between phenyl rings in the helical sequence. Hypochromism, therefore, is related to the conformational properties of the polystyrene chain in solution.

On the contrary, a copolymer sample characterized by a styrene content corresponding to the maximum hypochromism does not demonstrate any reversible change in optical density at 269.5 nm as a function of temperature. Moreover, the geometry and intensity of the whole band centered around 260 nm are quite different from those of polystyrene (Figure 2). For copolymers, hypochromism seems to be confined only to the peak at 269.5 nm. A more detailed discussion of these differences will be published elsewhere (*11*), but we can already report that the hypochromism of styrene–methyl methacrylate copolymers differs from that of isotactic polystyrene.

The maximum per cent hypochromism and the associated copolymer composition appear to be functions of the solvent used (*see* Table II); however, no simple correlation with solvent thermodynamic power was detected. To the contrary, a linear correlation with the solvent dielectric constant was demonstrated. In fact, the copolymer composition corresponding to the maximum hypochromism at 269.5 nm is lineally dependent on the solvent dielectric constant (*see* Figure 3). The intercept corresponding to the value dielectric constant = 1 (which means that there are no interactions of solvents) gives the copolymer composition really responsible for the maximum hypochromism, namely about 30 mole % methyl methacrylate. Furthermore, the maximum per cent hypochromism is a linear function of the solvent dielectric constant (*see* Figure 3); extrapolation gives a value of about 14%.

So far, we have proposed a correlation between hypochromic effect and copolymer composition. In order to explicate this dependence, it is convenient to analyze hypochromism in terms of copolymer structure—that is in terms of sequence distribution. In an approach suggested by Stützel *et al.* (*12*), the molar extinction coefficient (ε) of the random copolymer at a given wavelength can be written as follows (A = styrene, B = methyl methacrylate):

$$\varepsilon_{\text{copolymer}} = f_{AAA}\,\varepsilon_{AAA} + f_{BAA}\,\varepsilon_{BAA} + f_{BAB}\,\varepsilon_{BAB} \tag{1}$$

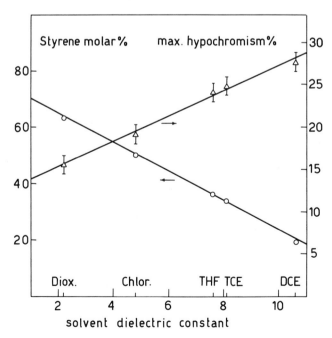

Figure 3. Correlation between hypochromism and solvent dielectric constant

where f_{AAA}, f_{BAA}, and f_{BAB} denote the fractions of A monomers found in the middle of triads AAA, BAA (and AAB) and BAB. ε_{AAA} can be assumed equal to the extinction coefficient of polystyrene, and ε_{BAB} to that of the alternating copolymer.

It follows that the difference between the optical density of polystyrene and that of a statistical copolymer is given by:

$$\Delta = \frac{10^{-2}}{(r_1 x + 1)\left(2 + r_1 x + \frac{r_2}{x}\right)}\left[2r_1 x\,(\varepsilon_{PS} - \varepsilon_{BAA}) + (\varepsilon_{PS} - \varepsilon_{ALT})\right] \qquad (2)$$

where r_1 and r_2 are the reactivity ratios and x is the molar ratio in the feed. In terms of run number (21) and styrene content in the copolymer, Equation 2 becomes:

$$\Delta \cdot 10^4 = \left(R - \frac{R^2/2}{\%St}\right)(\varepsilon_{PS} - \varepsilon_{BAA}) + \frac{R^2/4}{\%St}(\varepsilon_{PS} - \varepsilon_{ALT}) = ak_1 + bk_2 \qquad (3)$$

In Figure 4, coefficients a and b of Equation 3 are plotted as functions of styrene concentration in the statistical copolymer. The data for calculation refer to copolymers synthesized in bulk at 60°C ($r_1 = 0.50$; $r_2 = 0.47$) (15).

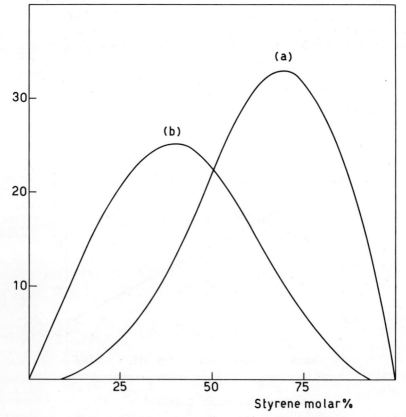

Figure 4. *Effect of styrene content in the copolymer on coefficients* a *(related to ε_{BAA}) and* b *(related to ε_{ALT})*

It is evident that the coefficient of ε_{BAA} (a) is higher at copolymer compositions characterized by larger amounts of styrene whereas the opposite is valid for the coefficient of ε_{BAB} (b). Solvent effects on the hypochromism of styrene–methyl methacrylate random copolymers seem to be correlated, at least qualitatively, to the variation in a and b as functions of styrene content in the copolymer. Specific solvent effects on the extinction coefficients ε_{AAA}, ε_{BAA}, and ε_{BAB} account for the quantitative aspects of hypochromism. Full details will be given (11).

Conclusions

Solvent effects on the optical densities of alternating and random copolymers of styrene and methyl methacrylate can explain the variation in hypochromism at 269.5 nm (*see* Figure 1 and Table II). Actually, the hypochromism seems to depend on specific interactions between methyl methacrylate mers and styrene mers. Good correlations between hypochromism and chemical structure of the copolymers can be deduced on the basis of the sequence length distribution.

Anyhow, careful choice of solvent can confine hypochromism to copolymer compositions characterized by very low or very high styrene content. For this purpose, solvents of low dielectric constant seem preferable for copolymers containing less than 50% styrene, and solvents with relatively high dielectric constants would be best for high styrene copolymers (*see* Figures 1 and 3). This can overcome criticism of the use of a UV spectrometer as a second detector for GPC analyses of styrene copolymers.

Literature Cited

1. Meehan, E. J., *J. Polym. Sci.* (1946) **1**, 175.
2. Funt, B. L., Collins, E., *J. Polym. Sci.* (1958) **28**, 359.
3. Tobolsky, A. V., Eisenberg, A., O'Driscoll, K. F., *Anal. Chem.* (1959) **31**, 203.
4. Gruber, U., Elias, H.-G., *Makromol. Chem.* (1965) **84**, 168.
5. Cantow, H.-J., Probst, J., Stojanov, C., *Kaut. Gummi Kunstst.* (1968) **21**, 609.
6. Runyon, J. R., Barnes, D. E., Rudd, J. F., Tung, L. H., *J. Appl. Polym. Sci.* (1969) **13**, 2359.
7. Adams, H. E., *Separ. Sci.* (1971) **6**, 259.
8. Brüssau, R. J., Stein, D. J., *Angew. Makromol. Chem.* (1970) **12**, 59.
9. O'Driscoll, K. F., Werz, W., Husar, A., *J. Polym. Sci. Part A–1* (1967) **5**, 2159.
10. Gallo, B. M., Russo, S., Part 1, *J. Macromol. Sci. Chem.* (1974) **8**, 521.
11. Gallo, B. M., Russo, S., Part 2, *J. Macromol. Sci. Chem.*, in press.
12. Stützel, B., Miyamoto, T., Cantow, H.-J., IUPAC Intern. Symp. Macromol., Helsinki (1972) *Prepr.* **II–59**, 337.
13. Bontá, G., Gallo, B. M., Russo, S., *J. Chem. Soc. Faraday Trans. I* (1973) **69**, 328.
14. Gallo, B. M., Pallesi, B., Pedemonte, E., Russo, S., unpublished data.
15. Russo, S., Gallo, B. M., Bontá, G., *Chim. Ind. Milan* (1972) **54**, 521.
16. Hirooka, M., Yabuuchi, H., Iseki, J., Nakai, Y., *J. Polym. Sci. Part A–1* (1968) **6**, 1381.
17. Ohnuma, H., Kotaka, T., Inagaki, H., *Polymer* (1969) **10**, 501.
18. Loux, G., Weill, G., *J. Chim. Phys.* (1964) **61**, 484.
19. Vala, Jr., M. T., Rice, S. A., *J. Chem. Phys.* (1963) **39**, 2348.
20. Reiss, C., *J. Chim. Phys.* (1966) **63**, 1307.
21. Harwood, H. J., Ritchey, W. M., *J. Polym. Sci. Part B* (1964) **2**, 601.

RECEIVED April 3, 1974.

9

Vinylene Carbonate—A Study of Its Polymerization and Copolymerization Behavior

M. KREBS and C. SCHNEIDER

Institute of Physical Chemistry, University of Cologne, Cologne, Germany

Bulk polymerization of vinylene carbonate (VCA) initiated by ^{60}Co γ-rays was studied at 30°–110°C at a constant dose rate of $1 \cdot 10^5$ rad/hr. An overall activation energy of 5.0 kcal/mole and a maximum reaction rate of $1 \cdot 10^{-3}$ mole/l-sec were obtained. As has been reported, purification of the monomer is a crucial point because inhibiting impurities are formed during the synthesis. From experiments with chlorine-substituted ethylene and vinylene carbonates, we tentatively conclude that, in addition to mono- and dichloroethylene carbonate, dichlorovinylene carbonate is mainly responsible for the inhibition. The copolymerization behavior of VCA with some chlorine-substituted olefins was studied. Chlorotrifluoroethylene (CTFE) is an especially suitable comonomer; the reactivity ratios found were $r_{VCA} = 0.42$ and $r_{CTFE} = 0.48$.

Generally, 1,2-disubstituted ethylene derivatives have only a small tendency for radical homopolymerization. An exception is vinylene carbonate (VCA) which can be easily polymerized by chemical as well as radiation initiation. However, the reaction is strongly affected by traces of impurities formed during the synthesis. Inhibition experiments are discussed with regard to the nature of the inhibiting impurities. The copolymerization behavior of VCA with some halo-substituted olefins was studied; with chlorotrifluoroethylene (CTFE), a statistical copolymer with a slight tendency for alternation was obtained.

VCA, the cyclic carbonic acid ester of an enediol is a remarkable

exception to the well known rule that 1,2-disubstituted ethylene derivatives show only a small tendency for radical homopolymerization. Since the synthesis of VCA was first described by Newman and Addor (1), its polymerization behavior has been studied by various research groups (*2, 3, 4, 5, 6, 7, 8*). The purification of the monomer, however, turned out to be a crucial point, and kinetic results were considerably affected by unknown inhibiting impurities. Although various methods have been developed to improve monomer quality (*5, 9, 10, 11*), the nature of the inhibiting impurities is not yet clear.

The copolymerization behavior of VCA has been the subject of several investigations including some in industry (*12–20*). Comonomers of different types have been used for copolymerization experiments; however, with the exception of N-vinylpyrrolidone (*14*) and isobutyl vinyl ether (*20*), VCA was incorporated into the copolymer only slightly.

Experimental

VCA was prepared according to the procedure described by Field and Schaefgen (*5*). For bulk polymerization, the monomer was sealed in glass ampoules and irradiated at 31°–111°C and a dose rate of $1 \cdot 10^5$ rad/hr using ^{60}Co γ-rays. The polymer obtained was dissolved in dimethylformamide (DMF) and then precipitated with methanol.

All copolymerization experiments were carried out in DMF (1.5 ml) at 65°C using azobis(isobutyronitrile) (AIBN) as initiator. The concentration of AIBN was kept constant (0.01 mole/kg monomer mixture), and the total weight of both monomers was always 3 g. The products were precipitated in a water/methanol mixture, and they were characterized by elemental analysis and IR spectrometry.

Results and Discussion

Homopolymerization of VCA. Radiation-induced homopolymerization of VCA was investigated by Hardy and Nyitrai (*8*). They showed that the rates obtained with radiation initiation were extremely low (maximum rate, 0.2% conversion/hr at 50°C and $9.3 \cdot 10^4$ R/hr) and much smaller than with conventional initiation. However, our findings did not confirm these results. As can be seen from Figure 1, a plot of conversion *vs.* irradiation time for dif-

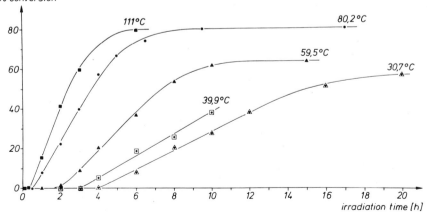

Figure 1. Dependence of conversion on irradiation time at temperature of 30.7°–111°C at a dose rate of $1 \cdot 10^5$ rad/hr

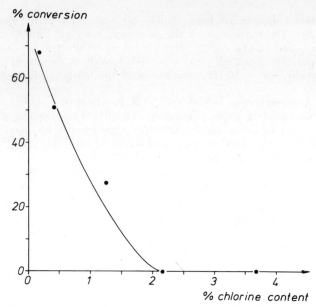

Figure 2. Dependence of conversion on the chlorine content of the monomer (values taken from Ref. 11)

ferent temperatures at constant dose rate, radiation-induced polymerization of VCA is a fast process. From an Arrhenius plot, an overall activation energy of 5.0 kcal/mole was found; this is in the normal range for radiation-induced radical polymerization processes. At higher temperatures, the rate decreased. Even small conversions produced an appreciable increase in viscosity, and, at conversions above 10%, the sample was a transparent, glassy solid. The polymers obtained were colorless, hard, and brittle; the molecular weight was in the order of 10^5.

A typical feature of the conversion–time curves is the presence of induction periods, the lengths of which depend on temperature and dose rate. We assume that they were caused by traces of impurities which, despite the very careful purification process, were still present in the monomer. It is very likely that the low rates obtained by Hardy and Nyitrai were caused by impurities, and consequently the molecular weights of their polymers were about two orders of magnitude smaller than ours.

The Nature of the Inhibiting Impurities. The problem of suitable purification techniques and synthesis conditions has been discussed before (*5, 9, 10, 11*). Thomas, who failed to get high molecular weight polymer, suggested (*10*) that this was attributable to traces of chlorinated or other-wise impure polymer which could not be removed by fractional distillation. Zief and Schramm (*11*) demonstrated that the conversion of monomer depends on its chlorine content (Figure 2). The chlorine content may therefore be considered a criterion of purity.

Since the synthesis of VCA occurs *via* the following reaction, chlorine-containing substances may be formed as intermediates or byproducts or by rearrangement during the chlorination reaction, and they may be responsible for the inhibition periods.

Field and Schaefgen (5) demonstrated that a high molecular weight poly(vinylene carbonate) (PVCA) was obtained if the first step of the synthesis, photochlorination of ethylene carbonate, was carried out in CCl_4 instead of bulk reaction and if the monomer was distilled from $NaBH_4$ shortly before polymerization. On the other hand, bulk chlorination led invariably to poor quality VCA.

The monomer used for the homopolymerization experiments reported here was synthesized with strict adherence to the method of Field and Schaefgen, and it was of very good quality. A second sample of VCA, synthesized under more vigorous conditions using less CCl_4 and a higher chlorination rate, could not be polymerized by γ-rays under the given conditions. Nevertheless, the monomer had the exact melting and boiling points and the same properties as the first sample with only two exceptions: it failed to polymerize with γ-rays, and its chlorine content was about 0.1% whereas the other's was only about 0.01%.

It has been reported (11, 21) that the presence of mono- and dichloroethylene carbonate decreases the yield and molecular weight of PVCA. This can be confirmed for the radiation-induced polymerization when these compounds are present at levels above 1%. However, as inhibition experiments demonstrated, the inhibition effectiveness of mono- and dichloroethylene carbonate was too small to account for the induction period observed. Furthermore, even after careful separation of these two substances, an unpolymerizable monomer with a chlorine content of about 0.1% was obtained. It should be noted that radiation-initiated polymerization of VCA is especially sensitive to inhibiting impurities, and that the induction period can be easily overrun by using conventional initiators.

We attribute the observed inhibition to dichlorovinylene carbonate. There are good arguments for this assumption: the photochlorination of ethylene carbonate is not restricted to monosubstitution, but it leads to a whole spectrum of chlorinated products. Separation by rectification is difficult since the chlorine-substituted derivatives of ethylene and vinylene carbonate show a marked anomaly of boiling points (1). One byproduct of the chlorination reaction is trichloroethylene carbonate which, in spite of careful distillation, remains in traces with the monochloro compound and is dehydrohalogenated in the second step of the synthesis to give dichlorovinylene carbonate.

To test our assumption, we conducted polymerization experiments with small amounts of dichlorovinylene carbonate added to VCA. The inhibition effect was marked; addition of even less than 0.5% dichlorovinylene carbonate completely suppressed radiation-induced polymerization of VCA under the given conditions. Although this may not be considered convincing proof, it is supported by the effect of reaction conditions on monomer quality: the amount of trichloroethylene carbonate increases with more drastic chlorination condi-

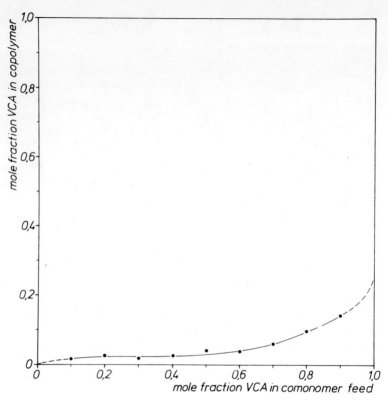

Figure 3. Copolymerization diagram of the system vinylene carbonate/ vinylidene chloride at 65°C

tions and a rather poor monomer should be obtained if the chlorination is carried out in bulk or at higher temperatures. This agrees with the findings of Field and Schaefgen and with our own experiments. In order to obtain a polymerization-grade monomer, VCA should therefore be synthesized in an excess of CCl_4 under mild reaction conditions, and the raw monomer should be distilled from $NaBH_4$ at least twice.

It is very likely that chain transfer by abstraction of a chlorine atom from the impurity leading to a less reactive radical is responsible for the observed inhibition. Smets and Hayashi (2) observed that adding small amounts of acetone or benzene considerably decreased the rate of homopolymerization. The low resonance stabilization of the VCA macroradical obviously favors chain transfer reactions.

Copolymerization of Vinylene Carbonate with Some Halo-Substituted Olefins. Copolymerization experiments were conducted using *trans*-dichloroethylene, vinylidene chloride, and CTFE since these monomers have a structural relation to the inhibiting impurities discussed above. With *trans*-dichloroethylene, no polymerization occurred, and only oligomers of VCA with a molecular weight of ∼300 were formed. Like dichlorovinylene carbonate, *trans*-dichloroethylene acts as an inhibitor, probably through degradative chain transfer by abstraction of a chlorine atom.

With vinylidene chloride, VCA was incorporated into the copolymer only to a small extent (Figure 3). It is assumed that the unfavorable difference between the Q values of the monomers ($Q_{VCl_2} = 0.2$ and $Q_{VCA} = 0.007$) as well as the sensitivity of the VCA radical to chain transfer with vinylidene chloride hinder a higher incorporation of VCA. Since the copolymer precipitated during the polymerization, this system could not be used for evaluation of copolymerization characteristics.

Quite different results were obtained when CTFE was used as a comonomer. A statistical copolymer with a slight tendency for alternation was formed (Figure 4). Reactivity ratios were $r_{VCA} = 0.42$ and $r_{CTFE} = 0.48$. The copolymer obtained was a white solid which was soluble in various solvents, *e.g.* acetone and DMF. The molecular weight was rather low ($M_n \approx 2 \cdot 10^3$) which indicates that chain transfer occurred in this system as in the others mentioned above. In addition to the VCA/N-vinylpyrrolidone (*14*) and VCA/isobutyl vinyl ether (*20*) systems, this is a third system which permits incorporation of VCA at high levels. There was, however, one remarkable difference: whereas with the other two systems the rate decreased with increasing VCA content in the feed, this did not occur with the VCA/CTFE system. As can be seen in Figure 5, the reaction rate passed through a maximum at

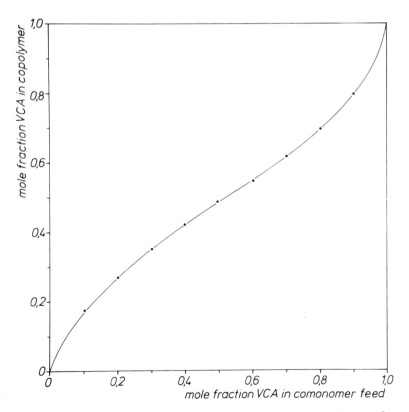

Figure 4. Copolymerization diagram of the system vinylene carbonate/chlorotrifluoroethylene at 65°C

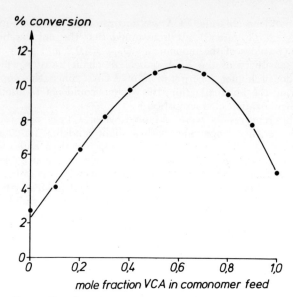

Figure 5. Plot of conversion vs. mole fraction of VCA in comonomer feed for the system vinylene carbonate/ chlorotrifluoroethylene at 65°C (reaction time, 45 min)

~60 mole % VCA in the monomer mixture, a maximum rate that is more than fourfold that of CTFE and more than twice that of pure VCA.

Acknowledgments

We are indebted to H. D. Scharf and G. Hesse for generous gifts of reagents and to R. C. Schulz for informing us about the unpublished results of his studies on vinylene carbonate.

Literature Cited

1. Newman, M. S., Addor, R. W., *J. Amer. Chem. Soc.* (1953) **75**, 1263.
2. Smets, G., Hayashi, K., *J. Polym. Sci.* (1958) **29**, 257.
3. Haas, H. C., Schuler, N. W., *J. Polym. Sci.* (1958) **31**, 237.
4. Ham, G. E., Zief, M., U.S. Patent **2,993,030** (1961).
5. Field, N. D., Schaefgen, J. R., *J. Polym. Sci.* (1962) **58**, 533.
6. Kazanskaya, V. F., Klimova, O. M., *Zh. Prikl. Khim. Leningrad* (1965) **38**, 432.
7. Vollkommer, N., Thesis, Mainz, 1967.
8. Hardy, G., Nyitrai, K., *Acta Chim. A. Sc. Hung.* (1968) **56**, 39.
9. Nesbitt, B. F., Goodman, I., British Patent **899,205** (1962).
10. Thomas, R. M., U.S. Patent **2,873,230** (1959).
11. Zief, M., Schramm, C. H., U.S. Patent **3,041,353** (1962).
12. Price, J. A., Padbury, J. J., U.S. Patent **2,722,525** (1955).
13. Gluesenkamp, E. W., Calfee, J. D., U.S. Patents **2,847,398** (1958), **2,847,401** (1958), and **2,847,402** (1958).
14. Hayashi, K., Smets, G., *J. Polym. Sci.* (1958) **27**, 275.
15. Overberger, C. G., Biletch, H., Nickerson, R. G., *J. Polym. Sci.* (1958) **27**, 381.
16. Judge, J. M., Price, C. C., *J. Polym. Sci.* (1959) **41**, 435.
17. Marder, H. L., Schuerch, C., *J. Polym. Sci.* (1960) **44**, 129.
18. Klimova, O. M., Kazanskaya, V. F., *Zh. Prikl. Khim. Leningrad* (1965) **38**, 434.
19. Wolf, R., Thesis, Mainz, 1965.
20. Schulz, R. C., Wolf, R., *Kolloid Z. Z. Polym.* (1967) **220**, 148.
21. Cantor, P. A., Kesting, R. E., U.S. Patent **3,332,894** (1967).

RECEIVED May 7, 1974. Work supported by the Deutsche Forschungsgemeinschaft.

10

Process for the Agglomeration of Latex Particles Based on Electrolyte Sensitization

HERBERT SCHLUETER

Chemische Werke Huels AG, Production Department 4, 4370 Marl, Kreis Recklinghausen, West Germany

Agglomeration rate studies were made with SBR base latex and oxidized poly(ethylene oxide)s (PEO). Included in the investigation were a number of factors which demonstrated that the colloidal basis of this agglomeration is a sensitization to electrolytes. Small amounts of PEO accelerated the rate 10^7-fold compared with unsensitized electrolyte agglomeration. This rate difference is attributable to a cooperative action of PEO and anionic emulsifier; there is no necessity to postulate a bridging mechanism. The amount of electrolyte and emulsifier normally needed for the polymerization of SBR latexes may be used for a fast agglomeration, and some technologically important factors may be changed at will without any resultant coagulum formation. This makes the process versatile for large scale operations.

Reactions based on interactions between hydrophilic polymers and hydrophobic colloids are common in nature and in industry. In biological systems, many reactions and processes have been interpreted in terms of polymer–colloid interactions (*1, 2, 3*). In various branches of industry, hydrophilic polymers are used as stabilizers or destabilizers (*4, 5, 6*). Especially in latex technology, hydrophilic polymers are used as agglomeration activators (*7, 8*). In this case, agglomeration means aggregation and coalescence of small particles into larger ones, as required for fluidity at high latex concentrations.

Often stabilization and destabilization may occur in the same system. At high polymer concentrations the system may be strongly stabilized whereas at low concentrations the dispersion may flocculate. In the latter case, a distinction can be made between (*a*) adsorption flocculation (*i.e.*, flocculation by polymer only) and (*b*) sensitization (*i.e.*, in addition to polymer some electrolyte is needed to ensure complete flocculation).

The mechanism of sensitization is far from clear. Basically, there are different concepts regarding the roles of polymer and electrolyte in this process. In one case, for example, the effective charge density on particle surface is lowered by polymer adsorption so that the colloidal particle is made sensitive to electrolyte, and it can be flocculated even at lower electrolyte concentrations than in the absence of polymer (*9*). In the other case, bridging between the particles through segments of the polymer molecule is the deciding factor, and a certain critical amount of electrolyte is needed (10). The electrolyte reduces the thickness of the diffuse double layer, and this could make the polymer more effective since additional lengths of polymer molecule would then be available for bridging. Along these lines, the agglomeration of latex particles promoted by poly(methyl vinyl ether) (PVM) has been discussed (11).

In this connection, the question arises how synthetic rubber particle agglomeration carried out with oxidized poly(ethylene oxide) (PEO) (12) can be interpreted in colloidal terms. This agglomeration may also be defined in terms of polymer–colloid reactions as a flocculation occurring at low polymer concentrations which, however, is restricted to the colloidal particle size range (13). The reason for termination of agglomeration, short of latex particle coagulation, is not only of interest because of the common occurrence of polymer–colloid interactions in nature and industry, as has already been mentioned, but because it is of major importance especially in latex technology for further developments in the field of controlled agglomeration with hydrophilic polymers. Attempts to answer these questions by quantitative rate studies are reported here.

Experimental

Materials. STYRENE–BUTADIENE RUBBER (SBR) LATEX. SBR latex was prepared by redox emulsion polymerization using (in parts): butadiene (69) and styrene (31) at 6°–40°C (pinane hydroperoxide/sodium formaldehyde sulfoxylate/Fe^{++} as initiator) in the presence of potassium oleate (2.7) inorganic electrolytes (0.45) as polymerization aids, and demineralized water (135) until a conversion of 70% was achieved. Residual monomers were then removed.

The latex had the following physical properties: specific surface, 106 m^2/g; soap coverage, 26%; and particle concentration, 5.3 × 10^{15}/ml [corresponding to 32.5% latex total solids (TS)]. The size distributions of the latex particles are plotted in Figure 1. The size scale in all figures showing particle size distributions is logarithmic for convenience in presentation since the agglomerated latexes have relatively large size ranges.

In addition, SBR latexes with less emulsifier and inorganic electrolyte as well as some with larger particles and modified particle size distributions were also prepared for controlled variations of these factors.

CARBOXYLIC STYRENE–BUTADIENE (SB) LATEX. The carboxylic latex was prepared by emulsion polymerization at 60°C using (in parts): butadiene (40), styrene (57.5), and acrylic acid (2.5) in the presence of demineralized water (138), ^{14}C-sulfonate (0.5) as emulsifier, and tertiary dodecyl mercaptan (0.5) and ammonium persulfate (0.5) as initiator.

POLY(ETHYLENE OXIDE). PEO was prepared using polyethylene glycols of commercial grade (12).

POTASSIUM OLEATE. Pamolyn 100, a product of Hercules Powder Ltd., was used.

INORGANIC ELECTROLYTE. The concentration of potassium ions, [K$^+$], in the aqueous latex phase was varied using potassium chloride and potassium sulfate (analytical reagent grade) in a 2:1 ratio.

Methods. DETERMINATION OF AGGLOMERATION RATE. The kinetics of PEO agglomeration may be described in a wide initial range (up to 50–70% of total agglomeration) by all three forms of Smoluchowski's equation, in which the reciprocal particle number $(1/N)$, the mean particle volume $(\bar{v})$, and the reciprocal cube of the particle surface $(1/\Sigma^3)$ vary linearly with time (t) (14,

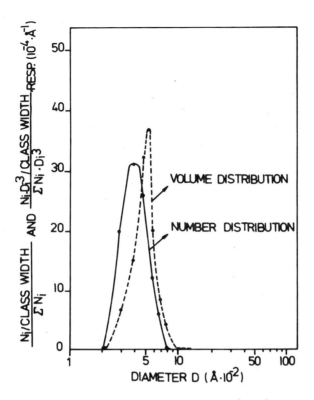

Figure 1. Particle size distributions of base latex
N_i, number of particles with diameter D_i; $N_i \cdot D_i^3$, volume of particles with diameter D_i; ΣN_i, sum of total N_i; and $\Sigma N_i \cdot D_i^3$, sum of total $N_i \cdot D_i^3$

15, 16). $1/N_o$, $\bar{v}_o$, and $1/\Sigma_o^3$ are the corresponding values of the base latex when $t = 0$. Φ represents the volume fraction of the latex particles. The

$$1/N - 1/N_o = k_N \cdot t \qquad (1)$$

$$\bar{v} - \bar{v}_o = k_{\bar{v}} \cdot t \qquad (2)$$

$$\frac{18\pi\Phi^3}{\Sigma^3} - \frac{18\pi\Phi^3}{\Sigma_o^3} = k_\Sigma \cdot t \qquad (3)$$

equations can be used to determine the agglomeration rate constants k_N, $k_{\bar{v}}$, and k_Σ. Determination of $1/N$ or $\bar{v}$ by electron microscopy, ultracentrifugation (17), or an alginate creaming method (18, 19) is very time-consuming. Therefore, soap titration to minimum surface tension (20) was used almost exclusively for the determination of Σ, and Equation 3 was used in these experiments. Figure 2 shows that $18\pi\Phi^3\Sigma^{-3}$ increased linearly with time over a relatively wide initial range.

The stability factor is defined as:

$$W_{expt} = \frac{k_{\Sigma o}}{k_\Sigma} \qquad (4)$$

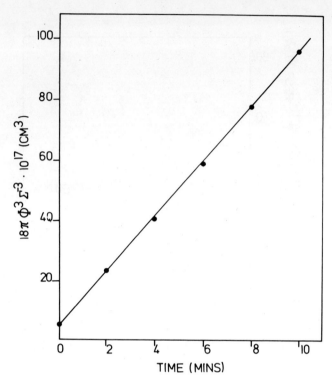

*Figure 2. Variation of reciprocal cube of particle surface (Σ^{-3})
with time at 40°C*

Particle surface (Σ) in cm²/ml latex, and volume of dispersed phase (Φ)
in cm³/ml latex

where k_{Σ_0} is the fast Smoluchowski rate constant assuming that each particle collision leads to agglomeration because of diffusion. This can be derived from:

$$k_{\Sigma_0} = \frac{2\,k\,T\,\Phi}{3\eta} \tag{5}$$

where k = Boltzmann constant, T = absolute temperature, and η = viscosity of water.

Unless stated otherwise, the agglomeration experiments were carried out in a temperature chamber at 40°C and 32.5% TS using 0.1% PEO/TS. Samples were collected at fixed intervals, and the reaction was terminated by adding the samples to sufficient potassium oleate solution to decrease the surface tension of the latex to about 40 dyne/cm or by diluting with water to about 6% TS. The particle surface (Σ) was then determined by soap titration so that rate constant (k_{Σ}) and stability factor could be calculated from Equations 3, 4, and 5.

High agglomeration rates in the region of log $W_{expt} < 4.3$ were determined by a rapid flow method, referred to as quenched flow mode (21), using an apparatus specially constructed for this purpose (22).

SURFACE TENSION. Surface tension as a function of potassium oleate concentration in the presence and absence of PEO was measured by the Ring detachment method at 25°C. The values were corrected by the method of Harkins and Jordan (23).

DETERMINATION OF PARTICLE SIZE DISTRIBUTION. Particle size distributions of base latex and agglomerated latex were determined by evaluating electron micrographs (24).

Results

Particle Size Distribution Changes. The number frequency distribution of particles for the base and end latexes are plotted in Figure 3. No particles in the end latex formed by agglomeration had a diameter larger than 9600 A. Furthermore, particles of 1300–9600 A constituted only about 1.2% of the total number of particles. The small particle size (200–400 A) that had constituted more than 5% in the base latex disappeared completely. The peak of the distribution curve shifted from about 380 A to 700 A.

The volume frequency distribution of latex particles after agglomeration and that of the base latex are plotted in Figure 4. Large particles 1300–9600 A in diameter constituted 63% of the total particle volume (*cf.* Figure 3).

Particle size distributions of samples collected during agglomeration demonstrate that mean particle size increased steadily, though it is interesting that many of the larger particles were already formed during early stages of

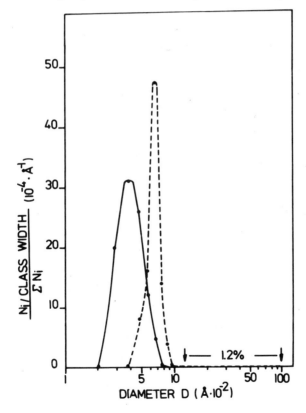

*Figure 3.   Number distributions of particles before and
after agglomeration*
———, *base latex; and* – – –, *end latex*

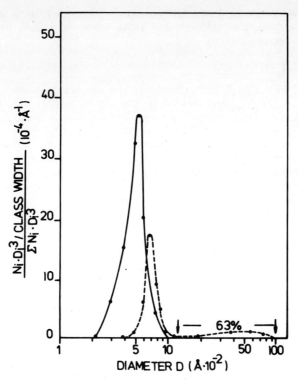

Figure 4.   Volume distributions of particles
—————, *base latex; and* – – –, *end latex*

agglomeration. Thus, the larger particles may serve as targets for the smaller ones that act as bullets.

Factors Affecting PEO Agglomeration. By using Equation 3, it is easily possible within a relatively short time to study the effects of the many factors which usually play a part in polymer–colloid reactions and which have to be considered in PEO agglomeration, too. These factors are tabulated in Table I according to reactants and reaction conditions.

Table I.   Factors in PEO Agglomeration

SBR Base Latex
    Inorganic electrolyte (amount and type)
    Anionic emulsifier (amount and type)
    Particles (concentration, size, and size distribution)
    Polymer (butadiene/styrene ratio, molecular wt, residual monomers, etc.)

PEO
    Concentration
    Polymer (molecular wt, structure [*e.g.* degree of branching], hydrophobic groups, etc.)

Reaction Conditions
    Temperature
    Agitation
    Operation (batchwise or continuous)
    Method of mixing PEO and latex

Inorganic Electrolyte and Anionic Emulsifier. The plot of log W_{expt} *vs.* log $[K^+]$ is a straight line (Figure 5). The slope of the line indicates that the rate of agglomeration varied with the 5th to 6th power of $[K^+]$. The dependence, however, could be followed experimentally only as far as log $W_{expt} \approx$ 4.3 because at this point agglomeration time was already reduced to a few seconds.

A special flow apparatus was used in order to follow agglomeration over a wider $[K^+]$ range, including the range where agglomeration rate becomes independent of $[K^+]$ (*see* Figure 6). This occurs when each particle collision leads to an agglomeration, and the time of agglomeration, which at this point was already less than 10^{-3} sec, cannot be reduced further. Nevertheless, the rate constant (k_{Σ_0}) can still be determined experimentally. It was found that $k_{\Sigma_0} = 0.8 \cdot 10^{-12}$ cm^3/sec whereas the calculated value is $1.4 \cdot 10^{-12}$/sec. Consequently, as a rule the values of log W_{expt} would be 0.24 lower if we had used the experimentally determined value of k_{Σ_0}. If the straight line shown in Figure 6 is extrapolated to $[K^+] = 0$, agglomeration time increases to about 200,000 years. It is very impressive that this huge time can be reduced to about 10^{-3} sec by increasing $[K^+]$ in the aqueous phase from 0 to 2.4%.

These findings agree with the relations which were determined in electrolyte coagulation studies, and they can be substantiated by the Derjuin–Landau–Vervey–Overbeek (DLVO) theory (25, 26, 27). Particle agglomeration of a commercial SBR latex using electrolyte only is also possible under certain conditions without coagulum formation (28). The emulsifier content was increased from 4.41 to 6.4%/TS in order to avoid coagulum formation and to cover a wider $[K^+]$ range. The findings for agglomeration obtained with and

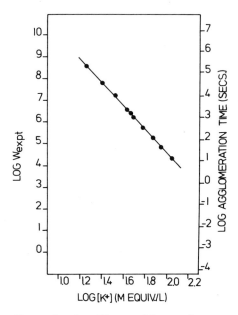

Figure 5.   Log W_{expt} and log agglomeration time vs. log $[K^+]$

$[K^+]$ = *potassium ion concentration of neutral electrolyte in mequiv/l*

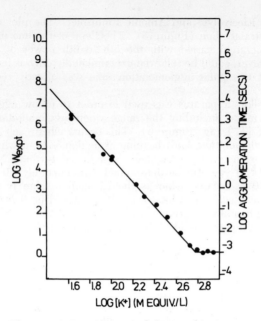

Figure 6. Log W$_{expt}$ *and log agglomeration time* vs. *log* [K$^+$] *in the range of lower stability factors* W$_{expt}$ *and shorter agglomeration times*

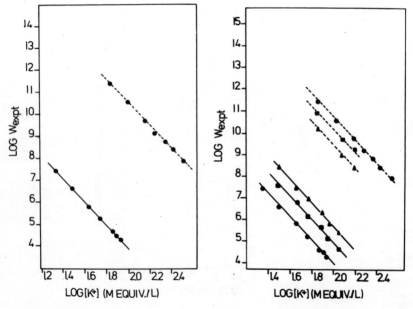

Figure 7. Log W$_{expt}$ vs. *log* [K$^+$]
———,*with PEO; and* - - -, *without PEO*

Figure 8. Log W$_{expt}$ vs. *log* [K$^+$]
———, *with PEO; and* - - -, *without PEO.*
% K oleate/TS latex: ▲, *2.4;* ■, *4.46; and*
●, *6.4.*

without PEO are presented in Figure 7. The distance between the two straight lines indicates that PEO accelerated the agglomeration rate 10^7-fold, and that at constant rates [K⁺] can be reduced to almost 1/20.

When the emulsifier content of latex was changed systematically, a fundamental difference between agglomerations with and without PEO was apparent (*see* Figure 8). The distance between the straight lines, *i.e.* the acceleration by PEO, decreased considerably as emulsifier content was reduced.

Log W_{expt} is plotted against log K oleate concentration in Figure 9. Dependence was not linear. In the range of greatest dependence, the agglomeration rate varied even with the 7th power of [K oleate]. Therefore, in this range the dependence on oleate concentration is even more pronounced than that on inorganic electrolyte concentration. The plot also suggests that the curvature tends to become sigmoidal with increasing range of potassium oleate concentration.

Regarding the type of inorganic electrolyte, there was no marked difference in effectiveness between the monovalent cations, but preliminary experiments using SBR latexes with emulsifiers which, in contrast to oleate, do not form insoluble salts with divalent cations, demonstrated that in this case PEO was less effective. Findings in earlier sensitization experiments were similar (*29, 30*). As to the type of anionic emulsifier, a slight correlation between emulsifier surface activity and agglomeration rate was observed.

Particle Concentration. The effect of particle concentration was essentially that which can be expected from a second order reaction (time for half-agglomeration was inversely proportional to the number of particles originally present). However, at TS > 35% and higher viscosities and TS < 15% of base latex, the reaction was faster or slower, respectively, than expected.

Particle Size. The effect of particle size on agglomeration rate at constant electrolyte concentration corresponded roughly to that which would be expected from Smoluchowski's rate equations, *i.e.*, no effect was observed. Figure 10 shows that in the plot of log W_{expt} *vs.* log [K⁺], the slope of the straight line was not dependent on particle size. Such dependence could not be confirmed although it is predicted by theoretical calculations (*25*). Interestingly, this finding agrees with that in electrolyte flocculation experiments using polystyrene latex dispersions of various particle sizes (*31*), *i.e.*, an increase in slope with increase in particle size was not observed experimentally. It should be mentioned that attempts were made recently to eliminate this discrepancy by more refined calculations (*32*). Nevertheless, the effect of base latex particle size on the TS/viscosity relation of the end latex was marked. As base latex particle size increased, TS decreased when the latex was concentrated after agglomeration to a viscosity of 1200 cP.

Particle Size Distribution. The effect of particle size distribution was roughly in accordance with the theory of Mueller who extended Smoluchowski's theory to polydisperse systems (*33*). According to this theory, a particle diameter ratio of 1:10 in the polydisperse system reduces the time of half-coagulation by the ratio 1.0 to 0.4.

In contrast to other polymers proposed for agglomeration, it is unimportant whether or not the base latex has a fairly uniform particle size in order to avoid coagulum formation (*34*). With base latexes containing large amounts (*e.g.*, 40–50 wt %) of large particles (> 1200 A), no coagulum at all was obtained after agglomeration.

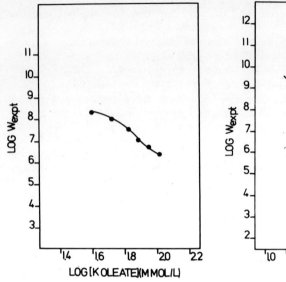

Figure 9.   Log W$_{expt}$ *vs. log* [K *oleated*]
[K *oleate*] *in mmol/l;* [K$^+$] = 34.8 *mequiv/l*

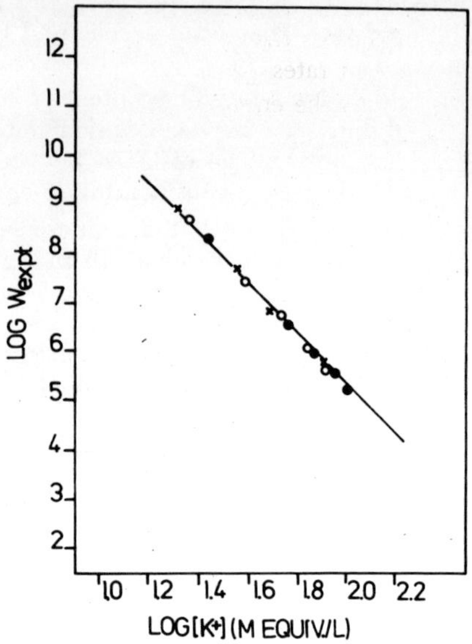

Figure 10.   Log W$_{expt}$ *vs. log* [K$^+$]
Particle diameter in angstroms: ×, 560; O, 760;
and ●, 820

SBR Latex Polymer. All factors which make particle coalescence easier should also affect the agglomeration rate. Therefore, the second order transition temperature of SBR latex polymer should be low. Thus it is understandable that various factors (*e.g.*, the ratio of butadiene to styrene, residual monomer, and molecular weight of polymer) are of importance in agglomeration.

PEO Concentration. A plot of log W$_{expt}$ against log PEO concentration is presented in Figure 11. The slope of the straight line indicates that at low [PEO] the rate varied with the 1.5th power of PEO concentration. After reaching a minimum, the curve rose sharply when PEO concentration was increased further.

PEO Polymer. There can be little doubt that polymer properties—*e.g.* molecular weight, structure (degree of branching and slight crosslinking), and content of hydrophobic groups—are extremely critical factors, and they have to be adjusted optimally. However, much additional work is needed in order to describe these factors quantitatively. This also applies to the role of helix conformations which were recently detected by calorimetric measurements using aqueous polyethylene glycol solutions (*35*).

Temperature. As regards dependence on temperature, there was an interesting parallel to the anomalous temperature effects (*36*) observed in many enzyme reactions (above 35°C, the rate increased by 40%/10°C and below 35°C by 80–100%/10°C).

Agitation. In contrast to other polymers proposed for agglomeration, in this system agitation conditions did not have any effect on agglomeration rate and coagulum formation. This agrees with Smoluchowski's calculations, namely that agitation can affect coagulation rate only when particle radius is larger than 10^4 A. As the latex particle radius before and after agglomeration was far smaller than this value, agitation conditions had practically no effect on agglomeration.

Batchwise or Continuous Operation. It is also unimportant whether the agglomeration is carried out batchwise or continuously. This can be demonstrated by conducting the agglomeration in an agitated vessel or in a special flow apparatus like that mentioned above. There was good agreement of agglomeration rate in both cases. It was, of course, possible to make the comparison only in the higher W_{expt} range where agglomeration in an agitated vessel can still be followed and measured.

Method of Mixing PEO and Latex. Again in contrast to other polymers (*10, 34*), no effect of mixing method on either agglomeration rate or coagulum formation was detectable in this system. Whether or not it has noticeable effect on particle size distribution remains to be elucidated by evaluating electron micrographs.

Discussion

Mechanism. There can be no doubt from the above findings that an intense sensitization does exist in PEO agglomeration. This is also illustrated by the plot of arbitrarily defined agglomeration concentration of monovalent cations *vs.* PEO concentration (*see* Figure 12). The curvature is characteristic of sensitization. A minimum of agglomeration concentration and a change from sensitization to protection at higher PEO concentration are apparent.

The $\log W_{expt}/\log [K^+]$ relations reveal quite clearly that, now as before, the reaction is governed by inorganic electrolyte or, more exactly, by counter ions in an unadulterated manner though at considerably lower concentrations. Bridging as a decisive factor, as was assumed in PVM agglomeration (*11*), can therefore be excluded. Obviously the role of PEO is to destabilize the particle by adsorption on the particle surface and to make the particle sensitive

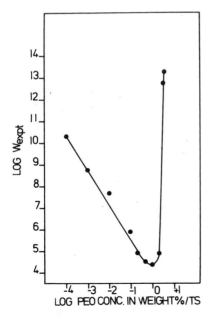

Figure 11.   Log W_{expt} *vs. log PEO concentration in wt % /TS latex*
$[K^+] = 56$ *mequiv/l; 6.4% K oleate/TS latex*

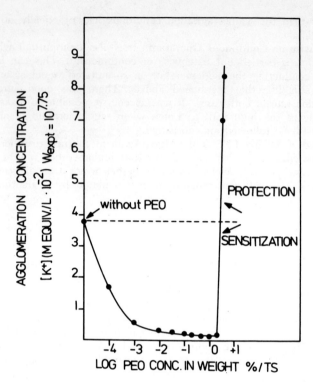

Figure 12. Agglomeration concentration of K^+ in mequiv./l vs. log PEO concentration in wt %/TS latex

Agglomeration concentration of K^+ without PEO is defined arbitrarily as 380 mequiv/l which corresponds to $W_{expt} = 10^{7.75}$

to electrolyte. According to the experimental findings, this is caused by a cooperative action of soap and PEO molecules. In fact, an interaction between K oleate and PEO molecules can be detected *in vitro* by plotting the surface tension of potassium oleate against its concentration in the presence of PEO. This becomes especially evident if PEO is fractionated by a special precipitation method (37) into active and inactive components and if these components are used in measuring surface tension as a function of oleate concentration. The active component forms a distinct plateau whereas the inactive one does not (Figure 13). The plot of surface tension *vs.* oleate concentration is included for comparison. No plateau-forming component could be obtained when non-oxidized polyethylene glycol was fractionated. According to Saito (38, 39, 40, 41), the formation of a plateau suggests complex formation between polymer and anionic soap molecules.

The schematic representation in Figure 14 depicts the mechanism of PEO agglomeration in light of the experimental findings and, in particular, the soap–polymer interaction. By adding PEO, a relatively large contact surface on which the particles are adsorbed is created in the latex (stage 1, adsorption). The PEO molecule is strongly branched and may be somewhat cross-linked. It may be a globular coil with a diameter twice that of the base latex particle. The latex particles, however, are present in excess. As was demon-

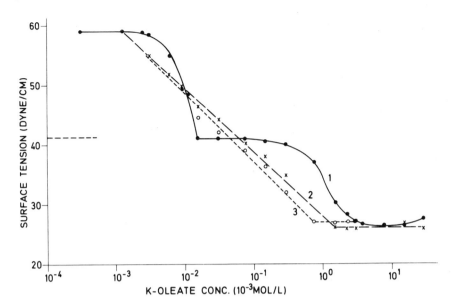

Figure 13. Surface tension vs. [K oleate] in the presence of PEO
Curve 1 (●), active component of PEO, 2.5 g/l; Curve 2 (×), inactive component of PEO, 2.5 g/l; and Curve 3 (O), no PEO

strated, the tendency for complex formation between PEO and oleate molecules is relatively great so that complex formation on the particle surface may also be expected, thus making the particle sensitive to electrolyte. Additional work is needed to determine how this electrostatic mechanism (that is based on DLVO theory) proceeds in detail; some possibilities have been suggested (*42, 43*). A sensitized particle may also aggregate directly with a non-sensitized particle since it has been reported (*44, 45*) that repulsion-free energy between two particles of unequal stern potential is determined largely by the lower one, *i.e.*, in this case by that of the sensitized particle (stage 2, aggregation). The coalescence process occurs spontaneously because of the free surface

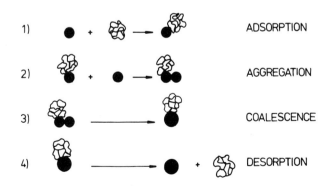

Figure 14. Schematic representation of PEO agglomeration
●, base latex particle; and   , PEO molecule

energy (stage 3, coalescence). However, the scheme of PEO agglomeration would be incomplete and contrary to all experimental findings (*e.g.*, the slight effect of the method of polymer admixture) without emphasizing the adsorption reversibility of PEO (stage 4, desorption). Undoubtedly, the real reaction cycle is not as ideal as depicted in Figure 14, but, when compared with other agglomerations using hydrophilic polymers, PEO agglomeration appears closest to ideality.

The antagonistic behavior of soap molecules, *i.e.*, their ability to accelerate and even stop agglomeration, may also be interpreted in terms of soap–polymer interaction. According to Lange (46), the formation of soap–polymer complexes is a cooperative process, *i.e.*, the more soap the polymer already contains, the more pronounced the tendency to complexation. However, the polymer can bind only a limited number of soap molecules. Consequently, PEO agglomeration should also stop when the PEO molecule is saturated with oleate molecules and the tendency to complexation expires. Thus, even under favorable agglomeration conditions (*e.g.*, at high electrolyte concentrations), agglomeration remains restricted to the colloidal particle size range as the surface tension of latex cannot drop below a certain level. The final surface tension observed experimentally is roughly the plateau surface tension marked in Figure 13 by the dashed line on the left. But not every hydrophilic polymer which is prone to complex formation should be an ideal agglomeration activator. Aside from soap–polymer interaction, polymer adsorption should be reversible in order to obtain fast agglomeration without any coagulum formation.

Technical Application. The experimental finding that with PEO, in contrast to other polymers, many factors have little or no effect on agglomeration is unquestionably the main reason for the great versatility that the agglomeration process offers in the manufacture of SBR latexes. It is noteworthy that the agglomeration result is the same irrespective of whether agglomeration is carried out in a 1-l beaker or in a 200-m³ reactor, even if base latex viscosity is extremely high as the result of short-time polymerization techniques.

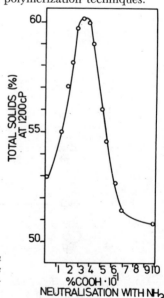

Figure 15.   Relation between TS content of a carboxylic SB latex at 1200 cP viscosity and its degree of partial neutralization before agglomeration

According to the experimental findings, PEO agglomeration is in principle a four component process (hydrophobic dispersion, hydrophilic polymer, anionic emulsifier, and inorganic electrolyte with monovalent cations). Polymerization recipes for commercial SBR latexes indicate that these latexes contain sufficient anionic emulsifier and inorganic electrolyte with monovalent cations to allow agglomeration to proceed in a relatively short time. Consequently, the scope of this process for the manufacture of high concentration latexes is by no means restricted as compared with other commercial processes using mechanical means (47, 48).

An interesting case is the application of this process in the manufacture of carboxylic SB latexes which are generally polymerized in the presence of relatively small amounts of anionic emulsifier and inorganic electrolyte. It is possible to agglomerate the particles of these latexes, prepared by short-time polymerization techniques, by neutralizing part of the carboxyl groups on the particle surface before agglomeration (49); obviously, the carboxylate group on the particle surface behaves as anionic emulsifier. Improvement of the TS/viscosity relation of a carboxylic SB latex can be determined from Figure 15. Thus, carboxylic SB latexes with new property combinations (50) have become available. The described agglomeration process can minimize capital expenditure and operating costs of plants producing high concentration SBR latexes and carboxylic types.

Acknowledgments

The author is indebted to K. v. Bassewitz for determination of particle size distributions by electron microscopy, to G. Schreier for measurements of surface tension as a function of soap concentration, and H. O. Dopp for his conscientious execution of the experiments. The permission of Chemische Werke Huels AG to present this paper is also appreciated.

Literature Cited

1. Goldberg, R. J., *J. Amer. Chem. Soc.* (1952) **74**, 5715.
2. Pettica, B. A., *Exper. Cell. Res. Suppl.* (1961) **8**, 123.
3. Busch, P. L., Stumm, W., *Environ. Sci. Technol.* (1968) **2**, 49.
4. Kuzkin, S. K., Nebera, V. P., "Synthetic Flocculants in Dewatering Processes," Moscow, 1963 (Transl. Nat. Lend. Libr., Boston, 1966).
5. Tajoma, S., *Technol. Rep. Kyushu Univ.* (1957) **30**, 2.
6. Williams, B. G., Greenland, R. J., Quirk, J. P., *Austral. J. Soil Res.* (1968) **6**, 59.
7. Howland, L. H., Aleksa, E. J., Brown, R. W., Borg, E. L., *Rubber Plast. Age* (1961) **42**, 868.
8. Schlueter, H., Kraenzlein, P., Chemische Werke Huels AG, German Patent **1,208,879** (1961).
9. Nemeth, R., Matijevic̆, E., *Kolloid Z. Z. Polym.* (1968) **225**, 155.
10. Fleer, G. J., Lyklema, J., *J. Colloid Interface Sci.* (1974) **46**, 1.
11. White, W. W., Reynolds, J. A., Gilbert, R. D., *J. Appl. Polym. Sci.* (1964) **8**, 2049.
12. Schlueter, H., Chemische Werke Huels AG, German Patent **1,213,984** (1963).
13. Schlueter, H., *Kaut. Gummi Kunstst.* (1966) **10**, 608.
14. Smoluchowski, v. M., *Phys. Z.* (1916) **17**, 557,585.
15. Lawrence, A. S. C., Mills, O. S., *Disc. Faraday Soc.* (1954) **18**, 98.
16. Hill, R. A. W., Knight, J. T., *Trans. Faraday Soc.* (1965) **61**, 170.
17. Cantow, H. J., *Makromol. Chem.* (1964) **70**, 130.
18. Schmidt, E., Biddison, P. H., *Rubber Age* (1960) **88**, 484.
19. Schmidt, E., Biddison, P. H., *Rubber Chem. Technol.* (1961) **34**, 433.
20. Maron, S. H., Elder, M. E., Ulevitch, J. N., *J. Colloid Sci.* (1954) **9**, 89.

21. Weissberger, A., "Techniques of Chemistry," 3rd ed., Vol. VI, Part II, p. 8, G. G. Hammes, ed., Wiley–Interscience, New York, 1974.
22. Pflug, G., unpublished data.
23. Harkins, W. D., Jordan, H. F., *J. Amer. Chem. Soc.* (1930) **52**, 1751.
24. Bassewitz v., K., *Microsc. Acta* (1975) in press.
25. Reerink, H., Overbeek, J. N., *Disc. Faraday Soc.* (1954) **18**, 74.
26. Paine, H., *Kolloid Z.* (1912) **11**, 145.
27. Sherman, P., "Emulsion Science," pp. 97–117, Academic, London-New York, 1968.
28. Selivanovskij, S. A., Ershova, N. M., *Colloid J. USSR* **21**, 659 (1959); *Kolloid Zh.* **21**, 686 (1959).
29. Troelsta, S. A., Thesis, Utrecht, 1941, p. 81.
30. Troelsta, S. A., Kruyt, H. R., *Kolloid Beih.* (1943) **54**, 251.
31. Ottewill, R. H., Shaw, J. N., *Disc. Faraday Soc.* (1966) **42**, 154.
32. Wiese, G. R., Healy, T. W., *Trans. Faraday Soc.* (1970) **66**, 490.
33. Mueller, H., *Kolloid Z.* (1926) **38**, 1.
34. Westerhoff, C. B., White, W. W., Gilbert, R. D., *Rubber Age* (1963) **94**, 446.
35. Maron, S. H., Filisko, F. E., *J. Macromol. Sci.* (1972) **B 6**, 79.
36. Kistiakowski, G. B., Lumry, R., *J. Amer. Chem. Soc.* (1949) **71**, 2006.
37. Ring, W., Cantow, H. J., Holtrup, W., *Eur. Polym. J.* (1966) **2**, 151.
38. Saito, S., *Kolloid Z. Z. Polym.* (1954) **137**, 98.
39. *Ibid.*, (1957) **154**, 19.
40. *Ibid.*, (1959) **165**, 162.
41. *Ibid.*, (1958) **158**, 120.
42. Hauser, E. A., Cines, M. R., *J. Phys. Chem.* (1942) **46**, 705.
43. Martinow, G. A., Smilga, V. P., *Kolloid Zh.* (1967) **27**, 250.
44. Devereux, O. F., de Bruyn, P. L., "Interaction of Plane-Parallel Double Layers," Massachusetts Institute of Technology, Cambridge, 1963.
45. Hogg, R., Healy, T. W., Fuerstenau, D. W., *Trans. Faraday Soc.* (1966) **62**, 1638.
46. Lange, H., *Kolloid Z. Z. Polym.* (1971) **243**, 101.
47. Talalay, L., *Proc. Rubber Technol. Conf. 4th*, London (1962) p. 442.
48. Jones, R. D., *Proc. Rubber Technol. Conf. 4th*, London (1962) p. 485.
49. Schlueter, H., Chemische Werke Huels AG, German Patent **1,770,934** (1968).
50. Schlueter, H., Chemische Werke Huels AG, U.S. Patent **3,842,025** (1974).

RECEIVED April 3, 1974.

11

Preparation of Acrylonitrile–Styrene Copolymers by Calorimetrically Controlled Monomer Feeding

B. N. HENDY

Research Department, Imperial Chemical Industries, Ltd., Plastics Division, Welwyn Garden City, Hertfordshire, England

A fairly general solution to the problem of composition drift during batch copolymerization permits making copolymer in sufficient quantities for evaluation as a thermoplastic. By polymerizing in a calorimeter, it was possible to control monomer feed accurately in relation to rate of polymerization. The apparatus has a 2-l reaction vessel which is basically a distillation calorimeter that controls the batch temperature and measures the reaction rate and the total integrated amount of reaction, or heat, to which is aligned the monomer feed. Acrylonitrile–styrene copolymers rich in acrylonitrile were prepared by feeding styrene. The heterogeneous product formed without feeding was useless as a thermoplastic, but the homogeneous product made by feeding was very good. The mechanical properties of these copolymers improved with increasing acrylonitrile content.

The composition of a copolymer produced by simple batch copolymerization of monomers with very different relative reactivities changes as the reaction proceeds so that a heterogeneous product is obtained. This is well illustrated by the copolymerization of acrylonitrile (AN) and styrene (S) because mixtures of these monomers consisting mainly of AN form instantaneous copolymers that contain much more styrene than the parent mixture. Thus, as such polymerizations proceed, the concentration of styrene in the remaining monomer mixture falls progressively, and consequently the composition of the copolymer changes accordingly. This tendency to styrene enrichment is tabulated in Table I which summarizes the whole composition range from acrylonitrile-rich mixtures to styrene-rich mixtures. In the former, styrene enrichment is pronounced. However, it is almost nonexistent in the latter, and thus styrene-rich copolymers can be made without the problem of composition drift. In light of this, it is noteworthy that, whereas styrene-rich copolymers have long been established commercially, acrylonitrile-rich copolymers have only recently been investigated in detail (1).

Table I. The Relation between the Compositions of the Monomer Mixture and the Resulting Copolymer

$S_m{}^a$, mole %	$S_p{}^b$, mole %	$S_p{:}S_m{}^c$
1.0	16.8	16.8
2.0	25.4	12.7
5.0	36.7	7.3
10.0	43.5	4.34
20.0	48.7	2.44
25.0	50.4	2.02
35.0	53.2	1.52
45.0	56.0	1.24
55.0	59.2	1.08
61.9	61.9	1.00
70.0	65.8	0.94
80.0	72.3	0.90
90.0	82.4	0.92
95.0	89.8	0.94

[a] S_m, styrene in monomer mixture.
[b] S_p, styrene in instantaneous copolymer.
[c] $S_p{:}S_m$, styrene enrichment ratio.

The drift in product composition during copolymerization can generally be avoided by feeding one or both of the monomers into the reacting mixture in such a way as to keep the ratio of comonomers constant at the appropriate value for formation of the required copolymer (2). In order to maintain this condition, it is necessary to measure the composition of the monomer mixture itself or of the copolymer as it forms; alternatively, the rate at which monomer should be replenished can be deduced from the rate at which the polymer is forming. Thus either an analytical technique or knowledge of the kinetics of the process is required.

The preparation of polymers for evaluation as plastic materials usually involves complicated procedures and the use of commercial grade materials. Consequently, each preparation tends to reflect different kinetic features, and, therefore, for kinetically controlled feeding to be accurate, one must know for certain the kinetic course of each separate reaction. Measurement of the heat evolved as the copolymerization proceeds provides a means of following the reaction. Thus, use of a calorimeter as the reaction vessel permits idiosyncrasies in the kinetics of each separate polymerization to be taken into account, and then the feeding can follow variations accurately in rate as the polymerization proceeds.

Experimental

The Apparatus. The apparatus was basically an isothermal distillation calorimeter consisting of three parts: a reaction vessel, a vapor tube plus condensate buret, and a monomer feed buret (*see* Figure 1 for diagram and key to parts). A 2.5-l glass reactor inside a vacuum flask was surrounded by liquid coolant. The reaction heat was transferred to the coolant which boiled out of the flask (**13**), up through a heated tube (**20**), and then into a condenser (**23**) and buret where the condensate was measured before being returned to the vacuum flask in aliquots. Monomer was fed from the feed buret (*via* 8) at a rate corresponding to the rate of condensate formation so that, throughout the polymerization, the total amount of feed added corresponded to the total condensate, or heat, measured.

A significant amount of reaction heat was lost from the vacuum flask to the surroundings, and the resulting error was corrected continuously by adding

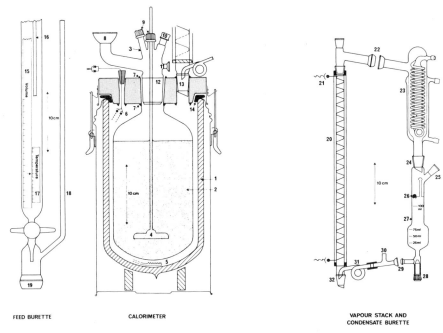

FEED BURETTE   CALORIMETER   VAPOUR STACK AND
                                                                     CONDENSATE BURETTE

Figure 1. The apparatus

1, Vacuum flask (height, 30 cm; internal diameter, 16 cm; and capacity, 4 l) from Thermos Lt. housed in a convenient steel canister, modified by openings cut into lid; 2, glass reaction vessel (22.5 cm from bottom to shoulder; external diameter, 14.0 cm; and working capacity, 2.5 l); 3, opening for thermometer (angled to avoid the vortex in the stirred reactants); 4, stirrer (collapsible paddle type, Quickfit and Quartz ST 1/4); 5, background heater (wire-wound ceramic resistor of 12.5 ohm); 6, plug and socket connecting electrical power to background heater (maximum power supply, 25 V at 2 amp); 7, epoxy cement seals which fix the glass parts to the lid and prevent vapor leakage; 8, spherical joint (MS 29) to which is connected the feed buret (19); 9, screw thread joints (bore, 7.5 mm; SQ 13); 10, vaccine cap (bore, 9 mm); 11, spherical joint (FS 13); 12, cone and socket (B40/38); 13, socket (B24/29) to which is connected the vapor take-off tube (32); 14, rubber sealing ring between the vacuum flask and its lid; 15, volume scale with fixed arrow (see text); 16, tube for purging with nitrogen; 17, temperature scale (see text); 18, pressure compensating tube; 19, joint which connects with the calorimeter head at 8 (FS29); 20, vapor take-off tube (bore, 24 mm); 21, electrical connection to heating wire on vapor take-off tube (power applied, ca. 37 W); 22, spherical joint (MS and FS 29); 23, condenser (ether type, Quickfit and Quartz C11/23 with minor modification); 24, cone and socket (B24/29); 25, vent; 26, valve (63 mm-diameter ball bearing) fitting into a glass seat; operated by an external magnet; 27, buret; 28, stopcock (Rotaflow TF6/18); 29, spherical joint (MS and FS 13); 30, side-arm for filling the calorimeter jacket with pre-heated coolant; 31, vapor trap; and 32, joint which connects with calorimeter jacket at 13

heat from an electrical heater (**5** and **6**) contained within the flask. The required amount of background heating was established by adjusting the power to that level at which trace distillation occurred in the absence of any reaction heat, or it could be determined from calibration experiments (*see* below). Typically, the heat loss from a reaction at 60°C required the heater to be set at about 8.5 W which is about one-sixth the heat evolution from a polymerization lasting 4–5 hrs.

It was necessary to correct the volume of distillate for any changes in batch temperature which occurred because such changes, which may be increases or decreases, represented significant amounts of heat. This situation arose because the reactor wall was a thermal barrier which was influenced as the reaction proceeded by changes in the viscosity of the reaction medium and by build-up formation; furthermore, changes in the reaction rate affected the batch temperature. Typically, for a polymerization lasting 4 or 5 hrs, the average differential across the reactor wall was about 4°C. This sensible heat correction was expressed as a volume of monomer feed by repositioning the volume scale (15) on the feed buret, the necessary adjustment being indicated by a temperature scale (17) fixed alongside. A change in heat content was thus expressed directly as the corresponding feed volume without need for a separate calculation.

The monomer feed buret had the following useful features. It permitted the monomer to be added continuously by regulation of the drip rate; it gave accurate feeding because of its length (70 cm); its volume was sufficient to hold the entire feed in one filling; and oxygen could be removed by bubbling nitrogen through. The feed was most conveniently controlled if the volume scale on the buret was marked in the corresponding volumes of coolant condensate rather than in milliliters of actual feed. The dimensions of the scale markings are given by:

$$\text{scale markings (cm/100 ml condensate)} = \frac{V \times L \times S}{\Delta H_{total} \times 10}$$

where V is the volume of feed for the entire reaction (ml), L is the latent heat of the coolant (cal/ml), S is a characteristic of the feed buret which expresses the length of the buret per unit volume of its contents (cm/ml), and ΔH_{total} is the expected heat evolution for the entire reaction (kcal). The dimensions of the smaller scale, which represents the batch temperature, are given by:

$$\text{scale markings (cm/°C temperature change)} = \frac{C \times V \times S}{\Delta H_{total}}$$

where C is the heat capacity of the batch (kcal/degree) calculated by summation of the values of the individual ingredients and inclusion of half the heat capacity of the reaction vessel itself.

These two scales were operated as follows. The volume scale was positioned prior to the start of polymerization so that its zero mark was level with the meniscus of the feed in the full buret; the batch temperature scale was positioned later at the instant polymerization started, and it was set so that the arrow on the volume scale corresponded to the batch temperature at that moment. Subsequently, as the reaction proceeded, the temperature scale was kept fixed while the volume scale was adjusted so that the arrow followed any change in batch temperature. In practice, during a polymerization, if the batch temperature was changing, the adjustment was made frequently so that the resulting alteration to the feeding rate was small and gradual.

The condensate from the condenser accumulated in the condensate buret where it was measured and returned to the vacuum flask in 100-ml aliquots. For simplicity, it was returned cold without reheating at a fairly reproducible temperature of 17°C, and, as a consequence, a polymerization occurring at 60°C was disturbed by about 0.2°C at each return. The latent heat used in calculations had to allow for the heat used in warming the recycled condensate to the boiling temperature of the coolant in the flask. In a polymerization lasting 4–5 hrs, a typical rate of distillation was about 11 ml/min, and the sensitivity of the measurement was such that about 5 ml condensate resulted from the formation of 1 g polymer.

In such an apparatus, maintenance of smooth boiling and adequate heat transfer was important. Mechanical agitation of the coolant, however, would have complicated the apparatus and was therefore not applied. It was found that the small disturbance caused by the background heater immersed in the coolant and a few boiling chips gave adequate boiling performance.

Calibration. The coolant used in this work was 4-methyl-1-pentene which boils at 54°C and is convenient for a polymerization temperature of 60°.

(Alternatively, 2,3-dimethylbutane may be used.) The latent heat, allowing for cold return, was 65.5 cal/ml as calculated from the latent heat of vaporization at 54°C, the mean specific heat in the 17°–54°C range, and the density at 17°C (*3, 4, 5*). This calculated latent heat used in conjunction with electrical heating provided a means of checking the working of the calorimeter; some calibration findings are presented in Table II.

The calibration procedure was as follows. A polymerization was simulated by electrical heating of a stirred charge of *o*-dichlorobenzene (800 ml). The experiment was continued for about 10–15 fillings of the condensate buret; measurements were started from a full buret so that the heat of rewarming was included in the first filling. The heater was supplied from a stabilized direct current supply, and power was determined by measuring voltage and current. The heater was placed inside the reactor rather than in the coolant jacket (as for background heating) because some experiments with high power heating in the jacket gave unusual results which suggested that liquid coolant may be swept into the heated vapor column as the result of the intense boiling around the small heater. The power used in these calibrations was 50 W which is the power generation of a polymerization taking 4–5 hrs. Each calibration also included an experiment at lower power (about 12 W); by applying a straight line relation, the heat loss, or power input for zero distillation rate, could be determined from the intercept as well as the power used to cause the measured amount of distillation.

In the calculation, it was necessary to consider any change in batch temperature which might have occurred. Losses of coolant vapor also had to be allowed for if necessary; such loss was best estimated separately by weighing the calorimeter containing its coolant at the start and after a sufficient number of fillings of the buret (transferring the coolant to another container for measurement considerably exaggerated the loss). The calorimeter used to obtain the data in Table II had a loss of 0.48 ml condensate per 100 ml distilled.

Polymerization Method. The copolymers were prepared by an emulsion method (*6, 7*). A convenient procedure for charging the calorimeter was as follows. The aqueous phase was preheated to about 20°C above the polymerization temperature (60°C); it was then added to the empty calorimeter and N_2 was bubbled through. Nitrogen-purged monomer mixture was added next, and the temperature was allowed to stabilize. Polymerization was started by successive injections of catalyst solution (through **10**) until a batch temperature change was noted. Boiling coolant was then added (through **30**) until an adequate heat transfer rate was attained.

The distribution of the monomers between the initial charge and the feed was calculated by the copolymerization equation with reactivity ratios of 0.04 for acrylonitrile and 0.41 for styrene (*8*).

Table II.   Electrical Calibration Experiments

Latent Heat, cal/ml

Observed[a]	Calculated[b]
65.3	
65.9	66.3 (15°C)
65.8	65.5 (17°C)
66.4	64.1 (20°C)
66.1	52.1 (54°C)[c]
64.9	
average, 65.7	

[a] Measured with cold return of 4-methyl-1-pentene from buret to vacuum flask.
[b] Calculated from published data (*3, 4, 5*); numbers in parentheses indicate temperature at return.
[c] Returned at 54°C, but measured in buret at 17°C.

In the homopolymerizations of acrylonitrile, the following recipe was used: water, 1750 ml; sodium dodecylbenzenesulfonate, 16.2 g; KH_2PO_4, 0.81 g; 10% aqueous H_2SO_4, 0.81 ml; acrylonitrile, 543 ml; and n-octyl mercaptan, 3.65 ml. The reaction was initiated at 60°C by injecting 5% aqueous solutions of ammonium persulfate (2 ml) and sodium metabisulfite (1 ml); additional ammonium persulfate (ca. 1 ml) was injected during the reaction. Polymerization proceeded to 70–85% conversion over 4–5 hrs; it was then terminated with sodium dimethyldithiocarbamate. Polymer yield was determined by coagulation of the latex in 1% aqueous $MgSO_4$ followed by washing, or by evaporation of the whole latex to dryness.

Results and Discussion

Monomer Feeding. There were two different ways of feeding monomer, and each met the requirements for composition control. In one, a monomer mixture with the same composition as the copolymer was fed to the reaction; in the other method, only the more rapidly consumed monomer was fed.

In the first method, the rate of feeding equalled the rate at which polymer was being produced. Therefore:

$$\text{rate of feeding (moles)} = \frac{1}{\Delta H_p} \times \text{rate of heat evolution} \qquad (1)$$

where ΔH_p is the heat of copolymerization per mole, and a mole of copolymer is taken as the appropriately weighted mean of the values for the homo units. Thus, at any point in the process:

$$\text{total feed added (moles)} = \frac{1}{\Delta H_p} \times \text{total heat evolved} \qquad (2)$$

In the second method, it can be shown that the relation between the rate of feeding and the rate of heat evolution was a simple linear expression:

$$\text{rate of feeding (moles)} = \text{rate of heat evolution} \times \frac{(N - n)}{\Delta H_{total}} \qquad (3)$$

where N is the total number of moles of the more active monomer for the whole process (i.e., charge plus feed), n is the number of moles of the more active monomer charged to the reactor initially, and ΔH_{total} is the total heat which would be evolved if all of the monomers polymerized. It follows that at any point in the process the accumulated amount of monomer added as feed, expressed as a fraction of the total feed for complete conversion $(N - n)$, must be equal to the accumulated heat evolved, expressed as a fraction of the total heat evolved at complete conversion (ΔH_{total}). For the copolymers described here, it was found that the single monomer feeding process (i.e., styrene only) led to better products; hence this procedure was used in preference to mixed monomer feeding.

The effect on the process and on the product of an error in the feeding of styrene was considered. The type of error investigated was that which would arise from a systematic error in the measurement of heat, or from the use in the calculations of an incorrect heat of copolymerization. A monomer mixture was simulated on a computer and polymerization was allowed to proceed in small steps; at the end of each step an increment of monomer was added in an amount that was deliberately too large, too small, or correct for the incremental amount of polymer formed. The findings from such a study

Table III. The Effect on Copolymer Composition of a 10% Error in Feeding Styrene to Make 75/25 mole % AN–S Copolymer[a]

Feeding Underestimated		Conversion[b], %	Feeding Overestimated	
Cumulative	Instantaneous		Instantaneous	Cumulative
24.98	24.98	0	24.98	24.98
24.69	24.41	5	25.54	25.27
24.35	23.79	10	26.11	25.60
23.83	23.01	20	26.89	26.11
23.44	22.61	30	27.25	26.43
23.20	22.47	40	27.48	26.71
23.04	22.43	50	27.57	26.88
22.92	22.42	60	27.61	27.02
22.84	22.42	70	27.62	27.12
22.78	22.42	80	27.63	27.20
22.73	22.42	90	27.63	27.26
22.71	22.42	95	27.63	27.28

[a] Copolymer composition is expressed as mole % styrene.
[b] $100 \times \frac{P}{P + M}$ where P is the amount of polymer formed, and M is the amount of monomer actually in the reactor.

of the making of a copolymer with high AN content are presented in Table III. The process did not go out of control, but it was self-stabilizing. The error caused the monomer mixture composition, which was in a kind of equilibrium, to shift to a new equilibrium, and then uniform copolymer of a different composition was produced. Thus, a 10% error in heat measurement led, after the transition period, to a 10% error in the composition of the instantaneous copolymer. Table IV presents the findings from a similar study of a copolymer with a higher styrene content; the pattern was different because the transition was drawn out over the whole process and the monomer mixture never reached a new equilibrium.

If the composition of the product changes because of a major feeding error of this type, then the magnitude of the heat of copolymerization ΔH_p must also change because the value for the new copolymer will be slightly different. Thus, the initial error will be altered by what may be described as a feedback effect in the error; this can be positive or negative depending on the direction in which ΔH_p changes in relation to the initial error, and its magnitude is

Table IV. The Effect on Copolymer Composition of a 10% Error in Feeding Styrene to Make 50/50 mole % AN–S Copolymer[a]

Feeding Underestimated		Conversion[b], %	Feeding Overestimated	
Cumulative	Instantaneous		Instantaneous	Cumulative
50.03	50.03	0	50.03	50.03
50.01	49.99	5	50.06	50.05
49.99	49.95	10	50.10	50.07
49.93	49.84	20	50.20	50.12
49.88	49.71	30	50.34	50.17
49.81	49.54	40	50.48	50.24
49.72	49.31	50	50.69	50.32
49.60	48.96	60	50.95	50.42
49.45	48.46	70	51.37	50.56
49.22	47.51	80	52.07	50.76
48.80	44.91	90	53.66	51.08
48.29	39.21	95	56.28	51.39

[a] Copolymer composition is expressed as mole % styrene.
[b] See Table III.

dependent on the sensitivity of ΔH_p to changes in composition. The computer simulation did not take this into account and so, in a sense, it is a simplification. Nevertheless, in actual experiments no impairment of calorimetric control which could be ascribed to this feedback was encountered, perhaps because ΔH_p is not very sensitive to compositional changes in this system. However, the feeding errors in Table III that resulted from measurement or calculation error were encountered; this indicated that the system behaved very much in accordance with the calculations.

Evaluation of feedback error could be quite important in the broader application of the process. In particular, it would be useful to examine systems in which the heat of copolymerization was very sensitive to changes in composition because then the error would be most noticeable. Copolymerizations of monosubstituted and disubstituted ethylenes could be useful because the large difference between the heats of homopolymerization suggests a highly sensitive heat of copolymerization.

Another feature of this process could impair its straightforward operation; this is best explained by restating Equations 1 and 3 as follows:

$$\text{polymerization rate (moles)} = \frac{1}{\Delta H_p + h} \times \text{rate of heat evolution} \qquad (4)$$

where the quantity h allows for the possibility that the polymerization heat is accompanied by incidental processes such as mixing, solution, or precipitation. Although the heat evolved in AN–S copolymerization is discussed below, it is useful to consider here how the quantity h may affect the feeding. Obviously, if h is significant but is ignored, the validity of Equation 4 is impaired and the feeding is in error. If h is constant throughout the process, there is no serious problem provided that the appropriate heat is used in the equations. However, if h changes as the polymerization proceeds, then the relation between the amount of feed required and the heat evolved is no longer linear, and consequently such a process would have to be calibrated. Serious complications in which polymer solution processes affect h are not generally expected because such heats of solution are reported to be small, commonly ranging from 0 to -1 kcal/mole (9), i.e. from negligible to 5% of the heat of polymerization. AN–S copolymers that contain about 13 mole % S were insoluble in the parent monomer mixture and would have precipitated whereas copolymers that contain 25 mole % S were soluble.

In the making of AN–S copolymers by feeding styrene, linear feeding was used, and there was no serious experimental evidence that it should be otherwise, at least up to 80% conversion. In a series of experiments in which samples were withdrawn during the process and the polymer was analyzed, it was found that the product at the end (ca. 80% conversion) was, on the average, slightly richer in styrene (by 2% of the styrene content) than the half-way product. This small difference was not necessarily a calorimetric effect because it could have been caused by the slight solubility of AN in water (10%) which would have affected the S:AN monomer ratio in the latex particle. This is understandable if one considers that, in the later stages of the process, the amount of AN in the aqueous phase becomes an increasing proportion of the total amount remaining.

Heat of Polymerization. The heat of copolymerization was expected to differ slightly from that calculated from the weighted mean of the heats of homopolymerization because of the contribution of the crosspropagation reactions, *i.e.* AN–S and S–AN. Reactivity ratio considerations do in fact suggest that for acrylonitrile-rich copolymers, styrene homopolymerization contributes nothing to the heat because styrene pairs or sequences are very rare; the heat arises almost entirely from acrylonitrile homopropagation and crosspropagation.

Heats of copolymerization for this system were reported by Suzuki, Miyama, and Fujimoto (*10*) and by Chiu (*11*). The first group studied the whole composition range at 20°C. They reported values higher than expected from homopolymerization which indicated that crosspropagation is more exothermic than homopolymerization. Two aspects of their findings, however, need to be considered; first, the value which they obtained for the homopolymerization of acrylonitrile was unusually low while that for styrene was normal, and second, the copolymerization in the acrylonitrile-rich mixtures was probably affected by the monomer reactivities illustrated in Table I so that one wonders about the validity of the correlation with Alfrey's equation. For copolymer containing 13.5 mole % AN, they determined a value of about 17 kcal/mole. Chiu used a differential thermal analyzer; he reported a value of 17.1 kcal/mole for styrene-rich copolymer made in bulk and cited the value of 17.2 kcal/mole reported for the same composition by Suzuki *et al.* Chiu's determination, however, was probably made at 140°C so that, allowing for the specific heats of monomer and polymer and adjusting to the lower temperatures used by other workers, it probably represents a significantly smaller figure.

We measured the heat of copolymerization in emulsion for copolymer containing 13.5 mole % styrene at 60°C. Measurements were made with a styrene feed at 20°C; therefore, some reaction heat was used to warm the styrene to 60°C. Experiments indicated that this effect was partially compensated for by the heat of mixing because the mixing of styrene with the reactants was very slightly exothermic. If the styrene were preheated to 60°C, the effect would be to raise the measured heat of copolymerization by about 0.09 kcal/mole. A value of 20.2 kcal/mole (*i.e.*, 337 cal/g) was obtained for the heat of copolymerization. In order to elucidate this unexpectedly high finding, we also measured the heat of homopolymerization of acrylonitrile under similar emulsion conditions in the same apparatus; a value of 22.1 kcal/mole was obtained.

The heat of polymerization of acrylonitrile has been reported as around 18 kcal/mole. Joshi (*12*), using an isothermal distillation calorimeter (*13*) at 74°C, determined values of 18.3 and 18.5 kcal/mole for bulk polymerization and for polymerization in benzene solution, respectively; these values are about 1 kcal/mole higher than those reported previously by Tong and Kenyon (*13*) for bulk polymerization. Baxendale and Madras (*14*) used a non-isothermal method to determine a value of 18.3 kcal/mole for the polymerization of 5% aqueous solutions at 25°C. The lowest reported value is that of Suzuki, Miyama, and Fujimoto (*10*): 16.0 kcal/mole at 20°C. The highest reported value is that of Chiu (*11*): 20.5 kcal/mole for bulk polymerization. This last result was apparently obtained at a temperature of 140°C so the equivalent value for a temperature of 60°C would probably be slightly lower.

Our findings for both the copolymerization and the homopolymerization of acrylonitrile were thus unusually high, and the reason must be the emulsion polymerization process. The work of Joshi (*12*) on acrylamide and methacrylamide, which are water soluble, may be relevant. He reported values of 13.8 and 19.5 kcal/mole for acrylamide in hydrocarbon and in aqueous solution; for methacrylamide the corresponding values were 8.4 and 13.4 kcal/mole

Table V.   Heats of Mixing for Acrylonitrile (AN) and Water (W)[a]

Phase	Composition, g	Heat Absorbed, kcal	Possible Effect on Heat of Polymerization
aqueous	AN, 172 W, 1728	3.3	none
acrylonitrile	AN, 263 W, 21.5	1.9	0.3 kcal/mole more exothermic

[a] 435 g acrylonitrile and 1750 g water formed two phases. Measurements were made at 60°C in an adiabatic calorimeter.

(monomer-to-water ratio: average, 1.3; range, 0.4–3.1). Clearly, monomer-solvent and polymer–solvent interactions contributed significantly to the heats of these polymerizations. As for interactions between acrylonitrile and water in these emulsion polymerizations, there were two saturated liquid phases to consider, but the heats of mixing (see Table V) did not indicate that any major effect on the heat of polymerization would be expected. Likewise, no exothermic effect could be detected by mixing dried polymer powder (12.5 mole % styrene) with acrylonitrile and water although such synthetic mixtures must be very different from the real reaction mixture.

Thus, additional work is necessary on the heat of polymerization. It is tempting to explain our unusual results by suggesting that there are interactions involving the polymer which are reversed at very high conversions when the residual monomer becomes very small. Thus the difference between the reported findings may be because we studied slow polymerizations with fairly uniform reaction rates which stopped at about 80% conversion, whereas others studied reactions which went to or nearly to completion.

Properties of the Copolymers. The properties of the homogeneous copolymer made with controlled monomer feeding were very different from those of the heterogeneous copolymer made without feeding, i.e. by charging all the monomers at the start (see Table VI). The heterogeneous copolymer was hazed and brittle, but the homogeneous copolymer was clear and strong. It was calculated that unfed copolymer was mainly a 50:50 mixture of copolymer formed early in the reaction and polyacryonitrile formed later with very little intermediate material of the intended composition (see Figure 2). The presence of polyacrylonitrile in the unfed polymer was evident from x-rays of polyacrylonitrile-type two-dimensional order. By contrast, copolymer made by

Table VI.   The Products of Unfed

Process	Appearance[b]	Vicat Softening Point, °C		Un-notched Impact Strength, kJ/m²
		1/10	Full	
Without feeding	very brittle yellow badly hazed	108	123	2.4
With feeding	very pale yellow clear strong	103	108	ca. 24

[a] Unfed polymerization: monomer mixture contained 80 mole % AN and 20 mole % S; final product actually contained 23.6 mole % S because the reaction did not go to complete conversion. Fed polymerization: copolymer had a uniform composition of 80 mole % AN and 20 mole % S.

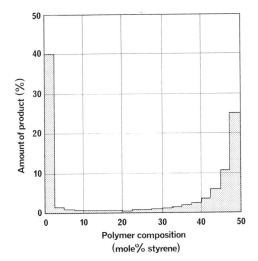

Figure 2. *Composition of the product from batch copolymerization of acrylonitrile (75 mole %) and styrene (25 mole %)*

feeding was amorphous because the styrene, which disrupts this order, was evenly distributed throughout the product.

The effect of composition on some physical properties of the homogeneous copolymer is plotted in Figures 3, 4, 5, and 6. The effects were determined on copolymers which had a reduced viscosity, as defined in footnote c to Table VI, of 0.8 to 0.9. This value of the reduced viscosity corresponded to the lowest molecular weight at which the polymers attained their limiting strength. It was thus a thermoplastic optimum, being the best balance between mechanical strength and viscosity in the melt.

Un-notched impact strengths and flexural strengths of the homogeneous copolymer (Figures 3 and 4) increased with increasing AN content to a maximum at *ca.* 87.5 mole % AN; at higher AN contents, these values decreased dramatically. X-rays of copolymers containing more than about 87.5 mole %

and Fed Polymerizations [a]

X-ray Observations	Reduced Viscosity [c]
appreciable two-dimensional order as made and a lot after annealing	2.28
amorphous as made and after annealing	0.8–0.9

[b] Composition-molded at 200°C.
[c] Reduced viscosity $= \dfrac{\text{relative viscosity} - 1}{\text{concentration}}$; concentration, 0.5 g polymer/100 ml solution; solvent, dimethylformamide; temperature, 25°C.

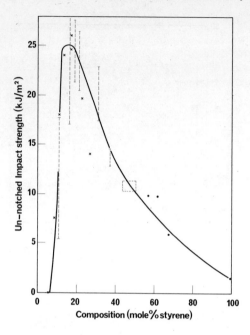

*Figure 3.   Effect of composition on un-notched
impact strength*
•, *commercial polymers;* ✕, *this work; and* |‑ ‑ ‑ ‑ ‑|,
the range when variation was large

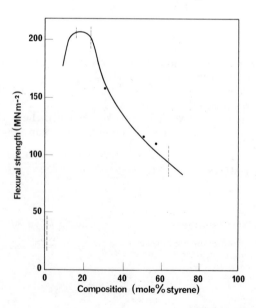

Figure 4.   Effect of composition on flexural strength
1 kpsi = 6.90 MNm⁻²; ‑ ‑ ‑ ‑, the range when variation was large

AN revealed polyacrylonitrile two-dimensional order, whereas copolymers containing less AN were amorphous. Thus the decrease in strength at very high AN contents was associated with polyacrylonitrile structure which developed because the styrene present was insufficient to disrupt the structure even though it was evenly distributed.

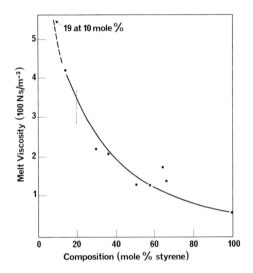

Figure 5. Effect of composition on melt viscosity

Measured at 260°C, 1000 sec⁻¹, and 100 Nsm⁻² = 1 kp; sample molecular weights gave reduced viscosities of approximately 0.8–0.9 (see text)

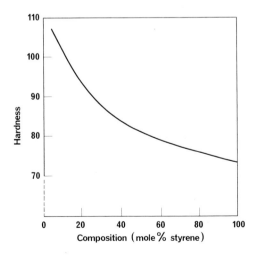

Figure 6.   Effect of composition on hardness
Measured on Rockwell M scale; sample thickness, 0.317 cm

Values for viscosity in the melt and hardness (Figures 5 and 6) increased continuously with AN content, and there was no maximum at 87.5 mole % AN. Softening points were affected very little by composition, the full Vicat softening points being *ca.* 105°–110°C except for compositions that contained more than 87.5 mole % AN; for these latter, the values were much higher. A high Vicat softening point was one of the characteristics of the copolymer made without monomer feeding (*see* Table VI), so high softening points were consistent with the polyacrylonitrile two-dimensional order.

It was observed that moisture uptake could affect the mechanical testing of these copolymers. Moisture caused lowered softening point and yield stress and increased notched impact strength, but it had no noticeable effect on un-notched impact strength.

The statements sometimes seen in the literature that S–AN copolymers have maximum strength when they contain 30% AN probably can be attributed to the effects of heterogeneity of composition (*15, 16*). The mechanical properties of these copolymers do in fact depend on the method by which they are prepared.

Acknowledgments

The author is grateful to numerous colleagues, in particular to J. B. Rose, E. Nield, D. D. J. Nightingale, and D. E. Meek.

Literature Cited

1. British Patent **663,268** (1951).
2. Billmeyer, F. W., "Textbook of Polymer Science," p. 310, Interscience, New York and London, 1964.
3. Dreisbach, R. R., "Physical Properties of Chemical Compounds," p. 237, American Chemical Society, Washington, 1959.
4. Gallant, R. W., "Physical Properties of Hydrocarbons," Vol. 1, Gulf, Houston, 1968.
5. "Physical Properties of Hydrocarbons C_1 to C_{10}," Spec. Tech. Publication **109A**, ASTM, Philadelphia, 1963.
6. British Patent **1,197,721** (1966).
7. British Patent **1,185,305** (1966).
8. Fordyce, R. G., Chapin, E. C., *J. Amer. Chem. Soc.* (1947) **69**, 581.
9. Dainton, F. S., Ivin, K. J., "A Dynamic Differential Colorimetric Technique for Measuring Heats of Polymerization," in "Experimental Thermochemistry," Vol. 2, p. 267, Interscience, New York and London, 1962.
10. Suzuki, M., Miyama, H., Fujimoto, S., *Bull. Chem. Soc. Jap.* (1962) **35**, 57.
11. Chiu, J., "Anal. Calorimetry," Vol. 2, pp. 171–83, Plenum, Chicago, 1970.
12. Joshi, R. M., *J. Polym. Sci.* (1962) **56**, 313.
13. Tong, L. K. J., Kenyon, W. O., *J. Amer. Chem. Soc.* (1947) **69**, 2245.
14. Baxendale, J. H., Madras, G. W., *J. Polym. Sci.* (1956) **19**, 171.
15. British Patent **590,247** (1947).
16. Muller, R. G., "ABS Resins," Process Economics Program, Report **20**, Stanford Research Institute, 1966.

RECEIVED April 1, 1974.

Aminimides XVI. Copolymerization Studies of Aminimides with 4-Vinylpyridine and N-Vinylpyrrolidone and Use of Copolymers in Adhesive Systems

H. J. LANGER and B. M. CULBERTSON

Research Center, Ashland Oil, Inc., Columbus, Ohio 43216

Copolymerization studies demonstrated that aminimide, 1,1-dimethyl-1-(2-hydroxypropyl)amine methacrylimide (DHA) copolymerizes readily with 4-vinylpyridine (4VP) and N-vinylpyrrolidone (NVP). These copolymers could be thermolyzed in solution to give soluble poly(4-vinylpyridine-co-isopropenyl isocyanate) and poly(N-vinylpyrrolidone-co-isopropenyl isocyanate) materials. The reactivity ratios of each monomer pair were determined, and the Alfrey-Price Q and e values for DHA were calculated: for DHA (M_1)–4VP (M_2), $r_1 = 0.41$, $r_2 = 0.77$, $Q = 0.68$, and $e = 0.58$; and for DHA (M_1)–NVP (M_2), $r_1 = 0.15$, $r_2 = 0.35$, $Q = 0.14$, and $e = 0.58$. The DHA–4VP copolymers quaternized readily to give a new family of water-soluble polyelectrolyte materials. The various copolymers were examined as adhesion promoters for rubber–tire cord composites.

It has been demonstrated that aminimide monomers, such as 1,1-dimethyl-1-(2-hydroxypropyl)amine methacrylimide (DHA), are very useful for preparing reactive polymers with pendent acyl aminimide ($-CO\overset{-}{N}-\overset{+}{N}\equiv$) residues (1). Further, when the materials are heated, the aminimide group on these polymers suffers a carbon–nitrogen migration reaction, giving tertiary amines and reactive polymers with pendent isocyanate moieties (1). Both the aminimide and isocyanate groups promote adhesion of polymers to various surfaces (1).

The literature offers numerous examples of investigations of copolymers of vinylpyridines and N-vinylpyrrolidones and of their use in adhesive applications. For example, 4-vinylpyridine (4VP) has been grafted onto rubber and used in copolymers blended with rubber to improve the adhesion of rubbers to metals and other tire cord materials (2, 3, 4). Poly(N-vinylpyrrolidone) has very good adhesion to glass, metals, and various plastics and can be used to give improved adhesion between glass fiber and other plastics in glass fiber-reinforced composites (5). Copolymers of N-vinylpyrrolidone (NVP) and vinyl acetate or acrylic esters are also useful as laminating and pressure-sensitive adhesives (6, 7).

This study was undertaken to see if improved tire cord adhesive systems could be developed using poly(DHA-co-4VP) and/or poly(DHA-co-NVP) materials. Since these copolymer systems were new and since previously published Q and e values for DHA (8), 4VP (9), and NVP (10) suggested poor copolymerizability between the monomer sets, a detailed copolymerization study of these monomer pairs was needed. In summary, this study was initiated to determine copolymerization characteristics, i.e. reactivity ratios, of DHA-co-4VP and DHA-co-NVP, to characterize briefly some of the physical and chemical properties of the resultant materials, and to evaluate select copolymers in tire cord adhesive systems.

Experimental

Monomers, Initiator, and Solvents. DHA was recrystallized from ethyl acetate (mp 146°–147°C). 4VP was dried over molecular sieves and fractionally distilled under nitrogen; a middle fraction with bp 64°C/18mm and $n_D^{25} = 1.5551$ was collected just prior to use. NVP was also dried and redistilled under nitrogen just before use (bp 90°C/10mm). Monomers, hydroxyethyl acrylate (HA), hydroxyethyl methacrylate (HEMA), hydroxypropyl methacrylate (HPMA), and glycidyl methacrylate (GMA) were used as received from commercial sources. Azobis(isobutyronitrile) (AIBN) was recrystallized from methanol (mp 102°–103°C). The solvents N,N-dimethylformamide (DMF) and cyclohexanone and the reagents methyl chloride and methyl chloroacetate were purified by standard techniques. All other solvents (AR grade) were used as received.

Analytical Procedures. The purity of all copolymers, i.e. absence of monomers, was checked by thin layer chromatography (TLC). Composition of the DHA-co-4VP copolymers was determined from elemental analysis data obtained by Micro-Analysis, Inc., Wilmington, Del. Compositions of the DHA-co-NVP copolymers were determined by non-aqueous titration, using 0.1N perchloric acid and gentian violet indicator in glacial acetic acid (11). Isocyanate was determined as reported previously (12).

The nuclear magnetic resonance (NMR) spectra were recorded on a Varian A-60 spectrometer, using deuterated dimethyl sulfoxide (DMSO) as solvent and tetramethylsilane as internal standard. IR spectra were obtained on a Perkin-Elmer 237B IR spectrophotometer.

A DuPont 900 differential thermal analyzer and a 950 thermal gravimetric analyzer unit were used for differential thermal analysis (DTA) and thermal gravimetric analysis (TGA). Alumina was used as reference, and the finely divided polymer samples were analyzed under nitrogen (0°–500°C) at heating rates of 10° or 20°C/min. Gel permeation chromatography (GPC) was run on a Waters model 200 GPC unit, using DMF as solvent. Inherent viscosities (η_{inh}) were determined in methanol at 30°C using Ubbelohde capillary viscometers (concentration, 0.5 g/dl).

Copolymerization Studies. All copolymerizations for determining reactivity ratios were carried out under nitrogen in sealed Wheaton pressure bottles fitted with a magnetic stir bar. The monomers, solvent (2-propanol, 40% of total monomer), and AIBN were combined; the mixtures were then degassed by conventional freeze–thaw techniques, flushed under nitrogen, sealed, and polymerized (see Tables III and IV for conditions). All copolymerizations were terminated by pouring the reaction mixtures into a large excess of non-solvent (acetone or diethyl ether). The crude polymer was redissolved, reprecipitated by a non-solvent, collected, washed, dried to constant weight *in vacuo*, weighed to determine conversion, and checked for purity by TLC.

Copolymerizations to high conversions were carried out in round-bottomed flasks fitted with stirrer, thermometer, gas inlet tube, and reflux condenser. All systems were deaerated, and the polymerizations were performed under nitrogen. Conversions were determined by weighing the isolated copolymers. The polymers were purified as described above.

Rearrangement Reaction. The copolymers were dissolved in freshly distilled DMF or cyclohexanone (*ca.* 10% solids) and heated under reflux while a dry nitrogen stream was passed through the solution. After 1.5 hrs of reaction, only a trace amount of amine was detectable in the distillate. The solution was allowed to cool, and the isopropenyl isocyanate copolymers were then isolated by pouring the thermolyzed polymer solution into a large excess of non-solvent. The polymers were filtered, washed, redissolved in chloroform, and then analyzed.

Quaternization Reaction. In a pressure bottle equipped with a magnetic stirrer, the copolymer was dissolved in methanol, and methyl chloride was then added to slight excess. For reaction conditions, *see* Table VI. After the specified reaction time, the excess methyl chloride was allowed to evaporate, and the polymer was isolated by precipitation into acetone. After purification, the quaternized copolymer was characterized by IR and chlorine elemental analysis. Similar experiments were conducted using equimolar amounts of methyl chloroacetate.

Adhesion Tests. The wire used for testing was National Standard single strand, brass-plated wire (diameter, 0.16"). Two polyester cord materials were used, DuPont T-68-1300/3 and Fiber Industries T-785-1000/3.

The general procedure for coating the steel wire consisted of cleaning the wire, coating it with the DHA–4VP copolymer, curing the coating 80 sec at 445°F, coating the wire a second time with an emulsion consisting of a resorcinol–formaldehyde (RF) resin and a styrene–butadiene–vinylpyridine latex (Gen-Tac FS, General Tire & Rubber Co.), and curing the second coating 80 sec at 445°F. For polyester cord, the procedure was modified so that each coating process gave correct pretensioning and stretch to the finished cord. The first dip consisted of quaternized DHA–4VP or poly(DHA–NVP) copolymer and a commercial epoxide resin, and it was cured 45 sec at 445°F.

In accordance with ASTM wire adhesion test D2229-635, the steel cord samples were embedded in a general purpose rubber stock, and the composite was vulcanized, aged, and then tested on an Instron tester at a head speed of 6 in./min. Eight samples per treatment were generally tested. Since the embedment length was 0.25 in., the static adhesion, *i.e.* the force required to extract the steel cord, is reported in pounds per 0.25 inch. The polyester cord H-pull tests were conducted according to ASTM D2138-67 procedure. Usually 18 samples were tested, and adhesion was reported as the number of pounds needed to pull a ¼-in. length of cord from the vulcanized rubber.

Results and Discussion

General Copolymerization Studies. DHA–4VP SYSTEM. In order to examine the copolymerizability of DHA with 4VP and to investigate some properties of the copolymer, equimolar amounts of DHA and 4VP were copolymerized in 2-propanol at 70°C. The powdery, off-white, purified polymer was soluble in DMF, slightly soluble in chloroform, and insoluble in water and tetrahydrofuran (THF). In contrast, poly(4VP) was soluble in THF, DMF, and chloroform.

Extraction of the copolymer failed to produce DHA or 4VP homopolymer. The GPC curve, with narrow molecular weight distribution and with only moderate amounts of molecular weight tails, demonstrated clearly that the copolymer was homogeneous.

Attempts were made to determine structure by both NMR and IR. The IR spectrum of the copolymer(s) (free film) had absorption bands at 1605 and 755 cm^{-1} (pyridine ring vibrations), 825 cm^{-1} (C–H deformation vibration), and 1580 cm^{-1} (aminimide carbonyl); this supported the structure of poly(DHA-co-4VP). The NMR spectrum showed DHA and 4VP moieties, but the resolution was too poor for quantitative assignments.

Thermal properties of the copolymer were investigated by DTA, TGA, and IR. The copolymer(s) had distinct thermograms, indicating transformation of the aminimide monomer segments to isopropenyl isocyanate residues and elimination of the tertiary amine, 1,1-dimethyl-1-(2-hydroxypropyl)amine. A typical DTA curve had a strong exotherm at 150°–225°C (midpoint ca. 180°C); the TGA curve indicated onset of weight loss at 150°C with an increase to a maximum rate in the region of the DTA exotherm. The weight loss of tertiary amine failed to match theoretical amounts because some amine was retained by the isocyanate–hydroxyl (urethane formation) reaction. When copolymer films were heated at 150°–200°C) for various times, the band in the spectra at 1580 cm^{-1} vanished and new bands appeared at 2260 cm^{-1} (—NCO) and 1710 cm^{-1} (urethane carbonyl).

Mixtures with various DHA–4VP compositions were copolymerized to study the effect of monomer composition upon copolymer conversion. The

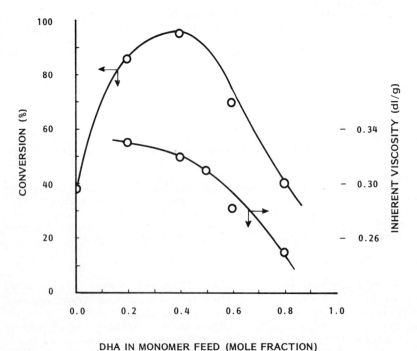

Figure 1.   Conversion and inherent viscosity as a function of monomer composition in the copolymerization of DHA with 4VP
Conditions: 70°C, 3 hrs, and AIBN 2 wt % of total monomer

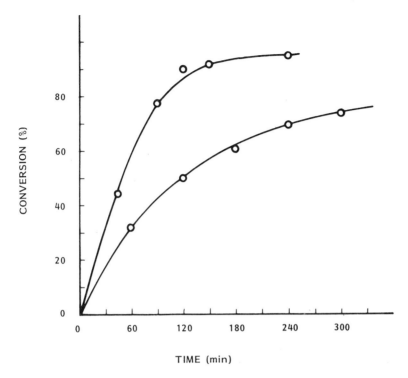

Figure 2.   Conversion vs. *time plots of the copolymerization of DHA with*
4VP

Conditions: 70°C, AIBN 2 wt % of total monomer, and monomer concentrations: upper
curve, 40 mole % DHA and lower curve, 60 mole % DHA

polymer yield increased until the content of DHA in the monomer mixture reached about 40 mole %; it then decreased when this concentration was exceeded (*see* Figure 1). Furthermore, the initial monomer feed had a strong effect on the inherent viscosity; increasing the DHA content in the feed lowered the molecular weight of the isolated polymer, probably by increased chain transfer reactions.

Typical examples of the relation between time and conversion for two different concentrations of DHA and 4VP are presented in Figure 2. Conversion increased linearly without any induction period, and the reaction was very fast in the early stages. The rate of copolymerization increased with the concentration of 4VP.

In the DHA–4VP studies, we found that copolymer yield and inherent viscosity varied considerably with the solvent used for polymerization (Table I). Solvents greatly affect free radical polymerization, although the mechanisms of some interactions is uncertain. For example, the effect of solvents on the rate of polymerization has been attributed to complex formation and to viscosity effects which alter rates of initiation, propagation, and termination (*13, 14, 15*). The data in Table I do not indicate any clear tendency, and no molecular weight-conversion correlation of the copolymers is apparent. The variation in copolymer yields is readily explained by differences in solubility of DHA.

Table I. Effect of Solvents on the Copolymerization of DHA with 4VP[a]

Solvent	Conversion, %	Viscosity, η_{inh}	Solubility at 25°C, g/dl
Benzene	3.7	0.34	1.0
2-Ethoxyethanol	19.0	0.32	23.0
2-Propanol	23.8	0.31	25.0
Dimethylformamide	13.0	0.48	18.0
Acetonitrile	0.6	0.28	8.4

[a] Polymerization conditions: 40 mole % DHA, 60°C, 3 hr, and AIBN 2 wt % of total monomer.

DHA–NVP System. We conducted several copolymerization experiments to high conversion in order to examine some characteristics of the systems and to determine some properties of the formed copolymers. The findings are summarized in Table II. The purified poly(DHA-co-NVP) materials were white, powdery solids completely soluble in water, alcohols, chloroform, DMF, and DMSO and insoluble in acetone, benzene, and ether. The copolymers had good film-forming properties, i.e. air-dried films were clear and hard with a high gloss.

The inherent viscosity of the DHA–NVP copolymer solutions varied with the feed composition (see Tables II and IV); the value was minimum for 80 mole % DHA in the monomer feed. Similar behavior, i.e. decrease in molecular weight with increase in DHA content in feed, was observed for the DHA–4VP systems. This is further evidence of DHA participation in lowering molecular weight by increased monomer-induced chain transfer reactions.

The IR spectra of the copolymers had strong absorption bands at 1710 cm^{-1} (pyrrolidone unit) and 1580 cm^{-1} (aminimide carbonyl). When films of the copolymers were heated (150°–200°C), the band at 1580 cm^{-1} vanished and a new strong band appeared at 2250 cm^{-1} which was indicative of the described aminimide–isocyanate rearrangement. Again, the NMR spectra had poor resolution but the broad bands indicated DHA and NVP moieties.

The DTA thermograms for DHA–NVP copolymers had endotherms near 120°C which could be associated with water loss. [Poly(NVP) is reported to have a very high affinity for water absorption.] However, these endotherms were also observed on carefully dried samples, and no weight loss was noted on the TGA curves. Unfortunately, the DTA samples could not be recycled since the copolymers underwent the rearrangement reaction. It may be argued that the endotherm was associated with the glass transition temperature of the materials. However, additional work is required to answer this question. Exothermic activity followed at 150°–240°C (midpoint, 185°C), which was also attributable to the aminimide–isocyanate rearrangement. The TGA curves,

Table II. Copolymerization of DHA with NVP to High Conversion[a]

Feed, mole % DHA	Time, hrs	Conversion, %	Copolymer, mole % DHA	Viscosity, η_{inh}	Adhesion H Values,[b] lbs
20	2.67	70	25	0.28	27.1
30	2.27	85	35	0.24	28.9
40	2.15	92	43	0.22	33.9
50	2.00	78	54	—	37.8
60	2.00	87	62	0.19	35.9

[a] Polymerization conditions: 70°C, and AIBN 2 wt % of total monomer.
[b] Obtained on DuPont T-68 1300/3 cord.

**Table III.   Copolymerization of DHA (M_1) with 4VP (M_2)
to Low Conversions** [a]

DHA, moles	4VP, moles	DHA in Monomer Feed, mole %	Con- version, %	0 Content of Copolymer, wt %	DHA in Polymer, mole %	Time, min	Copolymer Mole Ratio
0.01	0.19	5	4.80	3.48	12.54	60	0.143
0.02	0.18	10	9.00	4.48	16.60	70	0.199
0.03	0.17	15	5.30	5.44	20.73	80	0.262
0.04	0.16	20	6.25	5.98	23.15	80	0.301
0.06	0.14	30	8.00	7.11	28.27	90	0.394
0.08	0.12	40	3.50	7.80	31.95	60	0.469
0.10	0.10	50	1.00	9.03	38.48	40	0.625
0.12	0.08	60	7.15	10.51	47.08	90	0.889
0.14	0.06	70	3.14	12.24	58.34	60	1.400
0.16	0.04	80	9.60	13.91	70.62	90	2.404

[a] Polymerization conditions: 2-propanol, 70°C, and AIBN 1 wt % of total monomer.

with onset of weight loss at *ca.* 150°C, support this interpretation and show almost theoretical weight loss (150°–250°C) for amine.

Reactivity Ratio Studies. DHA–4VP System. A series of DHA–4VP (M_1–M_2) copolymerizations was carried out to low conversions (<10%) with the monomer pair ratio being varied. Table III summarizes the data. In the corresponding copolymerization diagram, the composition of the copolymer is plotted as a function of the composition of the initial monomer concentration

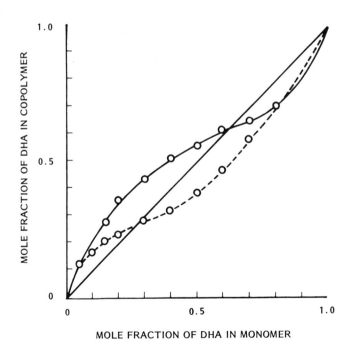

Figure 3.   Copolymer composition curves
- - -, DHA–4VP and ———, DHA–NVP

Table IV. Copolymerization of DHA (M_1) with NVP (M_2) to Low Conversions[a]

DHA, moles	NVP, moles	DHA in Monomer Feed, mole %	Conversion, %	Copolymer Analysis, meq. DHA	DHA in Polymer, mole %	Copolymer Mole Ratio	Viscosity, η_{inh}
0.03	0.17	15	8.60	2.09	27.55	0.380	—
0.04	0.16	20	9.40	2.59	35.82	0.558	—
0.06	0.14	30	10.00	3.03	43.66	0.775	0.45
0.08	0.12	40	10.00	3.41	50.94	1.038	0.41
0.10	0.10	50	5.40	3.63	55.40	1.215	0.36
0.12	0.08	60	7.40	3.90	61.50	1.597	0.29
0.14	0.06	70	2.67	4.05	64.68	1.831	0.27
0.16	0.04	80	4.92	4.25	69.40	2.268	0.19

[a] Polymerization conditions: 2-propanol, 60°C, and AIBN 1 wt % of total monomer.

(Figure 3). This monomer pair had an azeotropic copolymerization composition at *ca.* 26 mole % DHA.

Using the data in Table III, the general copolymer composition equations, and the Fineman–Ross procedure (*16*), the reactivity ratios were estimated at $r_1 = 0.40$ and 0.42 and $r_2 = 0.78$ and 0.75 with mean values of 0.41 and 0.77 respectively. Values of $r_1 = 0.406$ and $r_2 = 0.773$ were calculated by a computer program based on least squares.

The Alfrey–Price Q and e values (*17*) were calculated from the monomer reactivity ratios in the copolymerization with 4VP: $Q_1 = 0.68$ and $e_1 = 0.58$. The values used for 4VP in the calculations were $Q_2 = 0.90$ and $e_2 = -0.50$ (*9*).

DHA-NVP SYSTEM. The results from a similar series of copolymerizations of DHA (M_1) with NVP (M_2) are presented in Table IV (*see* Figure 3 for the copolymer composition curve). This system had an azeotropic copolymerization composition at a much higher DHA content (*ca.* 60 mole %). The differential form of the copolymer equation and the data in Table IV were used to calculate the reactivity ratios: $r_1 = 0.37$ and 0.33 and $r_2 = 0.16$ and 0.14 with average values of 0.35 and 0.15 respectively. Values of $r_1 = 0.352$ and $r_2 = 0.145$ were calculated by the computer program. Using the values $Q_2 = 0.14$ and $e_2 = -1.14$ for NVP (*10*), the calculated Q and e values for DHA in this system were $Q_1 = 0.14$ and $e_1 = 0.58$.

The data from these two studies of monomer reactivity with DHA, as well as data from a previous study in which M_2 was methyl methacrylate (MMA) and DMF the solvent, are summarized in Table V. When possible variations in the Q_2 and e_2 values estimated and used for NVP and 4VP are taken into account, and also that some deviation might be caused by the temperature difference, the agreement between the two Q_1 and e_1 values for DHA found in this study was satisfactory. However, there was no reasonable correlation between the Q_1 and e_1 values for DHA obtained with comonomer MMA. It is possible that some monomer–monomer or monomer–solvent interactions may account for this discrepancy, but we detected no proof of such interactions.

Table V. Reactivity Ratios and Q and e Values for DHA (M_1)

Comonomer (M_2)	r_1	r_2	Q_1	e_1
NVP	0.35	0.15	0.14	0.58
4VP	0.41	0.77	0.68	0.58
MMA	0.00	1.99	0.12	-2.45

$$\left[-(CH_2-CH-)_x-(CH_2-\underset{\underset{N}{\overset{CH_3}{|}}}{C}-)_y- \right]_n \quad \overset{CH_3}{\underset{\underset{CH_3\quad OH}{|}}{CONN-CH_2CHCH_3}} \qquad (1)$$

$$\xrightarrow{153°C}$$

$$\left[-(CH_2-CH-)_x-(CH_2-\underset{\underset{NCO}{\overset{CH_3}{|}}}{C}-)_y- \right]_n \quad + \quad \overset{CH_3}{\underset{\underset{CH_3\quad OH}{|}}{N-CH_2CHCH_3}}$$

Copolymer Reactions. REARRANGEMENT REACTIONS. Two poly(DHA-*co*-4VP) materials with 10–15 mole % DHA were thermolyzed in refluxing DMF (*see* Reaction 1). A strong nitrogen sparge was used to help remove the tertiary amine. The isolated poly(4-vinylpyridine-*co*-isopropenyl isocyanate) materials were insoluble in ether but soluble in chloroform, they had a strong band in the IR spectra at 2260 cm^{-1} and a weak band at 1710 cm^{-1}, and they consisted of 1.5–2.0 wt % NCO. A sample of the thermolysis solution had a shelf life, *i.e.* lack of gelation, of about five days.

Using the same procedure, a sample of poly(DHA-*co*-NVP) with 20 mole % DHA was rearranged in refluxing cyclohexanone (Reaction 2). A sample

$$\left[-(CH_2-CH-)_x-(CH_2-\underset{\overset{CH_3}{|}}{C}-)_y- \right]_n \quad \overset{CH_3}{\underset{\underset{CH_3\quad OH}{|}}{CONN-CH_2CHCH_3}} \qquad (2)$$

$$\xrightarrow{156°C}$$

$$\left[-(CH_2-CH-)_x-(CH_2-\underset{\underset{NCO}{\overset{CH_3}{|}}}{C}-)_y- \right]_n \quad + \quad \overset{CH_3}{\underset{\underset{CH_3\quad OH}{|}}{N-CH_2CHCH_3}}$$

of the isolated poly(N-vinylpyrrolidone-*co*-isopropenyl isocyanate) had a strong absorption band in the IR spectra at 2250 cm^{-1}. Titration of the isocyanate indicated that the NCO content was 3.25 wt %. The shelf life of the unstabilized polymer solution was also about five days.

QUATERNIZATION STUDIES. Quaternization of DHA–4VP copolymers with methyl chloride and methyl chloroacetate (Table VI) produced water-soluble, polycation-containing materials with the idealized structure:

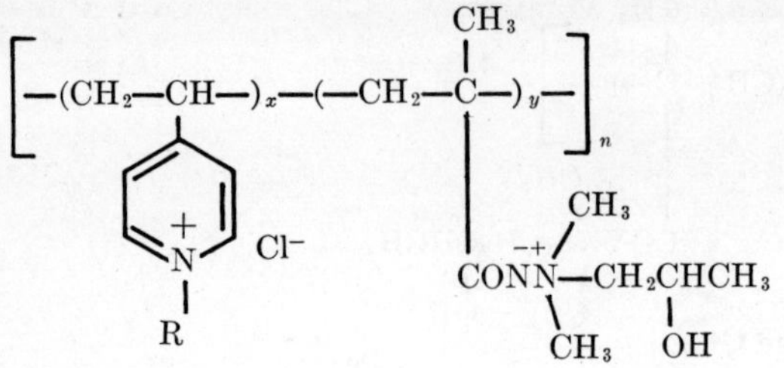

R = CH_3, $CH_2CO_2CH_3$

In methanol, the viscosities of the polycation materials resembled those of other polyelectrolytes (*18*), *i.e.* the inherent viscosity of a typical sample increased from 0.41 to 0.64 upon dilution from 0.5 to 0.125 g/dl. The IR spectra of the quaternized copolymers had aminimide absorption bands at 1585 cm^{-1} (as in the original polymer), a strong band at 1650 cm^{-1} (pyridine ring), and, with —$CH_2CO_2CH_3$, a strong band at 1750 cm^{-1}.

Table VI. Quaternization of DHA–4VP Copolymers

Copolymer, mole % 4VP	*Alkylating Agent*	T, °C	*Time, hrs*	*Quat.,[a] %*
40	CH_3Cl	60	3	68.5
40	$CH_2ClCO_2CH_3$	64	1	66.0
60	$CH_2ClCO_2CH_3$	25	1	—
60	$CH_2ClCO_2CH_3$	25	12	36.0
60	$CH_2ClCO_2CH_3$	64	3	91.5

[a] Calculated from Cl analysis and based on 4VP.

Complete quaternization (DHA included) could not be obtained, which may be explained by a low reactivity of DHA for alkylating agents or a decreased overall reaction rate. Fuoss and co-workers (*19*) observed strong, neighbor group effects in quaternization studies of poly(4VP). As the quaternization proceeded, the rate decreased as a result of the charge buildup on the polymer, and quaternization of the last half of the pyridine ring nitrogen occurred at about one-tenth the initial rate.

ADHESIVE STUDIES. The DHA–4VP copolymers had excellent properties as adhesion promoters for rubber–steel (brass-coated) composites (Table VII). The adhesion values of 44.8–47.6 lbs approach the limit of the test, *i.e.* rubber failure may occur at 45–50 lbs. The adhesion values listed in Table VII are the optima obtained in numerous tests in which pH, type of RF latex resin, curing temperatures, etc. were varied. Adhesion values ranged from the lower forties to the upper thirties, and coverages were in the 3+ range.

The quaternized DHA–4VP copolymers were very useful for promoting adhesion of rubber to polyester cord (Table VIII). The DHA–NVP copolymers were also tested as adhesion promoters for polyester–rubber composites. The

Table VII. Static Adhesion Data for DHA–4VP Copolymers on Brass-Plated Steel Wire[a]

Polymer	*Composition, mole %*	*Adhesion Value, lbs*	*Rubber Coverage, 0–5*
DHA–4VP	60/40	46.5	4.4
DHA–4VP	60/40	47.6	4.4

[a] RF performed *in situ*.

Table VIII.   Quaternized DHA–4VP Copolymers in Polyester Cord Adhesives

Copolymer, mole % 4VP	Quat.,[a] %	Adhesion H Values, lbs	Rubber Coverage, %
60	66	36.4[b]	90
40	89	38.0[c]	100
40	36	40.0[c]	100

[a] Alkylating agent $CH_2ClCO_2CH_3$; quaternization with respect to 4VP.
[b] Tested on polyester cord FI T-785-1000/3.
[c] Tested on polyester cord DuPont T-68-1300/3.

data (Table II) indicate that there had to be *ca.* 50 mole % DHA in the copolymer for the adhesion values to be comparable to those of the quaternized DHA–4VP copolymers.

Several terpolymers of DHA and NVP with HEA, HEMA, HPMA, or GMA were prepared and evaluated as film formers and adhesion promoters for polyester–rubber systems. All the copolymers were water soluble. However, cast films became water insoluble when heated (150°–200°C) because of the crosslinking reactions between isocyanate-hydroxyl or isocyanate-epoxide residues. Data from the terpolymerization experiments are presented in Table IX.

COATING STUDIES. Coatings were prepared from blends of poly(DHA-*co*-NVP) materials and expoxide resins. Typically, a 50:50 copolymer was dissolved in methanol and combined with Epon 812 (Shell Chemical) to give approximately 1:1 aminimide:epoxide. The solution was used to coat steel and glass panels to a thickness of *ca.* 1 mil after air drying and baking 30 min at 160°C. The coatings had high gloss, adhesion, H pencil hardness, 40-lb face impact, 20-lb reverse impact, and excellent mar resistance, and they passed the conical flex test. The good solvent resistance indicated that the films were crosslinked.

Summary

DHA monomer copolymerizes readily with both 4VP and NVP to form two new reactive polymers. The copolymers are easily obtained in high yield, with polymerization rates and molecular weights strongly affected by initial monomer feeds. Reactivity ratio studies, with $r_1 = 0.41$ and $r_2 = 0.77$ for DHA-*co*-4VP and $r_1 = 0.35$ and $r_2 = 0.15$ for DHA-*co*-NVP, clearly demonstrate that alternating copolymers are obtained.

Both DHA-*co*-4VP and DHA-*co*-NVP materials with low DHA content were thermolyzed in solution to give soluble, reactive 4VP-*co*-isopropenyl isocyanate and NVP-*co*-isopropenyl isocyanate copolymers. The DHA-*co*-4VP copolymers were quaternized to give new cationic polyelectrolytes with proper-

Table IX.   Terpolymers of DHA and NVP[a]

Monomers	Time, hrs	Conversion, %	Adhesion H Values, lbs	Rubber Coverage, %
DHA–NVP–HEA	5	50	33.6[b]	100
DHA–NVP–HEMA	3	51	37.7[c]	100
DHA–NVP–HPMA	5	60	33.2[c]	—
DHA–NVP–GMA	3	50	40.6[b]	97

[a] Polymerization conditions: 50:40:10 mole % monomer feed, methanol (50% solids), 70°C, and AIBN 1 wt % of total monomer.
[b] Tested on polyester cord DuPont T-68-1300/3.
[c] Tested on polyester cord FI-T-785-1000/3.

ties which could be greatly modified by adjusting copolymer composition, degree of quaternization, and type of alkylating reagent.

These copolymers have a wide range of possible applications. DHA-*co*-4VP and DHA-*co*-NVP copolymers are excellent for rubber–steel and rubber–polyester adhesive systems respectively. In addition, quaternized DHA-*co*-4VP copolymers promote strong adhesion of rubber to polyester tire cord in vulcanized composites.

Acknowledgment

The authors wish to acknowledge the help and encouragement of W. J. McKillip in this work, P. Menardi for NMR, M. Hallwachs for GPC, D. Gregerson for DTA and TGA experimental work, E. Luckman and R. E. Field for excellent technical assistance in the adhesive studies, and B. Bushong for general synthesis work.

Literature Cited

1. McKillip, W. J., Sedor, E. A., Culbertson, B. M., Wawzonek, S., *Chem. Rev.* (1973) **73**, 278.
2. Metallgesellschaft A.G., Brit. Patent **943,156** (1963); *Chem. Abstr.* (1964) **60**, 7012c.
3. Keskula, H., U.S. Patent **3,072,598** (1963); *Chem. Abstr.* (1963) **58**, 7000a.
4. Wolfe, W. D., U.S. Patent **2,817,616** (1957); *Chem. Abstr.* (1958) **52**, 5019d.
5. Lorenz, D. H., in "Encyclopedia of Polymer Science and Technology," N. M. Bikales, Ed., Vol. 14, p. 239, Interscience, New York, 1971.
6. Morner, R. R., Longley, R. I., U.S. Patent **2,667,473** (1954); *Chem. Abstr.* (1954) **48**, 6164.
7. Takemoto, K., *J. Macromol. Sci. Rev. Macromol. Chem.* (1970) **5**(1), 33.
8. Culbertson, B. M., Sedor, E. A., Slagel, R. C., *Macromolecules* (1968) **1**, 254.
9. Tamikado, T., *J. Polym. Sci.* (1960) **43**, 489.
10. Young, L. J., *J. Polym. Sci.* (1961) **54**, 411.
11. Culbertson, B. M., Slagel, R. C., *J. Polym. Sci. Part A-1* (1968) **6**, 363.
12. Culbertson, B. M., Freis, R. E., *Macromolecules* (1970) **3**, 715.
13. Schultz, G. V., Fischer, J. P., *Makromol. Chem.* (1967) **107**, 253.
14. Bengough, W. I., Henderson, N. K., *Chem. Ind. London* (1969) **20**, 657.
15. Burnett, G. M., Cameron, G. G., Parker, R. M., *Eur. Polym. J.* (1969) **5**, 231.
16. Fineman, M., Ross, S. D., *J. Polym. Sci.* (1950) **5**, 259.
17. Alfrey, Jr., T., Price, C. C., *J. Polym. Sci.* (1947) **2**, 101.
18. Flory, P. J., "Principles of Polymer Chemistry," p. 635, Cornell University, Ithaca, 1967.
19. Fuoss, R. M., Watanabe, M., Coleman, B. D., *J. Polym. Sci.* (1960) **48**, 5.

RECEIVED April 1, 1974.

Styrene–Polymer Interaction Parameters in High Impact Polystyrene

R. L. KRUSE

Monsanto Co., Polymers and Petrochemicals Div., Indian Orchard, Mass. 01051

The separation of polystyrene and polybutadiene into two phases, both containing styrene, was used to measure polymer-solvent interaction parameters. By this technique, the mean interaction parameter, χ_{12}, for polystyrene in styrene is 0.49 and that for polybutadiene in styrene, χ_{13}, is 0.29. Phase behavior at higher concentrations was calculated from data obtained at concentrations of less than 35%. As a result, phase behavior during polymerization of high impact polystyrene was interpreted.

The polystyrene/styrene/polybutadiene (PS/S/PBD) system occurs in the production of high impact polystyrene. The process for making toughened polystyrene as described by Moulau and Keskkula (1) starts with a rubber in styrene solution. As S is polymerized to PS, phase separation results in immediate formation of droplets of a PS phase. With further polymerization, the PS phase increases in volume until phase volumes are equal. At this point, phase inversion occurs—the dispersed PS phase becomes the continuous phase and the PBD phase becomes the disperse droplets. Complete conversion of S to PS yields the commercially important high impact polystyrene.

One method of quantifying phase behavior is to mix two polymers in a common solvent and observe the two liquid phase volumes (2, 3). The theoretical basis for the incompatibility of polymer solutions was discussed by Scott (4); however, complete phase relationships are rarely measured. The poly(methyl methacrylate)/benzene/rubber system was described by Bristow (5), but even he did not calculate solubility parameters from the data. Thus, measurement and data interpretation techniques need to be defined.

Experimental

Samples of linear PBD ($M_w = 149,000$; $M_w/M_n = 1.46$) and PS ($M_w = 290,000$; $M_w/M_n = 3.0$) were dissolved at six levels of total solids in S in jars on a roll mill. Styrene polymerization was prevented by adding 0.1% benzoquinone. The two-phase mixtures were separated by centrifugation 15 min at 20,000 rpm in a Beckman model 21 ultracentrifuge, and the solutions were then frozen in the metal centrifuge tubes. Frozen plugs were removed as needed by warming the outside of the tube. They were then sectioned at the interface, warmed to 25°C, and the solids level in each phase was measured by methanol

Table I.   Phase Data for the Styrene/Polystyrene/Polybutadiene System

Polystyrene in Its Phase, wt %	Polybutadiene in Its Phase, wt %
9	4
15	8
19	11
26	17
29	20
35	26

precipitation. IR measurements on the solids from the separated phases indicated an essentially complete separation of the two polymeric components. The data are tabulated in Table I. Duplicate samples agreed within 0.5%.

Monodisperse PS ($M_w = 160,000$; $M_w/M_n = 1.06$) and a narrow distribution PBD ($M_w = 78,500$; $M_w/M_n = 1.23$) were compared with their broad distribution equivalents at 10% PS and 6% PBD in S. Phase volumes were the same, which indicates that polymer molecular weight distribution is not a significant variable.

Solutions of PBD in S and PS in S, mixed and allowed to separate by gravity, gave the same results. Measurements at 0°–60°C also did not alter the phase relationships at concentrations over 4%. Phase separation of a 50/50 mixture of the two polymers at 25°C occurred at a solid concentration of 2–3%, but the solutions did not separate into equal volumes. Dilute solutions separate into equal volume phases if the PS concentration is approximately five times that of the PBD at total solids levels of 3–4%.

Theoretical

The change in chemical potential (4), $\Delta\mu_1$, of S in the S/PS/PBD system where the components are subscripted 1, 2, and 3, respectively, is:

$$\frac{\Delta\mu_1}{RT} = \log_e (1 - v_2 - v_3) + \left(1 - \frac{1}{x_2}\right) v_2 + \left(1 - \frac{1}{x_3}\right) v_3 + (\chi_{12} v_2 + \chi_{13} v_3) (v_2 + v_3)$$
$$+ \frac{\chi_{23}}{x_2} v_2 v_3 \qquad (1)$$

where v_i is the fraction of component i, x_i is its degree of polymerization, and χ_{ij} is its interaction parameter with a second component. The two high molecular weight polymers in S are segregated into separate phases—PS in S and PBD in S. The chemical potential, $\Delta\mu_1/RT$ for S with dissolved PS is:

$$\frac{\Delta\mu_1}{RT} = \log_e (1 - v_2) + \left(1 - \frac{1}{x_2}\right) v_2 + \chi_{12} v_2^2 \ (v_3 \approx 0) \qquad (2)$$

where v_2 is the volume fraction PS and χ_{12} is the S–PS interaction parameter. The relation for S with dissolved PBD is:

$$\frac{\Delta\mu_1}{RT} = \log_e (1 - v_3) + \left(1 - \frac{1}{x_3}\right) v_3 + \chi_{13} v_3^2 \ (v_2 \approx 0) \qquad (3)$$

where v_3 is the fraction PBD and χ_{13} is the S–PBD interaction parameter. At equilibrium, the chemical potentials of S in the two phases are equal; therefore:

$$\log_e \frac{(1 - v_3)}{(1 - v_2)} = v_2 - v_3 + \chi_{12} v_2^2 - \chi_{13} v_3^2 \ (x_2 \text{ and } x_3 \text{ large}) \qquad (4)$$

Since v_2 vs. v_3 are measured, the two unknown constants, χ_{12} and χ_{13} can be evaluated by plots of $[v_3 - v_2 + \log_e (1 - v_3) - \log_e (1 - v_2)]/v_3^2$ vs. $(v_2/v_3)^2$. Slope, χ_{12}; intercept, $-\chi_{13}$.

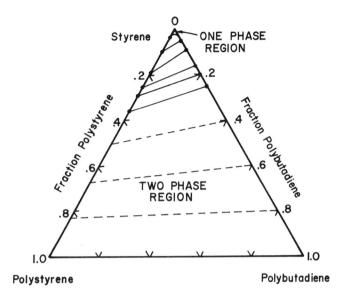

*Figure 1. Triangular phase diagram for the system
styrene/polystyrene/polybutadiene*
———, *measured tie line;* – – –, *calculated tie line*

The ratio $(1 - v_3)/(1 - v_2)$ is a partition coefficient that measures distribution of the solvent between the polymer phases. The partition coefficient is the fraction of S in the PBD phase divided by the fraction of S in the PS phase. For concentrated solutions, v_2 and v_3 approach one, and this ratio is a constant given by

$$\frac{1 - v_3}{1 - v_2} = e^{\chi_{12} - \chi_{13}} \tag{5}$$

The ratio is the distribution of residual styrene between the two phases.

Results

Data for the PS/S/PBD system in Figure 1 are replotted in Figure 2 according to Equation 4. The 0.49 slope by least squares is the interaction parameter of PS with S, χ_{12}. The negative 0.29 intercept by least squares is the interaction parameter for PBD with S, χ_{13}. The value 0.29 indicates that S is a better solvent for PBD than for PS. Least squares analysis indicated that the values of χ are precise to ±0.01 unit ($2\ \sigma$). Use of volume fractions instead of weight fractions does not change χ_{12}, but χ_{13} increases from 0.29 to 0.35.

Published values for the interaction parameters vary. Bristow and Watson (6) reported the value 0.43 for PS in toluene, and Boyer and Spencer (7) gave the value 0.424 for PS in S. Hild *et al.* (8) obtained the value 0.45 + 0.9 v_2 for model crosslinked PS networks in benzene. Our mean value for PS in S, 0.49, is in this range; our method, however, does not require an estimate of crosslink density. Scott and Magat (9) reported the value 0.30 for PBD with toluene, which is close to our mean value for PBD with S (0.29).

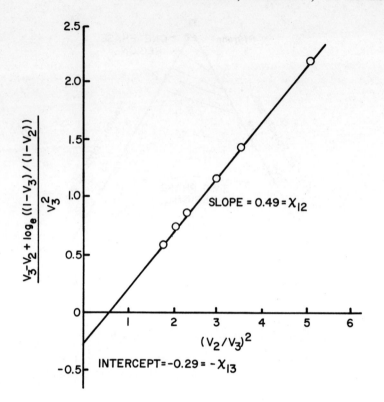

Figure 2. Calculation of solubility parameters from phase equilibrium data for the system styrene(1)/polystyrene(2)/polybutadiene(3)

$\chi_{12} = 0.49$, $\chi_{13} = 0.29$; *polymer concentrations in their respective phases are reported as* (v_2, v_3)

Phase relationships at polymer concentrations above 40% are difficult to measure. However, the concentrations of the phases can be calculated numerically using Equation 4, and they are used to position the additional tie lines (dashed) in Figure 1.

Values of the interaction parameters can be used to calculate the partition coefficient of residual styrene in the final polymer blend. The partition coefficient approaches 1.22 at complete conversion (from Equation 5). The approach to the limiting value with polymer concentration is illustrated in Figure 3.

Some assumptions in the theoretical development will now be examined. Ignoring the contribution of the PS–PBD interaction parameter (χ_{23}/x_2 or χ_{32}/x_3) to the numerical values of the ordinate in Figure 2 introduces an error of less than 1%. The magnitude of the interaction parameter, 0.02, was estimated from the concentration of polymer $(v_2 + v_3)_c$ at the point of initial phase separation using the procedure outlined by Scott (4), that is

$$\frac{\chi_{23}}{x_2} = \frac{(x_2^{-1/2} + x_3^{-1/2})^2}{2(v_2 + v_3)_c} \tag{6}$$

A similar value is calculated by using the solubility parameters of PS ($\delta_2 = 8.75$) and PBD ($\delta_3 = 8.4$) and the relation

$$\frac{\chi_{23}}{x_2} = \frac{(\delta_2 - \delta_3)^2 \ V}{RT} \tag{7}$$

where V is the molar volume of the repeat unit. Allen *et al.* (*10*) obtained the value 0.01 by measuring the phase compositions of high molecular polymers in carbon tetrachloride. Paxton's data (*11*) analyzed by Scott's method give the value 0.1–0.2 for the same two polymers in toluene, benzene, and carbon tetrachloride; however, his PBDs had a degree of polymerization of only 20. With the value 0.02, correction terms for the ordinate values are less than 1% at the concentrations measured. We assumed complete separation of the two polymers into their separate phases. Scott's dilution approximation for estimating the PS in the PBD phase v_2' is

$$v_2' = v_2 \ e^{-\chi_{23} \ (v_2 + v_3)} \tag{8}$$

For our experiments, $v_2 + v_3 > 0.13$ and χ_{23} is about 60. Therefore, the calculated fraction of PS in the PBD phase, v_2'/v_3, is less than 1% of the total polymer in that phase. Our measurements could not detect this small amount.

Discussion

Calculation of polymer concentrations at critical points in the process of polymerizing styrene in the presence of dissolved PBD is possible with use of

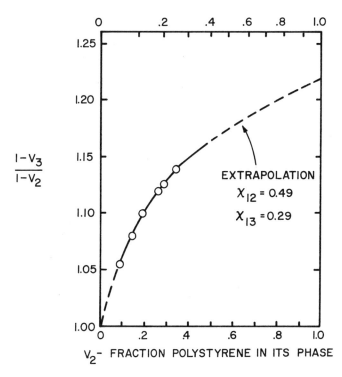

Figure 3.   Partition coefficients for styrene(1) between the poly-styrene(2) and the polybutadiene(3) phases

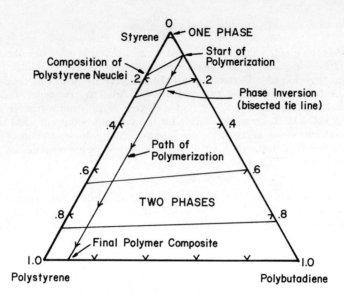

Figure 4. Phase concentrations during polymerization of styrene with 10% polybutadiene

The concentrations of styrene-v_1, polystyrene-v_2, and polybutadiene-v_3 are represented by (v_1, v_2, v_3)

the phase diagram. For example, the path of polymerization of a 10% PBD solution in styrene is illustrated in Figure 4. The separate PS phase is formed immediately, and concentration of PS in the initial droplets is 19%. The volume of dispersed PS phases increases with conversion of S to PS until phase volumes are equal at 13% PS. At this point, a tie line is bisected and the concentrations of PBD and PS in their respective phases are 19% and 28%. At higher conversions, the PBD phase is disperse. Although partitioning of the monomers at the higher conversions is difficult to measure experimentally, it can be calculated. For example, the concentrations of PBD and PS in their respective phases are 61% and 66% at 55% PS in the mixture and 83% and 86% at 75% PS. Thus, the polymer concentrations in their phases equalize at higher conversions. The partitioning of S, however, increases with conversion; near complete conversion, 12% residual S remains with the 10% PBD.

All the above calculations have assumed χ values independent of concentration, grafting, and crosslinking. In high impact polystyrene, all three factors can be important.

Acknowledgment

The interest and encouragement of Q. A. Trementozzi during this research is sincerely appreciated.

Literature Cited

1. Moulau, G. E., Keskkula, H., *J. Polym. Sci. Part A-1* (1966) 1595.
2. Dobry, A., Boyer-Kawenoki, F., *J. Polym. Sci.* (1947) **2**, 90.

3. Kern, R. J., Slocombe, R. J., *J. Polym. Sci.* (1955) **15**, 183.
4. Scott, R. L., *J. Chem. Phys.* (1949) **17**, 279.
5. Bristow, G. M., *J. Appl. Polym. Sci.* (1959) **2** (4), 120.
6. Bristow, G. M., Watson, W. F., *Trans. Faraday Soc.* (1958) **54**, 1742.
7. Boyer, R. F., Spencer, R. S., *J. Polym. Sci.* (1948) **3**, 97.
8. Hild, G., Haeringer, A., Rempp, P., Benoit, H., *Amer. Chem. Soc., Div. Polym. Chem., Prepr.* **14** (1), 352 (Detroit, May, 1973).
9. Scott, R. L., Magat, M., *J. Polym. Sci.* (1949) **4**, 555.
10. Allen, G., Gee, G., Nicholson, J. P., *Polymer* (1960) **1**, 56.
11. Paxton, T. R., *J. Appl. Polym. Sci.* (1963) **7**, 1499.

RECEIVED February 20, 1974.

14

Crosslinking Reactions in High Impact Polystyrene Rubber Particles

DIETER JOSEF STEIN, GERHARD FAHRBACH, and
HANSJÖRG ADLER

Kunststofflaboratorium, BASF Aktiengesellschaft,
Ludwigshafen/Rhine, West Germany

The crosslinking of rubber in the rubber particles of high impact polystyrene is investigated using suspension polymerization of rubber solutions in styrene as a model. Polybutadiene and butadiene copolymers with isoprene and styrene were used. The preferred crosslinking sites in butadiene and isoprene polymers are the 1,2- positions; 1,4-, and 3,4- are also involved in crosslinking reactions, but their reaction rate is significantly lower. The 1,2- reactivity is about the same in butadiene–styrene copolymers as in the homopolymers; in isoprene polymers it is less. For all investigated rubbers, the crosslinking reaction rate accelerates with increasing conversion of styrene; this is most striking in the conversion range above 98%. This phenomenon can be explained by copolymerization of the styrene monomer with the 1,2- units of the rubber backbone.

In general, high impact polystyrenes are multiphase systems consisting of a continuous rigid polystyrene phase and discrete rubber particles 0.5–10μ in diameter. The incorporated rubber particles are crosslinked and contain grafted polystyrene, and their inner structure is determined by the manufacturing process and can vary considerably. The principle polystyrene structures have been described in detail (1, 2).

According to more recent theories, the toughness of high impact polystyrene is caused by flow and energy dissipation processes in the continuous polystyrene phase. The rubber particles act as initiating elements. Considerable differences in the thermal expansion coefficients and in the moduli of the polystyrene phase on the one hand and of the rubber particles on the other lead to an inhomogeneous stress distribution in impact polystyrene. Stress maxima create zones of lower density, called crazes (3), in which the polystyrene molecules are extended parallel to the direction of stress. Macroscopically craze formation appears as whitening; the flow processes result in irreversible deformation (cold flow).

To trigger craze formation, rubber particles must be crosslinked (4). Qualitatively it is well known that crosslink density considerably influences the toughness of high impact polystyrene (*1, 2, 5, 6, 7*). However, little is known about the crosslink density of rubber particles because of the difficulty in determining it. This paper is an attempt to demonstrate the effect of rubber structure on its reactivity towards crosslinking to obtain a better understanding of this reaction during the manufacturing process.

Experimental

Rubbers. The rubbers investigated—polybutadiene, butadiene–styrene copolymers, and butadiene–isoprene copolymers—were prepared by anionic solution polymerization (initiator, *n*-butyllithium) in *n*-hexane on a laboratory scale. Rubber configurations were varied by adding tetrahydrofuran during the polymerization process. Polymerization was terminated by adding small amounts of methanol; precipitation of the rubber followed, then washing with large amounts of methanol containing 0.1% phenolic antioxidant. The rubbers were dried *in vacuo* at 50°C for 1 or 2 days (voltaile matter < 0.5%).

The molecular weights of these polymers, M_η, were 200,000–300,000 as estimated by the viscosity number $\eta_{\text{sp/c}}$ (0.5% in toluene at 25°C) and the $[\eta]$–M correlation for 1,4-polybutadiene (*8*) and 1,2-polybutadiene (*9*). The only exception, Taktene 1202 containing 2% 1,2-vinyl groups, had a viscosity number of 272 cm³/g which corresponds to an M_η of approximately 200,000.

Rubber compositions and configurations were determined by IR spectroscopy using the extinction coefficients of Simák and Fahrbach (*10*). The data are presented in Table I.

Preparation of High Impact Polystyrene. High impact polystyrenes were prepared according to the Stein and Walter method (*11*) except for the following alterations: 0.5 part paraffin oil, 0.1 part phenolic antioxidant, and suspension stabilization by 0.3 part of an organic suspension agent. To reach higher conversions, some of the experiments were run in suspension at 145°C two hours longer.

Table I.   Composition and Configuration of the Rubbers

	Butadiene, wt %				*Isoprene, wt %*				*Styrene,*	$\eta_{\text{sp/c}},$
No.	*Total*	*1,2*	*1,4-cis*	*1,4-trans*	*Total*	*3,4*	*1,4-cis*	*1,4-trans*	*wt %*	*cm³/g*
1	100	2	98	0	0	0	0	0	0	272
2	100	9	35	56	0	0	0	0	0	267
3	100	17	30	53	0	0	0	0	0	329
4	100	33	23	44	0	0	0	0	0	299
5	100	52	16	32	0	0	0	0	0	307
6	100	82	6	12	0	0	0	0	0	257
7	90	30	22	38	0	0	0	0	10	~330
8	90	40	19	31	0	0	0	0	10	~340
9	90	45	16	29	0	0	0	0	10	~330
10	80	34	17	29	0	0	0	0	20	~330
11	80	45	13	22	0	0	0	0	20	~290
12	0	0	0	0	100	5	81	14	0	~300
13	20	2	8	10	80	4	56	20	0	~290
14	21	2[a]	7	12	79	3	57	19	0	~350
15	20	2[a]	7	11	80	4	56	20	0	~350
16	40	4[a]	14	22	60	2	43	15	0	~300
17	40	4[a]	14	22	60	2	44	14	0	~310
18	40	3[a]	14	23	60	2	44	14	0	357
19	43	9[a]	8	26	57	13	26	18	0	368
20	65	7[a]	22	36	35	2	26	7	0	~270
21	82	8[a]	28	46	18	1	12	6	0	~270
22	90	9[a]	33	48	10	1	7	3	0	~280

[a] The 1,2-isoprene content, included in this figure, is negligible.

Determination of the Swelling Index (*SI*). A solution of high impact polystyrene in toluene (5:100) was centrifuged 1 hr at 25000 g. The supernatant liquid was decanted from the gel. Traces of solution remaining on the lip of the centrifuge cup were removed by filter paper. The gel was weighed before and after drying to constant weight, and from these two weights we calculated the swelling index.

$$SI = \frac{\text{wet weight of gel}}{\text{dry weight of gel}}$$

Determination of Residual Styrene Monomer Content. Residual styrene monomer content was determined by method II of Noffz and Pfab (*12, 13*).

Results

It is assumed that the crosslinking and chain propagation reactions are competitive.

Reaction paths 1a, 1b, and 1c illustrate the possible competitive reactions of polystyrene radicals and initiator radicals $R_n{}^*$. In the rubber phase, the subscript P indicates that the functional group is incorporated into a polymer chain. Double bonds and allylic hydrogen atoms will be the sites of attack favored by the radicals. The rubber radicals produced by Reactions 1b and 1c can cause crosslinking of the rubber by recombination or propagation steps; in the propagation reactions, copolymerization of the rubber double bonds with styrene must be considered. Since Reactions 1b and 1c precede the crosslinking and compete with the propagation reaction (Reaction 1a), one can expect that crosslinking of the rubber particles depends on conversion of the styrene monomer.

We tested this hypothesis with a polybutadiene containing 9% 1,2-vinyl units. At styrene conversions below 95%, the rubber crosslink density was too small for determining the *SI*. Therefore, measurements were limited to the range of very high styrene conversion. In Figure 1, the *SI* of the rubber particles is plotted against the amount of residual styrene monomer, *i.e.*,

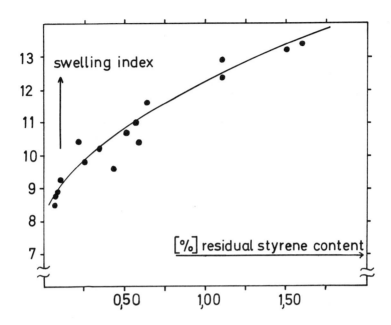

Figure 1. Swelling index vs. styrene conversion in polybutadiene

styrene conversion (wt %). The expected correlation was observed: the *SI* decreases with increasing styrene conversion, *i.e.*, low monomer concentration. It was therefore decided to include in the investigations only those experiments in which the styrene conversion scatter lay within the limits 99.50 ± 0.15%. With the aid of Figure 1, the effect of scatter was eliminated by correcting the SI's to the average styrene conversion, assuming that the curve in Figure 1 is applicable to the other systems. Experiments in which styrene conversion was outside the range of scatter were ignored.

To avoid misunderstandings about the importance and meaning of the swelling index in the context of this study, a brief discussion of the morphology of high impact polystyrene is necessary. Figure 2 shows electron micrographs of typical high impact polystyrenes; the dark lamellae represent the rubber. As was described by other authors (*1, 2*), multiple oil-in-oil emulsions are obtained; the rubber particles, discretely distributed in the continuous polystyrene phase, contain discrete areas of occluded polystyrene. The *SI* of these polystyrene-filled rubber particles can be measured after they are separated from the soluble continuous polystyrene phase. Under selected experimental conditions, it is not possible to extract the polystyrene occluded by the rubber lamellae. Consequently, one does not measure the *SI* of a uniformly crosslinked rubber but, indirectly, the osmotic pressure of the occluded polystyrene. This is evidenced by measuring the *SI* in methyl ethyl ketone (non-solvent for polybutadiene): for uncrosslinked polybutadiene, *SI* = 1.9; for the crosslinked rubber particles in high impact polystyrene on the other hand, *SI* values range between 4 and 8. This indicates that it is not possible to calculate network densities from swelling indexes by the usual methods, for example by the Flory-Rehner equation.

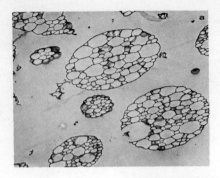 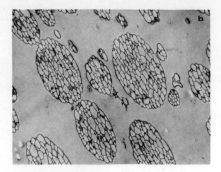

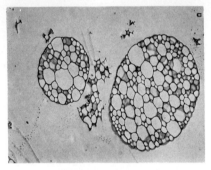

*Figure 2.   Morphology of high impact polystyrene pre-
pared from various rubbers ×5476*

Stained with OsO₄: dark areas, polybutadiene; and light areas,
polystyrene. a, Polybutadiene with 9% 1,2-; b, polybutadiene
with 82% 1,2-; and c, copolymer of butadiene (43%) and isoprene (57%) with 9% 1,2-

The osmotic pressure of the polystyrene solution is balanced by the elastic
retractive forces of the more or less densely crosslinked rubber lamellae; osmotic
pressures in the swollen rubber particles are rather high because of the high
positive virial coefficient of polystyrene in toluene ($A_2 = 5 \times 10^{-4}$ mole cm³ g⁻²)
(14). For $SI = 10$ (corresponding to $c = 0.1$ g × cm³) and $M_n = 5 \times 10^4$, one

$$\pi \sim c \cdot R \cdot T \left(\frac{1}{M} + A_2 c \right) \tag{2}$$

can estimate from Equation 2 that the osmotic pressure is 0.2 atm. Actual
values are probably even higher since higher virial coefficients were not taken
into account in the calculation.

Despite these limitations, one can qualitatively conclude from the results
depicted in Figure 1 that crosslinking increases with conversion. Furthermore,
the relative rate of crosslinking, *i.e.*, formation of crosslink sites as a function of
styrene conversion, probably also increases with conversion. However, the
second statement has yet to be proved in special experiments in which the
formation of network sites is measured directly.

According to reaction schemes 1a, 1b, and 1c, crosslinking is influenced
by the type and number of double bonds or allylic hydrogen atoms in the
rubber. Polybutadiene microstructure can be controlled easily by suitable
anionic polymerization conditions in which parameters like double bond num-

ber and molecular weight distribution remain constant. Figure 3 illustrates the effect on the crosslinking reaction of various concentrations of 1,2-vinyl units in the polybutadiene. The decrease in swelling indexes indicates clearly that crosslinking strongly increases with increasing 1,2- content. Since 1,2-polybutadiene is completely different from the 1,4- polymer, one might conclude that a change in phase inversion mechanism (*15*) occurs in going from 1,4- to 1,2-polybutadiene. Figure 2 illustrates that this is not the case; the differences in rubber particle size and internal structure in the various rubbers are very small. Therefore the large differences in swelling index must result primarily from different reactivity of the diene configurations towards crosslinking.

The data depicted in Figure 3 were supplemented by investigations of butadiene–styrene copolymers with styrene contents of 10 and 20 wt %. It is evident from Figure 4 that crosslinking of these rubbers is also essentially determined by the number of 1,2-vinyl units. The data for styrene copolymers coincide with those of the polybutadiene curve.

Measurements of butadiene–isoprene copolymers are summarized in a plot of swelling index *vs.* 1,2- content (Figure 5). These data were obtained at 99.85% styrene conversion rather than 99.50% (Figures 3 and 4), and this difference must be kept in mind. Crosslinking also increases with increasing 1,2- content in butadiene–isoprene copolymers although all points lie above the polybutadiene curve; this applies especially to copolymers with low 1,2-contents. The curvature in Figures 3 and 4 is not apparent in Figure 5 because of the limited range of the 1,2- content in the butadiene–isoprene copolymers.

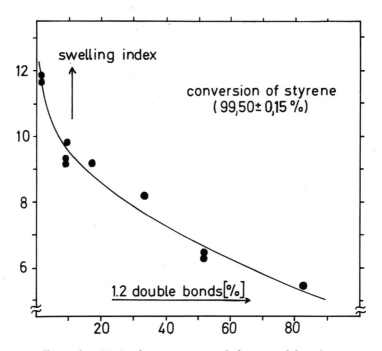

Figure 3.   1,2 Configuration vs. *crosslinking of polybutadiene*

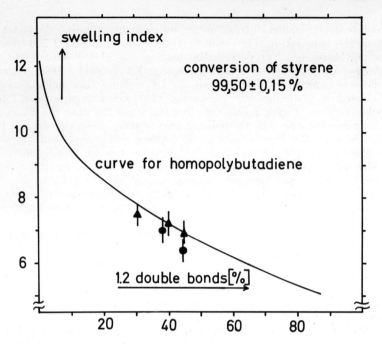

Figure 4. Swelling index of butadiene–styrene copolymers vs. 1,2 content in the rubber

△, 90/10 wt % butadiene/styrene, and ○, 80/20 wt % butadiene/styrene

Discussion

If one applies reaction schemes 1a, 1b, and 1c to polybutadiene, then the data in Figure 3 indicate that the double bonds and/or the allylic hydrogen atom of the 1,2- configuration must be primarily responsible for the crosslinking reaction. Van der Hoff (*16*) and recently Hergenrother (*17, 18*) were able to demonstrate that crosslinking of pure polybutadiene (*i.e.*, in the absence of polymerizing monomer) with dicumyl peroxide proceeds *via* a propagation step by Reactions 3 and 4. Hergenrother in particular demonstrated that the

$$
RO^* \;+\; \left|\begin{matrix} C{=}C \\ | \\ CH \end{matrix}\right|_P \;\longrightarrow\; ROH \;+\; \left|\begin{matrix} C{=}C \\ | \\ C^* \end{matrix}\right|_P \tag{3}
$$

$$
\left|\begin{matrix} C{=}C \\ | \\ C^* \end{matrix}\right|_P \;+\; \left|\begin{matrix} CH \\ | \\ C{=}C \end{matrix}\right|_P \;\longrightarrow\; \left|\begin{matrix} C{=}C \quad CH \\ | \qquad | \\ C{-}C{-}C^* \end{matrix}\right|_{2P} \quad \text{etc.} \tag{4}
$$

1,2-vinyl configuration is more reactive than the 1,4- configuration. These findings were confirmed by Reichenbach (*19*). Analogous results were obtained by Heusinger and co-workers (*20, 21*) by radiation crosslinking of polybutadienes and polyisoprenes of different configurations. They found that the crosslinking reaction of the 1,2- configuration proceeds *via* a chain reaction

with kinetic chain lengths of 2–5, whereas 3,4- and 1,4- configurations participate in the crosslinking reaction by recombination.

Fischer (22) on the other hand showed that the ratio of the rate constants
from reaction paths 1a and 1b for an attack by a polystyrene radical on the
1,2- double bond is

$$\frac{k_{12}}{k_{11}} = 1.5 \times 10^{-3} \qquad (T = 110°C)$$

In contrast to the grafting reactions (22), the crosslinking reaction during
preparation of high impact polystyrene becomes effective only at very high
styrene conversions (> 95%) at which the rubber double bond concentration
in the total reaction mixture approaches the monomer concentration. In rubber
lamellae, this shift in concentration in favor of the rubber double bonds is even
more pronounced because of the incompatibility of polybutadiene and polystyrene.

Styrene is distributed in moderately concentrated solutions (up to 30 wt
% polymer) between the polystyrene and the polybutadiene phases according
to Equation 5 (23):

$$\left(\frac{m}{p}\right)_B = 1.8 \left(\frac{m}{p}\right)_S \tag{5}$$

where $(m/p)_i$ is the weight ratio of styrene monomer to polymer in the individual phases and subscripts B and S indicate the rubber and polystyrene
phases respectively. From Equation 5 and the relative rate constant k_{12}/k_{11},
the concentration ratios z ($\equiv$ rubber double bonds/styrene double bonds)

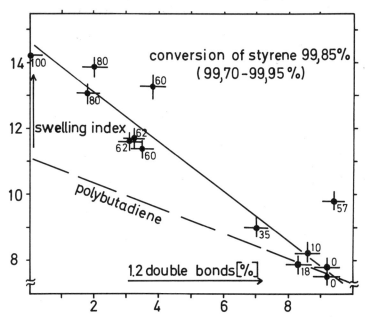

Figure 5.   1,2 Configuration vs. *crosslinking of butadiene–isoprene
copolymers*

The numbers indicate wt % isoprene in the rubber

Table II. Molar Concentration Ratio of Double Bonds (z) and Relative Rates (v_{12}/v_{11})

Rubber concentration, 5 wt % based on total charge; configuration, 100% 1,2 units

Styrene Conversion, %	$\bar{z}^a$	z^b	$v_{12}/v_{11}{}^b$
0	0.1	0.1	10^{-4}
50	0.2	1	0.0002
80	0.5	5	0.008
95	2	22	0.03
99	10	115	0.17

[a] Initial value.
[b] In the rubber phase.

and the relative rates v_{12}/v_{11} of competing reactions 1a and 1b in the rubber phase can be estimated. The values for various styrene conversions are presented in Table II.

Despite the small r values for copolymerization of styrene and the 1,2-vinyl units of polybutadiene, it is expected at styrene conversions over 95% that crosslinking copolymerization of styrene and rubber will occur. In addition, the allylic radicals formed in initiation Reaction 3 start graft copolymers of styrene monomer and polybutadiene double bonds. The existence of allylic radicals in our system is corroborated by the work of Fischer (*22*). In graft copolymerization, the 1,2- configuration is preferentially incorporated into the copolymer (*16, 17, 18, 19*) as in Structure 6. Networks are apparently formed

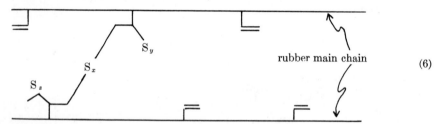

(6)

which contain the structural element shown in Figure 6. If the 1,4- configurations participated significantly in the crosslinking reaction, then the styrene–butadiene copolymer data would not lie on the curve shown in Figure 3. For example, rubber #10 contains only two-thirds the 1,4- configuration of rubber #4 because of the equal amount of 1,2- configuration in both rubbers.

S - styrene monomer units
x, y, z ≥ 1

Figure 6.   Schematic of a network of polybutadiene and polystyrene

This model of crosslinking polymerization cannot be applied directly to butadiene–isoprene copolymers. It is known that the polyisoprene crosslink mechanism differs from that of polybutadiene (*16, 17, 20, 21*). Crosslinking of polyisoprene occurs by transfer of a hydrogen atom from rubber to a free radical and ensuing recombination of two rubber radicals. As one might expect, introduction of 1,2- configurations by copolymerization with polybutadiene increases the reactivity towards crosslinking, although to a smaller extent. This effect is also observed when the 1,2-vinyl content is increased at a constant butadiene–isoprene ratio in the copolymer (compare rubber #19 with rubbers 16, 17, and 18). Steric blocking of the 1,2- double bonds by the methyl groups of the isoprene units could be implicated; however this argument is not supported by the fact that no decrease in reactivity is observed in butadiene and styrene copolymers although the bulky phenyl groups should already be effective, even at lower concentrations.

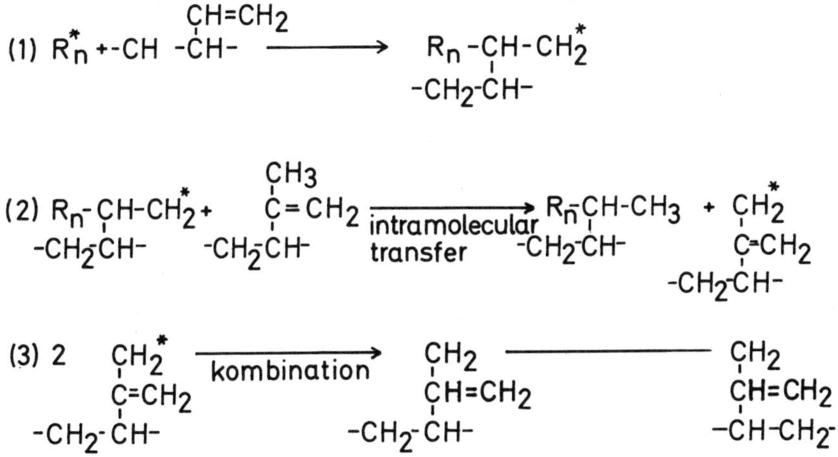

Figure 7. Intermolecular transfer reaction in butadiene–isoprene copolymers

We explain the apparently lower reactivity of the 1,2- units in isoprene copolymers by an intramolecular chain transfer reaction. Figure 7 shows the probable reaction sequence. In the first step, the radical of a growing chain adds to the 1,2- double bond of a butadiene unit. In the second step, an intramolecular chain transfer reaction of the 1,2-vinyl group will compete with the addition of styrene or rubber double bonds because of the high local concentration of allylic hydrogen atoms in the neighboring isoprene units. This transfer reaction corresponds to the attack by peroxide radicals on allylic hydrogen atoms of polyisoprene already postulated by van der Hoff (*16*). Rubber crosslinking probably also occurs by recombination (third step), but this path appears to be less efficient than crosslinking *via* the chain mechanism of the 1,2- double bonds.

Acknowledgments

We thank our colleagues E. G. Kastning, P. Simák, H. Hendus, and A. Echte for assistance and many valuable discussions.

Literature Cited

1. Willersinn, H., *Makromol. Chem.* (1967) **101**, 296.
2. Keskulla, H., *Appl. Polym. Symp.* (1970) **15**, 51.
3. Bucknall, C. B., Smith, R. R., *Polymer* (1965) **6**, 437.
4. Bohn, L., *Angew. Makromol. Chem.* (1971) **20**, 129.
5. Zitek, P., Mysik, S., Zelinger, J., *Angew. Makromol. Chem.* (1969) **6**, 116.
6. Dow Chemical Co., U.S. Patent **3,243,481** (1968).
7. Wagner, E. R., Robeson, L. M., *Rubber Chem. Technol.* (1970) **43**, 1129.
8. Kraus, G., Gruver, J. T., *J. Polym. Sci. Part A* (1965) **3**, 105.
9. Anderson, J. N., Barzan, M. L., Adams, H. E., *Rubber Chem. Technol.* (1972) **45**, 1270.
10. Šimák, P., Fahrbach, G., *Angew. Makromol. Chem.* (1971) **16/17**, 309.
11. Monsanto, U.S. Patent **2,862,906** (1956) example 1.
12. Noffz, D., Pfab, W., *Z. Anal. Chem.* (1967) **228**, 188.
13. Noffz, D., Pfab, W., "Polystyrol," in "Kunststoff-Handbuch," R. Vieweg, G. Daumiller, Eds., Vol. 5, p. 156, Carl Hanser, München, 1969.
14. Brandrup, J., Immergut, E. H., "Polymer Handbook," **4**, 126, Interscience, New York-London-Sidney, 1965.
15. Bender, B. W., *J. Appl. Polym. Sci. Part A* (1965) **3**, 2887.
16. van der Hoff, B. M. E., *Appl. Polym. Symp.* (1968) **7**, 21.
17. Hergenrother, W. L., *J. Appl. Polym. Sci.* (1972) **16**, 2611.
18. Hergenrother, W. L., *J. Polym. Sci. Part A-1* (1973) **11**, 1721.
19. Reichenbach, D., *Kaut. Gummi Kunstst.* (1965) **18**, 9.
20. Heusinger, H., Kaufmann, R., von Raven, A., Katzer, H., *IUPAC Internat. Symp. Macromol.*, Aberdeen, Sept., 1973.
21. Heusinger, H., Kaufmann, R., von Raven, A., Katzer, H., *Makromol. Chem.* (1973) **163**, 195.
22. Fischer, J. P., *Angew. Makromol. Chem.* (1973) **33**, 35.
23. Stein, D. J., unpublished data.

RECEIVED May 10, 1974.

15

Analytical Study of ABS Copolymers Using a Preparative Ultracentrifuge

B. CHAUVEL and J. C. DANIEL

Centre de Recherches Rhône-Progil, 93308 Aubervilliers, France

A quantitative method has been developed to separate free and graft copolymers in an ABS sample. The ABS powder is dispersed in MEK and then introduced into the cells of a preparative ultracentrifuge. After the reproducibility of the procedure was ascertained, the method was used to determine the grafting parameters of samples polymerized under specific conditions. This analytical technique is well suited to demonstrate how the grafting efficiency or grafting density is influenced by various polymerization conditions such as mercaptan content, monomer flow rate, emulsifier content, or polybutadiene content. The effects of other variables such as temperature, the initiator system, and characteristics of the polybutadiene latex can also be demonstrated.

When a mixture of styrene and acrylonitrile is polymerized in the presence of a polybutadiene latex by an emulsion radical process, an acrylonitrile–butadiene–styrene (ABS) copolymer is obtained. This ABS copolymer is actually a mixture of (a) a graft copolymer which contains some of the styrene/acrylonitrile (ST/AN) copolymer chemically bound to the polybutadiene backbone, and (b) a random copolymer, conventionally designated as a linear copolymer, which is not bound to the polybutadiene backbone but which consists of the portion of the styrene/acrylonitrile monomer that has polymerized separately.

The separation of these two components enables us to study their macromolecular characteristics and to determine the quantity of ST/AN copolymer which is bound to the substrate. If it is sufficiently accurate, a quantitative separation technique provides a valuable source of information for correlations between the properties of ABS copolymers and their structural characteristics. It can also demonstrate how the ABS structure depends on different polymerization variables.

Principle of the Analytical Method

As far as we know, the first application of ultracentrifugation to phase separation of graft copolymers was made by Shashoua and Van Holde (1).

Table I. Reproducibility of the Separation Technique as Shown by Values of L' [a]

Cell Number	Centrifugation Run		
	1	2	3
1	76.0	76.2	76.4
2	76.6	76.3	76.7
3	77.8	76.9	76.7
4	77.2	76.2	76.4
5	76.7	76.1	76.2
6	76.9	76.1	76.0
Mean	76.9	76.3	76.4

[a] In wt %.

This method was subsequently used by Gesner (2), by Moore and Frazer (3), and by Huguet and Paxton (4) for ABS copolymers.

The diameter of the polybutadiene latex particles used in ABS synthesis usually ranges from 0.1μ to 1μ, and the polymer is always crosslinked within the particles. Consequently, after grafting, every polybutadiene particle becomes a particle of grafted and crosslinked polymer which is much heavier than the linear copolymer molecules.

Table II. Analysis of Variance

Source of Variance	Sum of Squares	Degrees of Freedom	Variance
Between runs	1.0978	2	0.549
Between cells	1.6644	5	0.333
Residue	1.0089	10	0.101
Total	3.7711	17	

The separation technique consists of dispersing an ABS powder in an appropriate medium in which the linear copolymer is soluble and then centrifuging the dispersion. The graft copolymer microgels settle more rapidly than the free copolymer, despite the lower density of the ST/AN copolymer. S, the Svedberg sedimentation coefficient, depends on the molecular weight, according to the following equation (top of p. 162):

Table III. Effect of

	Antioxidant					
Sample	TNPP, %	DBPC, %	TDM, %	L', %	G, %	$L'+G$, %
1	1.25	5	0	52.1	47.1	99.2
2	1.25	5	0.1	67.8	36.8	104.6
3	—	2	0.1	64.3	37.1	101.4
4	1.25	5	0.2	70.3	30.2	100.5
5	0.5	—	0.2	72.1	30.3	102.4
6	—	2	0.3	74.5	27.5	102.0
7	0.5	—	0.3	74.3	28.0	102.3
8	1.25	5	0.4	76.0	23.5	99.5
9	0.5	—	0.4	74.7	26.6	101.3
10	1.25	5	0.5	79.8	21.6	101.4
11	0.5	—	0.5	76.5	25.4	101.9
12	0.5	—	0.6	80.3	23.8	104.2
13	1.25	5	0.6	80	19.5	99.5

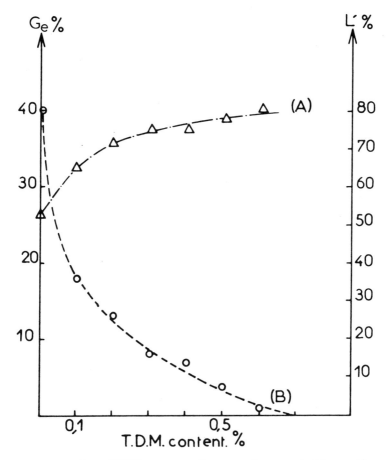

Figure 1.   Effect of TDM content on linear copolymer content (curve A) and on grafting efficiency (curve B)

the TDM Content

G_e, %	G_d, %	$\eta_r(L')$	$[\eta]$	$\overline{M}_w$, $\times 10^{-3}$	D_g, $\times 10^4$
40	160	—	—	—	—
19	75	92	83.9	270	1.5
18	68	98	88.9	295	1.25
16	65	74	68.7	195	1.8
10	39	78	72.1	210	1.0
9.5	38	62	58.2	150	1.35
7.2	29	59	55.6	140	1.1
8	32	46	43.9	95	1.8
6.5	27	52	49.3	112	1.3
2.5	10	46	43.9	95	0.55
4.5	18	46	43.9	95	1.0
0.1	1	43	41.1	85	0.06
2	9	46	43.9	95	0.6

$$S = \frac{MD\,(1-\overline{V}_s\rho_l)}{RT}$$

where M is the molecular weight, D the diffusion constant, ρ_l the solvent density, $\overline{V}_s$ the partial specific volume of the polymer, R the gas constant, and T the absolute temperature.

Separation Technique

An ABS sample prepared in our laboratory was used to develop the separation technique. Its composition was (wt %): butadiene, 20; acrylonitrile, 21; and styrene, 59. The ABS powder was dispersed in methyl ethyl ketone (MEK) at 1 wt % concentration. MEK was chosen because it is a good solvent for styrene/acrylonitrile copolymers, it is a poor solvent for polybutadiene, and its density is lower than that of polybutadiene and therefore lower than the density of the graft copolymer.

Experiments with an Analytical Ultracentrifuge. At first the sedimentation phenomenon was studied with an analytical ultracentrifuge (Spinco model E) equipped with a Philpott-Svenson optical device that enabled us to follow the moving boundaries of the different macromolecular species during a centrifugation run. We observed that with progressive increase of the rotor speed, the graft copolymer settled immediately at 30,000 g whereas sedimentation of the linear copolymer was negligible. With an acceleration of 200,000 g, the linear copolymer did not settle until 3 hrs later.

This observation agrees with those of Moore and Frazer (3) who reported good phase separation of an acetone suspension of ABS by centrifugation at 40,000 and 50,000 g.

Separation with a Preparative Ultracentrifuge. An MEK dispersion (20 ml) of ABS was centrifuged in a preparative ultracentrifuge (Spinco model L). The type 30 rotor operating at 25,000 rpm provides 35,000 g at the top of the sample cell and 70,000 g at the bottom. The run lasted 45 min. Afterwards, the graft copolymer was recovered as a deposit adhering to the bottom of the cell. Two successive, similar runs were made with the graft copolymer deposit being redispersed in fresh MEK before each run.

The two washing solutions were studied in the analytical ultracentrifuge, and no linear copolymer was detected in the second solution. Considering the sensitivity of the optical system, we estimate the quantity of linear copolymer collected in the second washing solution to be less than 1 wt % of the initial ABS. IR spectroscopy detected only traces of polybutadiene in the linear

Table IV. Effect of

Sample	Monomer Introduction, hrs	DBPC, %	L', %	G, %	L'+G, %
1	0	2	79.4	21.2	101.1
2	0	2	80.1	20.4	100.5
3	0	2	79	24.5	103.5
4	3	2	77.4	25.4	102.7
5	3	2	77.9	24.3	102.2
6	5	2	74.7	26.9	101.6
7	8	2	70	32.6	102.6
8	20	a	60.4	43.2	103.6

a Stabilized by a mixture of hydroquinone, 0.1% and β naphthylamine, 0.1%.

copolymer solution. The polybutadiene content of this solution was estimated at less than 0.5 wt % based on the initial ABS or less than 2 wt % based on the initial polybutadiene.

These data imply that all the sediment consists of microgels of graft copolymer. If a certain amount of polybutadiene were not grafted and not crosslinked, it would be insoluble in MEK and then appear when the ABS sample is dispersed in MEK. Although Gesner (2) detected soluble pure polybutadiene in ABS compounds, this might be the result of different preparation techniques.

Determination of the Grafting Parameters in ABS Copolymers

Our separation technique was used to determine the grafting parameters of an ABS sample prepared in our laboratory to check the reproducibility of the method.

Experimental. A 1% dispersion of ABS powder in MEK was prepared at 20°C and introduced into six stainless-steel cells fitted on a type 30 rotor of an ultracentrifuge (Spinco model L). Each cell contained 20 ml of the dispersion. The ultracentrifuge was run for 45 min at 25,000 rpm and 20°C. Afterwards, aliquots of the linear copolymer supernatant solutions were removed from the cells by syringe, placed in flasks, and stored at 20°C.

The graft copolymers, which were adhering to the bottom of the cells, were redispersed in 20 ml fresh MEK. After 15 hrs, these dispersions were centrifuged as above. The washing solution was then removed and replaced with fresh MEK for a second washing. After the third centrifugation, the washing solutions were added to the first linear copolymer solution. The graft copolymers were recovered by dispersing the remaining sediments in MEK.

Methanol was added to the various solutions or dispersions, and the contents were then dried *in vacuo* at 50°C under air. It was then easy to determine G (wt %), the graft copolymer content, and L' (wt %) $= L$, linear content, plus I, ingredient content. Emulsifiers and antioxidants which are soluble in MEK are among the ingredients.

Since I and B (butadiene content) are known, it is possible to calculate:

G_d, wt % of ST/AN grafts bound to 100 g polybutadiene

$$G_d = \frac{100 - (L'+B)}{B} \times 100$$

G_e, the grafting efficiency, *i.e.*, wt % of ST/AN grafts per 100 g of monomer

$$G_e = \frac{100 - (L'+B)}{100 - (B+I)} \times 100$$

the Monomer Flow Rate

G_e, %	G_d, %	$\eta_r(L')$	$[\eta]$	$\overline{M}_w$, $\times 10^{-3}$	D_g, $\times 10^4$
1.5	6.1	75	69.5	200	0.15
0.6	2.5	83	76.4	235	0.05
2.0	8.2	63	59.1	153	0.3
4.1	16.5	85	78.1	240	0.35
3.4	13.9	88	80.5	250	0.3
7.5	30	48	45.5	100	1.6
13	54	43	41.2	85	3.45
24	98	—	—	—	—

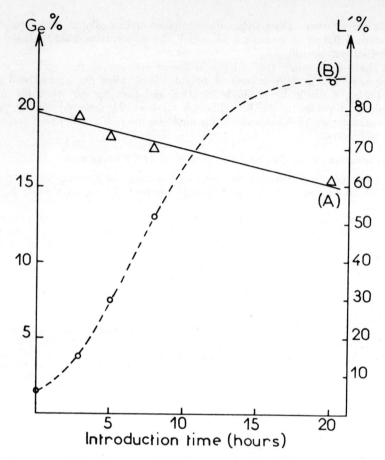

Figure 2. Effect of monomers flow rate on linear copolymer content (curve A) and on grafting efficiency (curve B)

η_r, the reduced viscosity of the linear copolymer. (Viscosity was measured at 30°C with 0.5 g copolymer/100 ml MEK.)

Discussion. The reproducibility of the experimental data were influenced by:

(a) Antioxidant volatility. An antioxidant is always added to the polymer to prevent oxidation during drying. Since most of the commonly used antioxidants are volatile, the material balance is disturbed. The importance of this phenomenon depends on the nature of the antioxidant and on drying conditions. To minimize this factor, we chose a drying temperature below 50°C.

(b) The tendency of the dry polymer to absorb moisture. Because of this, it is necessary to keep all copolymers in a dry atmosphere, over P_2O_5, after centrifugation.

(c) The copolymer's ability to retain solvent. It is impossible to obtain polymers by evaporating the solvent, and thus it is necessary to precipitate them with methanol and to dry them *in vacuo*.

Reproducibility. Three different centrifugation runs were carried out with the 1% ABS dispersion. Six cells were used in each experiment for a total of 18. The L' values are given in Table I. For the sample studied, analysis of variance is presented in Table II. The F ratio for the runs, $0.549/0.101 = 5.44$, exceeds the tabulated 5% critical of 4.1 but not the 1% value of 7.56. Therefore, there are differences among the runs. The F ratio for the cells does not exceed the 5% critical value.

If we consider the L' value with 15 degrees of freedom, we find an 0.18 variance. Hence the standard deviation of a single determination is 0.425, and the standard deviation of each $\overline{L'}$ is $0.425/\sqrt{6} = 0.175$.

Using the Fisher-Student function, the 95% confidence limits for $\overline{L'}$ (computed for 15 degrees of freedom) are: 76.86 ± 0.37; 76.30 ± 0.37; and 76.40 ± 0.37.

The best estimate for L' is 76.41 which is the average obtained from the data in Table I. Then 76.41 ± 0.26 are the 95% confidence limits for the average (17 degrees of freedom).

Application—Influence of Polymerization Conditions on Grafting Parameters

Sample Preparation—Standard Recipe. All samples were prepared in the same reactor with the same stirring conditions. A typical polymerization was performed according to the following recipe:

Backbone latex: commercial FRS 2004 polybutadiene latex (Firestone Co.).

Reactor feed (by weight): polybutadiene, 20; acrylonitrile, 21; styrene, 59; tertiary dodecyl mercaptan (TDM) (Aquitane-Total-Organico), 0.3; sodium Dresinate (Dresinate 731, Hercules Powder Co.), 1.

Polymerization temperature, 60°C.

Initiator system, potassium persulfate.

A mixture of monomers and TDM is added with mixing to the polybutadiene substrate for 5 hrs in the presence of the initiator system and the emulsifier. After polymerization, latexes were stabilized by adding 2,6-di-*tert*-butyl-*p*-cresol (DBPC) or tri(nonylphenyl) phosphite (TNPP) as antioxidants. They were then coagulated with a calcium chloride solution and finally dried at 50°C for 15 hrs.

Effect of Mercaptan Content. In the polymerization recipe, TDM is used as a transfer agent. It is interesting to know how each grafting parameter depends on this agent. The experimental results are presented in Table III and Figure 1. As the TDM content increased from 0.0 to 0.6, G decreased from 47 to 23%, G_d decreased from 160 to nearly 0%, G_e decreased from

Table V.   Effect of the Polybutadiene Content

Sample	B, %	Antioxidant, %		L', %	G, %	L'+G, %	G_d, %	G_e, %	Monomer Flow Rate, ml/hr
		DBPC	TNPP						
1	5	5	1.25	93.4	6.7	100.1	40	2	31.5
2	10	d°	d°	86.4	14.3	100.7	43	5	30
3	15	d°	d°	81.2	20.1	101.3	34	6	28
4	20	d°	d°	71.1	29.5	100.6	56	14	26.5
5	20	d°	d°	72.5	30.3	102.8	45	11	26.5
6	25	d°	d°	68.4	33.5	101.9	35	13	25
7	30	d°	d°	53.4	48	101.4	65	28	23.5
8	30	2	1	56.1	46.3	102.4	51	22	23.5
9	50	2	1	33.8	66.7	100.5	37	39	16.5

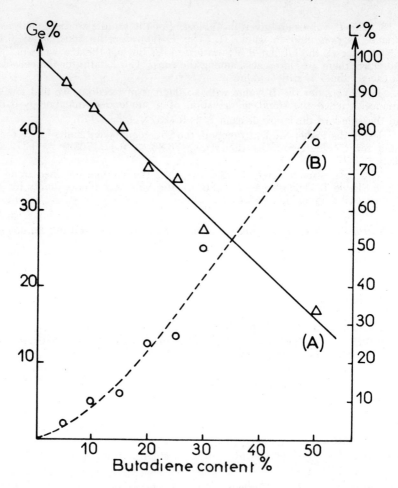

*Figure 3. Effect of butadiene content on linear copolymer content
(curve A) and on grafting efficiency (curve B)*

40 to nearly 0%, and η_r decreased from 1 to 0.4%. $[\eta]$, the intrinsic viscosity,
can be calculated from the value of η_r using the equation given by Gerrens
et al. (5):

$$[\eta] = \frac{\eta_r}{1 + 0.21\eta_{sp}}$$

Table VI. Effect of Temperature

Initiator: K$_2$S$_2$O$_8$

Sample	T, °C	DBPC, %	L', %	G, %	L'+G, %	G$_d$, %	G$_e$, %	η_r	$[\eta]$	$\overline{M}_w$, $\times 10^{-3}$	D$_g$, $\times 10^4$
1	50	2	78.3	23.8	102.1	8.5	2	77	71.3	200	0.2
2	60	2	73.8	28.3	102.1	31	8	53	50.2	120	1.4
3	70	2	67.9	34.3	102.2	60	16	42	40.2	83	3.9

Table VII.   Effect of Temperature

Initiator: Redox Catalyst

Sample	T, °C	TNPP, %	L', %	G, %	L'+G, %	G_d, %	G_e, %	η_r	$[\eta]$	$\overline{M}_w$, $\times 10^{-3}$	D_g, $\times 10^4$
1	60	0.5	59.3	43.3	102.6	90	24	44	42	89	5.5
2	50	0.5	59.3	43.4	102.7	90	24	44	42	89	5.5
3	40	0.5	63.2	38.9	102.1	72	19	69	64.3	178	2.2

From $[\eta]$ we can derive $\overline{M}_w$ using the relationship of Shimura *et al.* (6):

$$[\eta] = 3.6 \times 10^{-2}\,\overline{M}_w{}^{0.62}$$

It is then easy to calculate D_g, the grafting density, according to Dinges and Schuster (7). D_g is the average number of grafts per monomer unit of the substrate. Calculations have been made assuming that $\overline{M}_w$ is the same for the graft and the free copolymer. This assumption is based on the data of

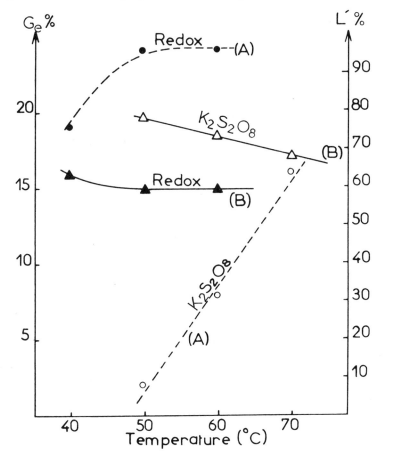

Figure 4.   Effect of temperature with different catalysts on grafting efficiency (curve A) and on linear copolymer content (curve B)

Dinges and Schuster and has been verified by separating the grafts from the backbone by an ozonolysis technique. By this method, we found that the reduced viscosity of the grafts is very close to that of the free copolymer. The data in Table III show that D_g is nearly constant within 0.1–0.4% TDM. This means that the TDM content has a strong influence on the molecular weight of the grafts but practically no influence on the number of grafts. When the TDM content is increased above 0.4%, however, the situation is apparently changed since D_g then decreases.

Effect of the Monomer Flow Rate. The rate at which the monomer mixture is introduced into the reactor greatly affects the grafting parameters. As the time for introduction was increased from 0 to 20 hrs, we observed (*see* Table IV and Figure 2) that: G increased from 21 to 43%, G_d increased from 5 to 100%, G_e increased from 1 to 25%, and η_r decreased from 0.75 to 0.43. When the monomer flow rate decreases, the monomer concentration in the polymer phase increases as does the probability for grafting. Thus, the variations in G, G_d, and G_e can be explained easily.

It is surprising that η_r is not constant when the monomer flow rate is varied. The decrease in η_r when decreasing monomer flow rate is probably related to a mercaptan influence; the mercaptan consumption rate is much higher than the monomer consumption rate. Thus, if the introduction rate of the monomer mixture is high compared with the polymerization rate, the mercaptan efficiency is poor, and the average molecular weight is high. On the other hand, if the monomer introduction rate is low, mercaptan is efficient throughout the reaction, and the macromolecular chains are shorter. When D_g is calculated, using the data of Table IV, it increases continuously when the monomer flow rate is decreased.

Effect of the Polybutadiene Content. A series of ABS samples was prepared using 0.1 part TDM and varying the polybutadiene content from 5 to 50%. The introduction time of the monomers was a constant 3 hrs. When the polybutadiene content increased, the grafting efficiency increased from 2 to 40% (*see* Table V and Figure 3). One must note, however, that an increase in the polybutadiene content automatically decreases the monomer flow rate. It ensues that the curves in Figure 3 result from the simultaneous effects of two parameters, the polybutadiene content and the monomer flow rate. Nevertheless, it seems reasonable to think that the concomitant variations of these two parameters produce effects in the same direction. The polymer/monomer ratio increases when the polybutadiene content is increased and when the monomer flow rate is decreased. The grafting efficiency probably depends on this ratio, but the relative importance of the two parameters cannot be ascertained from our data.

Effect of Temperature and the Initiator System. Polymerization temperature was varied from 50° to 70°C with $K_2S_2O_8$ as initiator and from 40° to 60°C using a redox catalyst. The redox catalyst was (wt %): cumene hydroperoxide, 0.75; ferrous sulfate ($FeSO_4 \cdot 7\ H_2O$), 0.01; dextrose, 1; and sodium pyrophosphate, 0.5. Cumene hydroperoxide was mixed with the monomers; the other ingredients were added into the emulsifier solution.

The experimental data (*see* Tables VI and VII and Figure 4) show that:

(a) The graft efficiency increases and the viscosity decreases with increase in temperature. These results were expected since the transfer reaction constants depend on temperature more than the propagation constants.

Table VIII.   Polymerization with γ-Rays[a]

Sample	Irradiation Time, hr	B, %	L', %	G, %	L'+G, %	G_d, %	G_e, %
A 1	6	35.8	56.8	45.2	102.0	21	12
A 2	7	28.2	63.1	39.1	102.2	31	12
A 3	8	26.6	67.6	34.7	102.3	22	8
A 4	9	21.5	71.6	31.2	102.8	32	9
A 5[b]	9.5	19.8	71.9	31.1	103.0	42	10

[a] Monomers were placed in the reactor for 9 hrs.
[b] η_r was 0.67.

(b) When the redox catalyst is used, the effect of temperature is lessened as it rises above 45°C. We assume that this phenomenon is associated with the shorter life time of the catalyst above this temperature. The very rapid decomposition of the catalyst yields a high instantaneous radical concentration; the radicals then destroy each other instead of attacking the polybutadiene backbone, according to Dinges and Schuster (7).

Table IX.   Effect of the Polybutadiene Backbone

Grafting onto LPF 1351

Sample	TNPP, %	L', %	G, %	L'+G, %	G_d, %	G_e, %	η_r	$[\eta]$	$\overline{M}_w$, $\times 10^{-3}$	D_g, $\times 10^4$
1	0.5	54.6	46.5	101.1	145	33	57	54	131	6.0
2	0.5	56.0	46.1	102.1	140	31	64	60	159	4.5

(c) As several workers have indicated (7, 8, 9, 10), grafting strongly depends on the nature of the initiator system. Under our polymerization conditions, grafting is promoted when persulfate is replaced by the redox catalyst. Thus, D_g increased by a factor of 3.5 when the operating temperature was 50°C.

Polymerization with γ-Rays. A few experiments were conducted using a ^{60}Co radiation source. The data obtained in one such experiment are presented in Table VIII. The radiation dose rate was 4000 rad/hr. The monomers were placed in the reactor for 9 hrs. From time to time, latex aliquots were removed to study the polymer characteristics as a function of time. The results did not show any important difference from a classical polymerization run carried out under the same conditions with $K_2S_2O_8$ used as catalyst.

Effect of the Substrate Latex Characteristics. Many authors (4, 11, 12, 13) have noted the great influence of the latex substrate on the structure and properties of ABS. Among the characteristics of the backbone latex, three have been reported as being very important. These are the degree of cross-linking of the polymer substrate (gel content, swelling index), the average

Table X.   Influence of the Emulsifier Content

Sample	Soap, %	Antioxidant TNPP, %	Antioxidant DBPC, %	L', %	G, %	L'+G, %	G_d, %	G_e, %
1	0	1.25	5	58.6	43.4	102	107	27
2	1	d°	d°	62.9	37.2	100.1	86	21.5
3	2	d°	d°	63	36.7	99.7	85	21
4	3	d°	d°	64.5	34.9	99.5	77	19
5	2	d°	d°	67.3	34.2	101.5	64	16
6	5	d°	d°	66.1	35.1	101.2	70	17.5

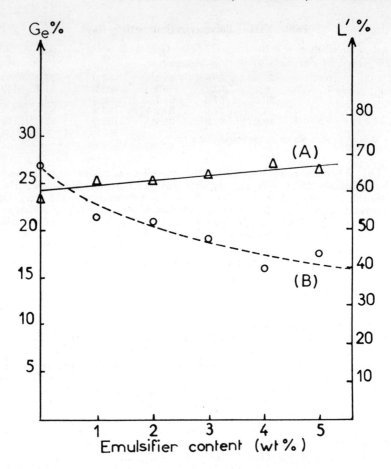

*Figure 5.   Effect of emulsifier content on linear copolymer content
(curve A) and on grafting efficiency (curve B)*

particle size of the rubber latex, and the nature of the polymer substrate
(chemical composition, glass transition temperature).

The data reported in Table IX were obtained in experiments in which
the polybutadiene latex used was Goodyear LPF 1351 rather than FRS 2004.
This is a fine particle latex; the particle diameter ranges from 500 to 5000 A,
the average being about 800 A. The degree of crosslinking of the polymer
is very close to that of FRS 2004. When using this latex with a higher specific
area, the standard recipe leads to G_d, G_e, and D_g values which are quadruple
those obtained with ABS grafted onto FRS 2004. The average molecular weight
of linear copolymer is not modified.

Effect of Emulsifier Content. Table X and Figure 5 show that grafting
efficiency decreases continuously when the emulsifier content varies from 0 to
5%. This phenomenon probably results from an increase in the micelle concentration which promotes the polymerization of a larger proportion of monomer
in the water phase.

Conclusions

Ultracentrifugation can be considered a sufficiently reproducible and efficient technique for separating the two components in an ABS graft copolymer. This method enabled us to show quantitatively how the grafting parameters are influenced by polymerization conditions. Consequently, this technique is not only a valuable means for analyzing unknown ABS samples, but it is also particularly convenient for studying and developing a new process, for controlling the sample characteristics when the process is extrapolated in a pilot or an industrial plant, and then for checking the reproducibility of different batches.

Literature Cited

1. Shashoua, V. E., Van Holde, K. E., *J. Polym. Sci.* (1958) **18**, 395.
2. Gesner, B. D., *J. Polym. Sci. Part A* (1965) **3**, 3, 825.
3. Moore, L. D., Frazer, W. J., *Amer. Chem. Soc., Div. Polym. Chem., Prepr.* **8**, (2), 1482 (Chicago, September, 1967).
4. Huguet, M. G., Paxton, T. R., *Amer. Chem. Soc., Div. Polym. Chem., Prepr.* **11** (2), 548 (Chicago, September, 1970).
5. Gerrens, H., Ohlinger, H., Fricker, R., *Makromol. Chem.* (1965) **87**, 209.
6. Shimura, Y., Mita, I., Kambe, H., *J. Polym. Sci. Part B* (1964) **2**, 403.
7. Dinges, K., Schuster, H., *Makromol. Chem.* (1967) **101** (2318), 200.
8. Allen, P. W., Merett, F. M., *J. Polym. Sci.* (1956) **22**, 193.
9. Allen, P. W., Ayrey, G., Moore, C. G., *J. Polym. Sci.* (1959) **36**, 55.
10. Locatelli, J. L., Thèse, Université de Mulhouse, 1973.
11. Frazer, W. J., *Chem. Ind.* (1966) **33**, 1399.
12. Farkas, G. Y., Crisan, T., Sirchis, I., *Mater. Plast.* (1970) **7** (7), 335.
13. Parsons, C. F., Suck, E. L., ADVAN. CHEM. SER. (1971) **99**, 340.

RECEIVED April 3, 1974.

16

ABS Resins: The Relation between Composition and Rheological Behavior

ANTONIO CASALE, ANTONINO MORONI, and CESARE SPREAFICO

Montedison S.p.A., Centro Ricerche, Divisione Petrolchimica,
21053 Castellanza (VA), Italy

The purpose of our study was a rheological evaluation of the effect of composition on the properties of ABS resins in the molten state. Steady-state viscosity was determined over a wide range of temperatures and shear rates. The shear modulus in the molten state was determined by measurement of the diameter of the extrudate. ABS resins in the molten state behaved as an amorphous homophase polymer. The effect of the elastomer phase on the viscoelastic properties which characterize the behavior of the continuous matrix, i.e. monomer friction coefficient and molecular weight between entanglements (M_e), *was calculated by the application of the molecular theories. The significance of these properties in heterophase systems is discussed.*

Acrylonitrile–butadiene–styrene (ABS) resins are typical heterophase systems. They consist of particles of crosslinked rubber, covered by a layer of graft resins (elastomer phase) embedded in a continuous matrix (resin phase). It is well known that the elastomer phase has a major effect on the ABS polymers. Varying the structure of the elastomer phase changes the properties of the resin. An increase in the elastomer phase induces the following changes in physical properties:

(*a*) in the solid state, blend rigidity decreases while toughness increases; and

(*b*) in the molten state, viscosity increases (*1, 2, 3, 4*) while the elastic component (defined as extrudate expansion) decreases (*5*), which agrees with findings for other heterophase systems (*3, 6*).

Relatively little information is available on the rheological behavior of ABS polymers (*4, 5, 7, 8, 9, 10*). Particularly little work of fundamental nature seems to have been done on the relation between ABS rheological properties and their composition, probably because of their complex structure. It is therefore difficult, on the basis of the published data, to develop a rational theory on the effect of the dispersed particles on the flow behavior of the blend. A number of papers were published on the viscoelastic properties of two-phase

172

systems with different morphology (blends of two homopolymers or of a polymer and a block copolymer).

Considering their morphology and the effect of the elastomer phase on the properties below T_g, ABS polymers may be expected to behave in the molten state (a) as a suspension with high solids concentration or as a colloidal dispersion (4), and (b) as a blend of a homopolymer and a diluent or of two homopolymers.

In our laboratory a systematic study was made with the aim of relating the composition of ABS polymers and their rheological properties. The findings enabled us to advance a hypothesis on ABS flow behavior and on the role of the elastomer phase. They also suggested a rheological criterion for polymer compatibility. Finally, on the basis of a method described previously (11), it was possible to use the rheological data to predict ABS processability.

This paper, the third in a rheotechnics series, reports on rheological studies performed at $\dot{\gamma}$ higher than 10^{-1} sec^{-1}. In two-phase polymer systems, Rosen and Rodriquez (3) hypothesized anomalous behavior (yield shear stress) at $\dot{\gamma} = 10^{-1}$. Recently, Zosel (9) verified this hypothesis experimentally for ABS systems. A study of the rheological behavior of ABS resins at very low $\dot{\gamma}$ is now in progress at our laboratories. The correlation between rheological behavior and moldability will be discussed in detail in a subsequent paper.

Experimental

ABS polymers were made by mechanically mixing the resin phase (SAN) and the elastomer phase. This last phase was prepared by grafting the same monomers onto a preformed rubber. By this procedure, the composition and molecular weight of the resins could be very accurately controlled before mixing. The following variables were included: an elastomer phase content of 0–40%, an acrylonitrile content of the SAN of 20–33%, and a weight average molecular weight of the SAN of 68,000–150,000. Table I lists the composition of the ABS resins used in this study.

The melt viscosity data were obtained in a conventional manner using an Instron capillary rheometer over the temperature range of 180–240°C. The capillary had a 90° entry angle, and it was 5 cm long and 0.125 cm in diameter. The well known equations were used to calculate apparent viscosity. Shear rates at the wall, calculated assuming a Newtonian fluid, were corrected for nonparabolic velocity profile using the Rabinowitch equation. No correction was made for entrance effect because of the length-to-diameter ratio of the capillary used.

The diameters of the frozen extrudates were measured by micrometer after annealing. The data are reported as the ratio of diameter of extrudate to diameter of capillary.

Basic Background

Non-Newtonian viscosity, η, is expressed in molecular theories for amorphous one-phase polymer as

$$\eta/\eta_0 = F\ (\dot{\gamma}\ \lambda) \tag{1}$$

where η_0 is the Newtonian viscosity, $\dot{\gamma}$ is the shear rate, and λ is a quantity proportional to the terminal relaxation time. The function, F, depends on the theory considered (Bueche or Graesslay) while its argument is still the same. λ, according to the modified Rouse theory, is a function of the structure and of the molecules, and it is expressed by the friction coefficient.

Table I.   Composition of the ABS Resins

ABS Sample	Elastomer Phase, %	Resin Phase, % AN	Resin Phase, M_w
1	0	26	68,000
2	20		
3	30		
4	40		
5	0	26	117,300
6	20		
7	30		
8	40		
9	0	26	145,700
10	20		
11	30		
12	40		
13	0[a]	26	226,000
14	20		
15	30		
16	40		
17	0	20	117,000
18	20		
19	30		
20	40		
21	0	33	119,000
22	20		
23	30		
24	40		

[a] SAN with broader molecular weight distribution.

As is well known, viscoelastic functions measured at different temperatures, molecular weights, and concentrations can be superposed in a single master curve reduced to a reference temperature, T_o, molecular weight M_o, and concentration C_o, by using appropriate shift factors a_T, a_M, and a_C (12). Equation 1 then becomes

$$\eta/\eta_o = F\ (\dot{\gamma}\ a_T a_M a_C) \tag{2}$$

Obviously, viscosity can be reduced not only to η_o, but also to a different value of viscosity. It is also possible to use the shear stress instead of the viscosity.

Molecular Meaning of a_T. The effect of temperature on viscosity is related to its effect on the friction coefficient, which, in turn, depends on the fractional free volume according to the equation:

$$\ln \zeta = \ln \zeta_{in} + \frac{B}{f} \tag{3}$$

where B is a numerical constant close to unity, and ζ is the inherent friction coefficient that is independent of temperature and molecular weight beyond a limit value. It is assumed that the fractional free volume at any temperature, f, increases linearly with temperature above the glass temperature T_g according to the equation:

$$f = f_g + \alpha\ (T - T_g) \tag{4}$$

where f_g is the fractional free volume at T_g, and α is the thermal coefficient of expansion of the fractional free volume above T_g. Both f_g and α can be calculated by the Williams–Landel–Ferry (WLF) equation if the shift factor a_T is known.

The applicability of the time–temperature superposition principle, not only to homopolymers but even to ABS polymers, has been demonstrated by

Scalco, Huseby, and Blyler (8), Zosel (9), and Bergen and Morris (10). Prest and Porter (13) applied the same principle to homopolymer blends [poly(2,6-dimethylphenylene oxide)–polystyrene]. Recently some papers were published on triblock copolymers of styrene–butadiene–styrene and on their blends with polybutadiene (14, 15). Triblock copolymers can be considered heterophase material as the different constituent blocks are thermodynamically incompatible with each other, and, consequently, polystyrene domains are enclosed in polybutadiene (continuous matrix). The findings indicate that these systems are in general thermorheologically complex, so that the shift factor a_T depends not only on temperature but also on time. These conclusions have been extrapolated to other two-phase systems.

Molecular Meaning of a_M. The dependence of the reduced viscosity of melt polymers on molecular weight is given by

$$a_M = \frac{[\lambda]_M}{[\lambda]_{M_o}} = \frac{[a^2]_M}{[a^2]_{M_o}} \cdot \left(\frac{M}{M_o}\right)^{4.4} \tag{5}$$

where a^2 is the square end-to-end distance per number of monomer units. Since a^2 is not dependent on molecular weight, a_M is only a function of the molecular weight ratio.

Molecular Meaning of a_C. With homophase polymers, the effect of the diluent on viscosity is attributable to: the change in the friction coefficient, ζ; a change in the entanglement factor, Q_e, due to a modification of the entanglement spacing; and a possible change in the a^2. The last factor is negligible in comparison with the others. The relative importance of the first two factors depends on test temperature; at temperatures much higher than T_g, the effect of the second factor is dominant (12).

Viscosity Dependence on Molecular Weight, Temperature, and Elastomer Phase Content

The shear stress-shear rate data were treated as follows.

(a) The curves obtained at different temperatures for each SAN of different molecular weight (distribution and acrylonitrile content being equal) and the relative blends were reduced to a reference temperature $T_o = 210°C$ by the reduced variables method (16). The log of the shift factor a_T plotted *versus* the reciprocal of the temperatures gave a straight line. The activation energy for viscous flow (ΔE_T) was calculated from the slope of the line (17). The activation energies for the different blends are listed in Table II. ΔE_T was independent of SAN molecular weight, and it decreased linearly as the elasto-

Table II.   Activation Energy, ΔE_T, and Shear Modulus, G,
of the Different ABS's[a]

	% Rubber							
	0		20		30		40	
SAN M_w	ΔE_T	G	ΔE_T	G	ΔE_T	G	ΔE_T	G
68,000	30.6	50.0	25.6	120.0	23.3	192.0	21.0	387.0
117,300	29.2	49.5	23.0	129.0	23.0	207.0	18.4	345.00
145,700	29.0	51.6	29.0	132.0	27.0	200.0	23.9	375.0
226,000[b]	30.3	35.0	26.7	82.0	24.8	135.0	—	266.0

[a] ΔE_T is given in kcal/mole, and G in dyne/cm² × 10⁻⁴.
[b] Sample with different molecular weight distribution.

mer phase content was increased. ΔE_T of the pure elastomer phase, obtained by extrapolation, was 5–10 kcal/mole, which is close to the value for rubber. This finding contrasts with that of Huguet and Paxton (4); they reported that the activation energy for the pure elastomer phase was 0 kcal/mole.

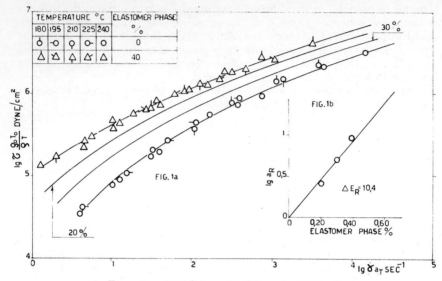

Figure 1.   SAN (M_w = 68,000) and ABS blends
a. Master curves reduced to 210°C; and b. shift factor vs. elastomer phase content

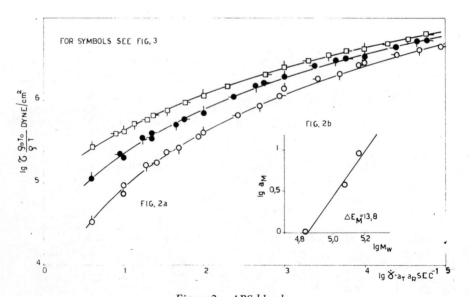

Figure 2.   ABS blends
a. Master curves reduced to 210°C and 0% elastomer phase content; and b. shift factor vs. molecular weight

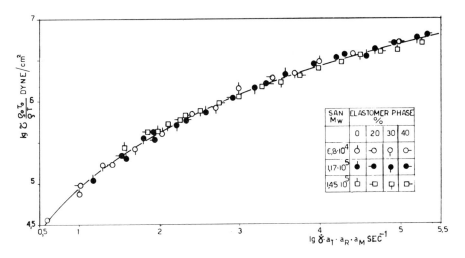

Figure 3. ABS master curves reduced to 210°C and 0% elastomer phase content
SAN Mw = 68,000

(b) It is well known that in polymer–plasticizers systems, flow curves for different concentrations can be superimposed (*16*). Consequently, the same procedure was applied to ABS polymers. The master curves at $T_o = 210°C$ for each of the resins and their ABS blends were reduced to a reference content (0%) by shifting the experimental curve along the abscissa (*see* Figure 1a). Again a straight line was obtained when the log of the shift factor a_R was plotted *versus* elastomer phase content. From the slope of the line, the effect of the rubber phase on polymer viscosity (defined ΔE_R by analogy with ΔE_T) was calculated (*see* Figure 1b).

(c) By superimposing the master curves at $T_o = 210°C$ and graft content 0% for SAN of three different molecular weights, a third shift factor a_M and the relative slope ΔE_M were obtained (Figure 2).

This procedure was very satisfactory, considering that in a single master curve 500 experimental data, referred to four rubber phase contents, three molecular weights (distribution being equal), and five temperatures were superimposed (Figure 3). The same procedure was applied to one SAN of different molecular weight distribution and to three samples with different acrylonitrile contents in order to determine the effect of these variables on the rheological behavior of ABS polymers. The following conclusions were drawn from the findings.

(a) ABS polymers in the molten state behave as one-phase polymers with regards to viscosity dependence on temperature, molecular weight, and elastomer phase content. Viscosity dependence on temperature agrees with the findings of Scalco *et al.* (*8*), Zosel (*9*), and Bergen and Morris (*10*). The first authors performed tensile stress relaxation experiments on a commercial ABS polymer, and they concluded that the temperature dependence on the viscoelastic behavior of polyblends was quite similar to that of many amorphous homopolymers above their T_g. On the other hand, recent evidence indicates that the time-temperature superposition principle cannot be applied without modification to block copolymers and their blends with homopolymers (*14*, *15*). According to our data, this concept cannot be extended to ABS polymers over our experimental range of graft content (0–40%).

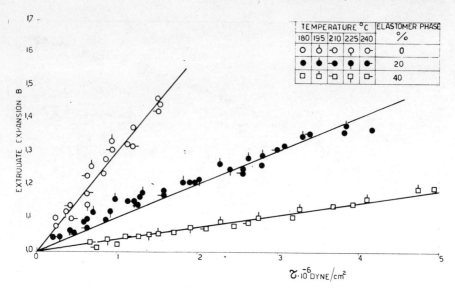

Figure 4. ABS extrudate expansion vs. τ at different temperatures and elastomer phase contents
SAN Mw = 145,000

(b) Consequently, the molecular theories developed for one-phase polymers (17, 18, 19) can be applied to polyblends. However, the reduced viscosity is a function not only of SAN molecular weight and of temperature, but also of rubber phase content and type.

(c) Non-Newtonian behavior at the same shear rate increases with molecular weight (as expected) and rubber content.

(d) The effect of the molecular weight distribution of the SAN is of fundamental importance. As expected, the master curve of a broad distribution sample could not be superimposed on the others. The effect of acrylonitrile content is very slight over the considered range.

Extrudate Expansion

Extrudate expansion data, B, are defined as the ratios of extrudate diameters to capillary diameters. B data obtained at different temperatures were plotted as a function of shear stress (*see* for example Figure 4 for SAN with $M_w = 145,000$). From B data, the shear modulus of the molten polymers, $G = \tau_w / \gamma_R$, can be calculated. τ_w is the shear stress at the wall and γ_R is the recoverable elastic strain. In fact, B was recently related quantitatively to the γ_R corresponding to τ_w (20) by the equation:

$$B^2 = \frac{2}{3} \gamma_R \left[\left(1 + \frac{1}{\gamma_R^2} \right)^{3/2} - \frac{1}{\gamma_R^3} \right] \tag{6}$$

It was shown that the theoretical values calculated by Equation 6 are in good agreement over the range of practical interest with the empirical relation:

$$B = 1 + 0.155 \, \gamma_R \tag{7}$$

These findings led to the following conclusions:

(a) Plots of B *vs.* τ_w were linear over the range $\tau_w = (1\text{--}40) \times 10^5$ dyne/cm², which agreed with the findings of Rosen and Rodriguez for a different heterophase system (6).

(b) Temperature did not effect extrudate expansion over the experimental range of 180–240°C, shear stresses being equal.

(c) The molecular weight of the SAN (distribution being equal) had no noticeable effect on the extrudate expansion of the resins and of the blends at the same elastomer phase content.

(d) As expected, B of the broader distribution sample was higher; the shear modulus, G, (*see* Table II and Figure 5) was not dependent on temperature, shear stress, or SAN molecular weight over the experimental ranges.

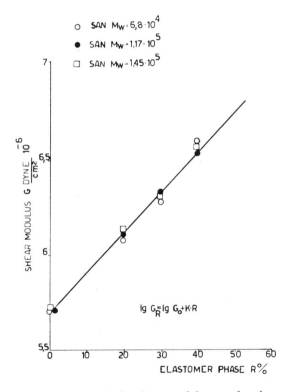

Figure 5. Log of the shear modulus vs. *the elastomer phase content*

(e) Progressive addition of elastomer phase definitely decreased the expansion, which agreed with previous reports on ABS resins (5) and high impact polystyrene (2). Han (5), on the basis of his previous data on blends of polystyrene and polypropylene, suggested that melt elasticity goes through a maximum at a certain blending ratio. We, however, have observed that the diameter of the elastomer phase extrudate was almost equal to the capillary diameter, even if it was difficult to collect precise data because of the poor consistency.

Discussion

As has already been stated, the verified possibility of extending the reduced variables principles to ABS resins makes it possible to treat these typical heterophase systems as blends of amorphous homophase polymers and plasticizers. One possible explanation is that over the experimental $\dot{\gamma}$ range it is not possible to separate the contributions of the two different phases, and the materials will behave as homophase polymer. In fact, long-time molten polymer rheology experiments measure viscoelastic processes over the entire molecule, and, as a consequence, molecular compatibility is evaluated (*13*). On the other hand, high frequency and/or low temperature tests involve the main chain as well as the side chains of the polymer system; the segmental miscibility of the polymer–polymer system is then evaluated. It is important in experimental measurements of polymer compatibility to evaluate the actual size of the volume subject to the test.

As was already stated (*see* Figure 6), the temperature dependence of the shift factor a_T is a function of the elastomer phase content. The strong effect of the rubber content on the temperature dependence of the shift factor a_T could be explained by an increase in free volume of the SAN resin induced by the elastomer phase, as was suggested by Prest and Porter (*13*) for polystyrene–poly(phenylene oxide) blends. In order to verify this hypothesis, log a_T experimental data for SAN and relative blends were used to calculate the WLF parameters and, in turn, the free volumes (f_o) at the reference temperature (T_o) and the thermal expansion coefficients (α) by the equation:

$$\log\ (a_T)\ =\ \frac{-C_1^\circ\ (T\ -\ T_o)}{(C_2^\circ\ +\ T\ -\ T_o)} \tag{8}$$

where $C_1^\circ = B/2.303\ f_o$ and $C_2^\circ = f_o/\alpha$ (*see* Figure 7).

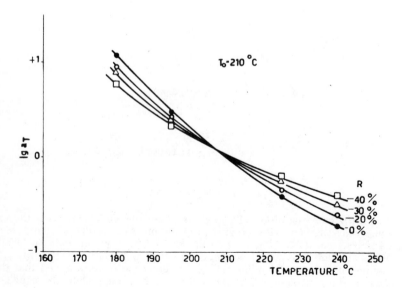

Figure 6.   Temperature dependence of the shift factor for the ABS blends

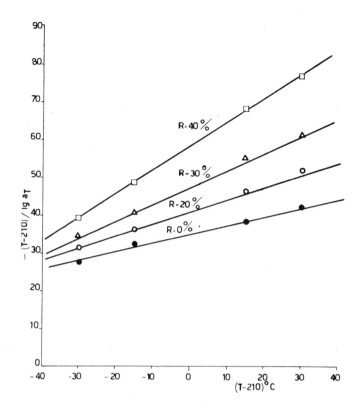

Figure 7. Determination of the WLF parameters at $T_o = 210°C$
for the ABS blends

The WLF and the free volume parameters are listed in Table III. As can be seen, the free volume of the blends increased as the elastomer phase content increased. Only with SAN resins has it been possible to determine $C_1{}^g$ and $C_2{}^g$ and, consequently, f_g/B and α/B. The findings $(f_g/B = 0.0313$ and $\alpha/B = 6.3 \times 10^{-4})$ agree with the values reported by Ferry for polystyrene (0.033 and 0.032; 6.9 and 6.3 $\times$ 10^{-4}) (21). The same parameters could be calculated for the different blends by assuming that:

$$f_R = f_{g_R} + \alpha_R (T - T_{g_R}) \tag{9}$$

where subscript R refers to the blends. This analysis was made by Prest and Porter (13).

Table III. WLF and Free-Volume Parameters

Rubber Content, %	$C_1°$	$C_2°$	f_o/B	$\alpha/B, \times 10^{-3}$	f_g/B
0	4.477	156.7	0.09699	0.619	0.0313
20	2.970	122.07	0.1463	1.20	—
30	2.222	105.1	0.1954	1.86	—
40	1.572	91.805	0.2762	3.01	—

With ABS resins, however, many difficulties arise in applying Equation 9 since dynamic measurements revealed clearly the T_g peaks of the two components. In particular one must decide: if f_{gR} is constant below T_{gSAN} or $T_{grubber}$, the value of T_{gR}, and the measurement system. For example, it must be decided if the resin phase transition must be considered as T_{gR} and if its value is changed by the elastomer phase. According to some authors, the SAN T_g could be modified by the presence of the elastomer phase. Scalco *et al.* reported that the T_g of their sample was 85°C which was 20°C lower than the SAN T_g (8). Bergen and Morris assigned T_g values of 84°C and 97°C respectively to two different ABS samples on the basis of findings in other thermal and mechanical experiments (10). However, as is well known, experimental results are a function of the system of measurement, and this also applies to T_g.

The study of ABS free volume is very important in relation to its role in the elastomer phase reinforcing effect; as suggested by Newmann and Strella (22, 23), the elastomer phase mainly induces a yielding in the matrix. A triaxial stress field in the environment of the dispersed particles causes a local increase in free volume which permits energy-consuming cold flowing phenomena.

The increase in free volume caused by the elastomer phase would in turn bring about a decrease in the friction coefficient and, consequently, a decrease in viscosity. Experimentally, we found an increase in ABS viscosity.

As we have already stated, the effect of the diluent on viscosity, in the case of one-phase polymers, is related to changes in both the friction coefficient and the entanglement factor. For one-phase polymers, the latter is predominant under our experimental conditions. We assume that the increase in blend viscosity could be explained by a change in the state of matrix entanglements induced by the elastomer phase. Bergen and Morris, on the basis of the number of grafted chains, suggested that the region surrounding each particle be considered a region of high chain entanglement density (10). The same idea was applied to branched polystyrene (24). The effect induced by the elastomer phase on the matrix entanglement state could be evaluated as a decrease in average molecular weight between entanglements, M_e^*. By this term we mean a molecular weight between entanglements averaged over the whole sample, taking into account both the regions with and without particles. This evaluation could be made by comparing the flow curves of the different blends at equal free volume. The free volume of the SAN at $T_0 = 210°C$ is 0.0913. Unfortunately, the corresponding temperatures of the blends are well beyond the experimental range (148°, 155°, and 168°C), and a longer procedure must be applied to calculate the effect of rubber on M_e^*.

The shift factor at a constant temperature, a_R, is the constant ratio between the relaxation times at two different elastomer phase concentrations and two different friction coefficients.

$$a_R(T) = \frac{\lambda_R}{\lambda_0} = \left(\frac{M_{e_0}}{M_{e_R}^*}\right)^{2.4} \cdot \frac{\zeta_R}{\zeta_0} \qquad (10)$$

The reference content, as already stated, is zero, that is $M_{e_0} = M_{e_{SAN}}$ and $\zeta_0 = \zeta_{SAN}$. By substituting Equation 3 in Equation 10, we obtain:

$$\log a_R(T) = 2.4 \log \frac{M_{e_0}}{M_{e_R}^*} + \frac{2.303 \ B}{f_R} - \frac{2.303 \ B}{f_0} \qquad (11)$$

Experimentally, we have found (*see* Figure 1) that

$$\log \, a_R(T_o) \; = \; m(T_o)R \tag{12}$$

where $m(T_o)$ is a function of the reference temperature, T_o. By combining Equations 11 and 12, we find that

$$\log M_{eR}^* = \log M_{e_o} - \left[\frac{m(T_o)}{2.4} \cdot R + \frac{2.303}{2.4} \left(\frac{B}{f_o} - \frac{B}{f_R} \right) \right] \tag{13}$$

In the iso-free-volume state, Equation 13 becomes:

$$\log M_{eR}^* = \log M_{e_o} - nR \tag{14}$$

This equation reveals the dependence of the ABS average molecular weight between entanglements (M_{eR}^*) on the molecular weight between entanglements of the pure resin (M_{e_o}), on the amount of the elastomer phase, and on factor $n = d\log a_R/dR$ which depends on the elastomer phase structure. The value of the constant n can be calculated by substituting in Equation 13 the experimental values of f_o, f_R, and m, calculated at $T_o = 210°C$ and three different rubber contents. With the elastomer phase used in this study,

$$\log M_{eR}^* = \log M_{e_o} - 2.0 \cdot R \tag{15}$$

However, experimentally it was calculated that the value of constant n would be changed 10% by increasing the elastomer phase content from 0% to 40%. This variation was considered reasonably acceptable, considering the complexity of the elaboration.

The hypothesis of the rubber effect on the change in M_{e_o} of the pure resin can be confirmed by the extrudate expansion data. According to the rubber elasticity theory:

$$G = \frac{\rho \, RT}{M_e} \tag{16}$$

From Equations 14 and 16, we obtain

$$\log G_R = \log \frac{\rho \, RT}{M_{e_o}} - nR = \log G_{e_o} - nR \tag{17}$$

With our graft polymer,

$$\log G_R = \log G_{e_o} + 2.0 \, R \tag{18}$$

Experimental findings (*see* Figure 5) indicated that G_R can be expressed by the equation:

$$\log G_R = \log G_{e_o} + 2.05 \, R \tag{19}$$

G_R is then independent of SAN molecular weight (which agrees with theory), and it is exponentially dependent on rubber content as predicted by Equation 17.

The agreement between the experimental and predicted values of the constant of the exponents was very good. The M_{e_o} of the SAN copolymer, as calculated by the extrudate expansion data, was 64,000. This value was higher than that for the polystyrene homopolymer.

From the value of the SAN M_{e_o}, it is possible to calculate the M_{eR}^* of the different blends by Equation 15. At an elastomer phase content of 40%, M_{eR}^* was equal to 10,000. As we have stated above, extrudate expansion of the pure elastomer phase was negligible. At higher graft contents, the relation between shear modulus and elastomer phase content probably could change. It is therefore impossible to derive the $M_{e_R}{}^*$ value for the pure graft phase by extrapolation.

From a scientific point of view, further studies must be performed in order to understand the physical meaning of the average molecular weight between entanglements of a two-phase polymer, considering that ABS polymers are composed of highly crosslinked polybutadiene particles surrounded by a region of high chain entanglement density and a continuous matrix. A variation in the friction coefficient probably superposes on the state of entanglement variation. In any case, the constant n permits characterization of the elastomer phase independently of SAN molecular weight and content. The lower its value, the closer the viscoelastic behavior of the ABS polymers is to that of the pure resin.

Our method, then, provides a rheological criterion for measuring the effect of a graft copolymer on a continuous matrix. As was discussed above, the master curves relative to a reference temperature and rubber content:

$$\frac{\tau_w}{\tau'_w} = F\,[\dot{\gamma}\,(a_T)_R\,a_R]$$

of the three SAN samples of different molecular weights were shifted onto the curve of the sample with lowest molecular weight. The shift factor a_M (Figure 2b) is related to the molecular weight according to the equation

$$\log a_M = 3.0 \log M - 3.0 \log M_o \qquad (20)$$

Equation 20 is similar to Equation 5. Its slope, however, is 3.0 instead of 4.4. The cause of this discrepancy is unknown. However, it should be remembered that a factor of 3.4 was found in stress relaxation experiments with polystyrene and poly(α-methylstyrene) (25).

Conclusions

ABS polymers in the molten state behaved as one-phase amorphous polymers in shear modulus and viscosity.

The elastomer phase increased the average free volume of the system as well as the value of the thermal expansion coefficient of the free volume, α. The α value could be related to the reinforcing effect of the graft polymers.

The elastomer phase caused a change in the entanglement state of the pure resin which could be evaluated as a change in the average molecular weight between entanglements, $M_{e_R}{}^*$. The compositional dependence on $M_{e_R}{}^*$ is given by

$$\log M_{e_R}{}^* = \log M_{e_o} - nR$$

The effect on $M_{e_R}{}^*$ was greater than that on free volume; consequently the viscosity of the system was increased by increasing the graft content. The change in $M_{e_R}{}^*$ also induced an increase in shear modulus.

The constant n is a graft property dependent on its structure, but, more important, it was independent of SAN content and molecular weight. Its value was closely related to ABS viscosity and viscoelastic behavior.

Acknowledgments

The authors wish to express their sincere appreciation to S. Baldoni for his assistance in the experimental work and in the elaboration of experimental data. Helpful discussions with G. Ajroldi are also gratefully acknowledged.

Literature Cited

1. Ebneth, H., Beohm, K., *La Nuova Chim.* (1969) **10**, 35.
2. Hagan, R. S., Davis, D. A., *J. Polym. Sci. Part B* (1964) **2**, 909.
3. Rosen, S. L., Rodriguez, F., *J. Appl. Polym. Sci.* (1965) **9**, 1601.
4. Huguet, M. G., Paxton, T. R., *Amer. Chem. Soc., Div. Polym. Chem., Prepr.* **11** (2), 548 (Chicago, September, 1970).
5. Han, C. D., *J. Appl. Polym. Sci.* (1971) **15**, 2591.
6. Rosen, S. L., Rodriguez, F., *J. Appl. Polym. Sci.* (1965) **9**, 1615.
7. Dzhagarova, E. K., *Sov. Plast.* (1971) **12**, 23.
8. Scalco, E., Huseby, T. W., Blyler, L. L., *J. Appl. Polym. Sci.* (1968) **12**, 1343.
9. Zosel, A., *Rheol. Acta* (1972) **11**, 229.
10. Bergen, R. L., Morris, H. L., *Proc. Int. Rheol. Cong., 5th,* Tokyo, Japan, 1970, p. 433.
11. Casale, A., Moroni, A., Ronzoni, I., *Polym. Eng. Sci.* (1974) **14**, 651.
12. Ferry, J. D., "Viscoelastic Properties of Polymers," p. 524, Wiley, New York, 1970.
13. Prest, Jr., W. M., Porter, R. S., *J. Polym. Sci. Part A-2* (1972) **10**, 1639.
14. Fesko, D. G., Tschoegl, N. W., *J. Polym. Sci. Part C* (1971) **35**, 51.
15. Choi, G., Kaya, A., Shen, M., *Polym. Eng. Sci.* (1973) **13**, 231.
16. Ferry, J. D., "Viscoelastic Properties of Polymers," p. 292, Wiley, New York, 1970.
17. Casale, A., Moroni, A., Civardi, D., *Int. Cong. Plast. Elast.,* Milan, October, 1972.
18. Ferry, J. D., "Viscoelastic Properties of Polymers," pp. 195 and 247, Wiley, New York, 1970.
19. Middleman, S., "The Flow of High Polymers," Interscience, New York, 1968.
20. Cogswell, F. N., *Plast. Polym.* (1970) **38**, 391.
21. Ferry, J. D., "Viscoelastic Properties of Polymers," p. 316, Wiley, New York, 1970.
22. Newman, S., Strella, S., *J. Appl. Polym. Sci.* (1965) **9**, 2297.
23. Strella, S., *Appl. Polym. Symp.* (1968) **7**, 165.
24. Fujimoto, T., Narukawa, H., Nagasawa, M., *Macromolecules* (1970) **3** (1), 56.
25. Ferry, J. D., "Viscoelastic Properties of Polymers," p. 415, Wiley, New York, 1970.

RECEIVED April 4, 1974.

17

Grafting Kinetics in the Case of ABS

G. RIESS and J. L. LOCATELLI

Ecole Supérieure de Chimie, 3 rue A. Werner, 68093 Mulhouse Cédex, France

During copolymerization of styrene (S) and acrylonitrile (AN) in the presence of polybutadiene, graft copolymer and free SAN were formed. After separation by a reversible crosslinking technique, the AN content and the molecular weight of grafted SAN and free SAN were determined. The difference in composition of these two species results from the preferential solvation of polybutadiene by styrene. The molecular weight of grafted SAN was higher than that of free SAN. This occurred even before macroscopic phase separation, and it can be attributed to the lower termination rate in the polybutadiene medium and to the preferential solvation of polybutadiene by the peroxide. Polymerization rate and grafting efficiency are given as functions of the 1,2-vinyl content and the type of initiator.

In rubber-modified polymers like high impact polystyrene or acrylonitrile–butadiene–styrene (ABS) resins, the toughening effect of the dispersed rubber particles appears only in the presence of block or graft copolymers. These copolymers regulate the particle size of the rubber dispersion and achieve adhesion of the two phases. Hence, graft copolymers are of practical importance in polymer alloys.

We achieved a systematic kinetic study of ABS. ABS resins, which are formed by copolymerization of styrene (S) and acrylonitrile (AN) in the presence of polybutadiene (PB), consist essentially of a mixture of SAN graft copolymer on PB and ungrafted SAN (styrene-co-acrylonitrile). The grafting kinetics and characteristics of the graft copolymer were studied in relation to the preferential solvation effects as functions of different variables: type and concentration of PB, type and concentration of initiator, monomer concentration, conversion degree, etc.

Separation and Characterization of the Graft Copolymer

The grafting reaction occurs in benzene solution. Since the common separation techniques (*e.g.*, selective extraction and fractional precipitation) were difficult to apply and were not always reproducible in separating small amounts of graft copolymers, we developed a new separation method based on reversible crosslinking. Formation of reversible gels greatly enhanced the solubility difference between the grafted and non-grafted species so that separation by solubility difference became very easy (*1, 2*).

Reversible crosslinking of the polymer backbone, *e.g.* polybutadiene, and the graft copolymer was effected by fixing COONa groups selectively on the PB part (*3, 4*). The dipole–dipole interaction between COONa groups in nonpolar solvents like benzene led to practically complete crosslinking of the polymers that contained COONa groups, *e.g.* the graft copolymer and eventually the polybutadiene that had not been grafted. The non-grafted SAN, however, remained soluble and was easily removed by separation of the two phases: soluble and gel.

Reversible crosslinking can be represented schematically:

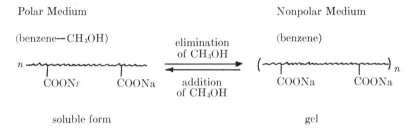

Polar Medium   Nonpolar Medium

(benzene—CH₃OH)   elimination of CH₃OH   (benzene)

soluble form   gel

The reversibility of the crosslinking made it possible to change easily from a crosslinked state to a soluble one and *vice versa*. Rupture of the crosslinks was simply effected by adding a solvating agent, like CH_3OH, for the COONa groups; in this way the soluble chains (non-grafted SAN) retained in the crosslinked network were released. After a new crosslinking, the soluble chains, that were now accessible, were separated from the PB fraction that contained the graft copolymer. This cycle of gel and soluble forms was repeated until pure graft copolymer was obtained. The use of ^{14}C-labeled SAN as tracer revealed that, after three successive separation steps, the remaining ABS was free of non-grafted SAN.

The weight and SAN content of the different fractions are noted on the diagram depicting the purification procedure. Methanol was eliminated from this system at room temperature by simple azeotropic distillation under light vacuum.

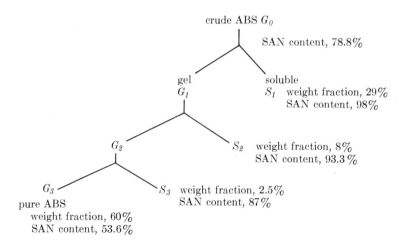

After separation of the pure graft copolymer, its SAN content was determined. In order to compare the characteristics of grafted and non-grafted SAN, the PB was selectively oxidized (5). The molecular weight and the acrylonitrile content of the graft could then be determined.

Characteristics of the Graft Copolymer

Composition of the Grafts—Preferential Solvation. A systematic study of different parameters has revealed that there could be an important difference in composition between grafted and non-grafted SAN. Especially at low conversion, this difference in AN content could be much greater than 4%, with resultant incompatibility of the two types of SAN prepared in the same batch (6).

Some of the data, which have been published elsewhere (7), are presented in Table I. The difference in acrylonitrile content of the grafted and the non-grafted SAN is apparent.

Table I.   Composition of Grafted and Non-Grafted SAN[a]

Experiment	Initiator[b]	Conversion Degree, %	AN in Non-Grafted SAN, %	AN in Grafted SAN, %	R[c]
8	Bz_2O_2	7.35	24.0	22.0	1.090
7	AIBN	10.6	23.70	22.2	1.075
3	Bz_2O_2	11.5	24.20	22.5	1.075
9–1	Bz_2O_2	21.5	24.70	23.2	1.060
2	Bz_2O_2	23.6	24.45	23.3	1.055
1	AIBN	33.0	24.35	22.2	1.095
5	AIBN	43.8	24.05	23.15	1.035
6	AIBN	43.8	23.85	22.8	1.045
4	Bz_2O_2	49.5	24.10	23.1	1.040

[a] Experimental conditions: PB concentration, 54 g/l; and monomer concentration (styrene + acrylonitrile), 20 wt % in solution.
[b] Bz_2O_2 = benzoyl peroxide, AIBN = azodi(isobutyronitrile), $[Bz_2O_2]$ = (1.45–3.1) × 10^{-3} moles/l, and [AIBN] = (0.55–1.6) × 10^{-3} moles/l.
[c] $R = \dfrac{\% \text{ AN in non-grafted SAN}}{\% \text{ AN in grafted SAN}}$.

By different techniques, especially by viscosimetry, it can be shown that the higher styrene content of the grafts results from the preferential solvation of PB by styrene. According to Dondos and Benoit (8), the preferential solvation can be expressed by λ', which is the excess volume of styrene near or in the polymer coil per gram of PB. Therefore, in the close vicinity of the PB, the concentration of styrene is higher than its concentration far from any macromolecule.

This effect correlated with the interaction parameters, χ, and with the solubility parameters, δ, of the different reagents (9, 10). For example, in Figure 1, the evolution of λ', which is characteristic of the preferential solvation, and the evolution of ΔC, which is the difference in composition between grafted and non-grafted SAN, are plotted as functions of the acrylonitrile volume fraction (11).

These observations also confirmed the hypothesis of two independent polymerizations even before visible phase separation, the one producing grafted SAN and the other non-grafted SAN.

It was also demonstrated that at low conversion internal free SAN is formed in the polymer coil; this non-grafted SAN therefore had the same composition as the grafted SAN (12).

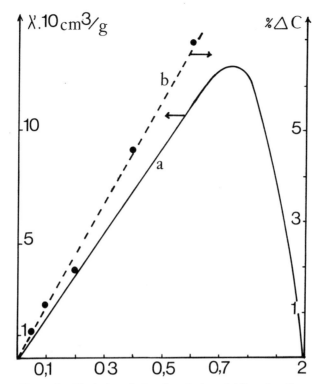

Figure 1. Variation of λ' and evolution of ΔC as functions of the acrylonitrile volume fraction

λ', *preferential solvation of polybutadiene by the monomers; and ΔC, difference in composition of grafted and non-grafted SAN*

Molecular Weights. Furthermore, it appeared that the molecular weight of grafted SAN was systematically higher than that of free SAN, regardless of whether Bz_2O_2 or AIBN was initiator (*see* Table II).

The study of this molecular weight difference as a function of other parameters (*e.g.*, polybutadiene and monomer) corroborated the fact that even before visible phase separation there were two independent polymerization reactions: one in the PB medium that produced grafted SAN, the other that produced free SAN. The difference in molecular weight results from this

Table II. Molecular Weight of Grafted and Non-Grafted SAN [a]

Run	Initiator I	[I], mole/l × 10^3	$\overline{M}_w$ Grafted SAN	$\overline{M}_w$ Non-Grafted SAN
B–1	Bz_2O_2	1.0	138,000	85,000
8		1.45	129,000	82,500
3		2.0	109,000	80,000
B–2		2.35	108,000	80,000
7	AIBN	0.55	166,000	85,000
A–1		1.35	125,000	79,500

[a] Experimental conditions: conversion degree, 10%; concentration PB, 54 g/l; total monomer concentration, 2.05 moles/l; and temperature, 70°C.

Table III. Preferential Solvation of PB by Bz_2O_2[a]

Experiment	$[PB]$, g/200 ml	$[SAN]$, g/200 ml	$[Bz_2O_2]$ in PB phase, mole/ml $\times 10^6$	$[Bz_2O_2]$ in SAN phase, mole/ml $\times 10^6$	S[b]
2A	6.65	6.65	1.44	0.63	2.28
2B	4.45	4.45	1.46	0.65	2.24
2E	6.65	2.22	1.35	0.59	2.30
2F	4.45	6.65	1.65	0.52	3.18

[a] Experimental conditions: polybutadiene: Cariflex BR 1202, $\overline{M}_n$ = 100,000; initial Bz_2O_2 concentration, 1.3 $\times$ 10^{-3} mole/l; and SAN: random copolymer of azeotropic composition, 24% AN, $\overline{M}_n$ = 100,000.

[b] $S = \dfrac{[Bz_2O_2] \text{ in PB}}{[Bz_2O_2] \text{ in SAN}}$.

heterogeneous polymerization with a lower termination rate in the PB medium where the grafted SAN was produced. However this effect was partially counterbalanced by an increase in the initiator concentration in the PB medium because of preferential solvation of PB by initiators like Bz_2O_2 (12).

Kinetic Study

Preferential Solvation by Peroxides. In a way similar to that used for the study of preferential solvation of PB by styrene, we also examined that solvation effect by different types of peroxides. The amount of peroxide in the PB was maximum when their solubility parameters were the same. Furthermore, the solvation effect was increased by adding to the system nonsolvents of PB and of the initiator. Data on the preferential solvation of PB by Bz_2O_2 as well as data on the peroxide surconcentration, S, are presented in Table III.

Rate of Polymerization—Retardation Effect. By studying the rate of polymerization in ABS formation, it was possible to demonstrate the importance of the concentration and the structure of PB, especially its 1,2-vinyl content (see Table IV).

By systematic study, we correlated the retardation effect of PB with the 1,2-vinyl content, the efficiency of the free radicals, and the preferential solvation of PB by the peroxides. This retardation effect, which disappeared in the presence of PB with high 1,2-vinyl content, was the consequence of two coincident phenomena:

Table IV. Effect of 1,2-Vinyl Content of Polybutadiene on Rate of Polymerization[a]

Run	1,2-Vinyl Content, %	R_p/R_{po}[b] AIBN	R_p/R_{po}[b] Bz_2O_2
A	without PB	1	1
17	0	0.86	0.84
20	4	0.63	—
16	18.5	0.705	0.68
18	70	1.23	1.20
19	95	1.45	1.45

[a] Experimental conditions: polybutadiene concentration, 26.9 g/l; temperature, 70°C; and monomer concentration, 1.90 moles/l.

[b] R_p/R_{po} = degree of conversion in presence of polybutadiene/degree of conversion in absence of polybutadiene; [AIBN] = 1.1 $\times$ 10^{-3} mole/l; and $[Bz_2O_2]$ = 1.6 $\times$ 10^{-3} mole/l.

(a) a non-homogeneous distribution of the peroxide and therefore of the free radicals with a higher concentration in the rubber phase, which resulted from the preferential solvation of PB by the initiator, and

(b) a reduced efficiency of the radicals in the PB medium by formation of more stable macroradicals and by a cage effect resulting from an increased viscosity in this medium.

Grafting Degree. The variation in degree of grafting was correlated with the 1,2-vinyl content of PB, the monomer concentration (effect of preferential solvation by the monomer), the type and the concentration of initiator (preferential solvation by initiators), and the degree of conversion. The effect of the structure of PB on the grafting degree is tabulated in Table V.

Table V. Effect of Polybutadiene Structure on Degree of Grafting[a]

Type of PB	1,2-Vinyl Content, %	Grafting Degree[b]	
		AIBN	Bz_2O_2
EQB 6 alt.	0	8.5	23.7
Cariflex BR 1220	4	17.2	—
J.L. 15000	18.5	11.5	21.0
EG 62/4	70	29.0	34.4
PVB 40	95	46.5	44.8

[a] Experimental conditions: polybutadiene concentration, 25 g/l; and monomer concentration, 2.05 moles/l.
[b] Grafting degree = grafted SAN/total SAN; [AIBN] = 1.1×10^{-3} mole/l; and [Bz_2O_2] = 1.6×10^{-3} mole/l.

The lower grafting degree in the presence of AIBN, compared with that obtained with Bz_2O_2, resulted from the greater stability of the primary free radicals; stable radicals cannot transfer hydrogen atoms in allylic position on 1,4-*cis*-polybutadiene. In the presence of a 1,2-vinyl structure, the tertiary hydrogen atoms were easier to remove, and in this way the AIBN could initiate grafting. Consequently, the presence of 1,2-vinyl structures increased the grafting efficiency (*13, 14, 15*).

A mechanism based on two simultaneous polymerizations can be proposed. Such a mechanism would account for the preferential solvation and the heterogeneity of the system (*4, 13*).

Literature Cited

1. Llauro-Darricades, M. F., Banderet, A., Riess, G., *Makromol. Chem.* (1973) **174**, 105.
2. *Ibid.* (1973) **174**, 117.
3. Locatelli, J. L., Riess, G., *Eur. Polym. J.* (1974) **10**, 545.
4. Locatelli, J. L., Thesis, Mulhouse, 1973.
5. Locatelli, J. L., Riess, G., *Angew. Makromol. Chem.* (1972) **26**, 117.
6. Molau, G. E., *J. Polym. Sci. Part A* (1965) **3**, 4235.
7. Locatelli, J. L., Riess, G., *Angew. Makromol. Chem.* (1972) **27**, 201.
8. Dondos, A., Benoit, H., *Makromol. Chem.* (1970) **133**, 119.
9. Locatelli, J. L., Riess, G., *Angew. Makromol. Chem.* (1973) **32**, 101.
10. Locatelli, J. L., Riess, G., *J. Polym. Sci. Part A-1* (1973) **11**, 3309.
11. *Ibid., Part B* (1973) **11**, 257.
12. Locatelli, J. L., Riess, G., *Makromol. Chem.* (1974) **175**, 3523.
13. Locatelli, J. L., Riess, G., *Angew. Makromol. Chem.* (1973) **28**, 161.
14. *Ibid.* (1973) **32**, 117.
15. *Ibid.* (1974) **35**, 57.

RECEIVED May 7, 1974. This work was supported by the Société Nationale des Pétroles d'Aquitaine.

18

Properties-to-Composition Relation of ABS Resins by Statistical Analysis

MARIO CATONI, GIUSEPPE PIZZIGONI, and ISIDORO RONZONI

Montedison S.p.A., Divisione Petrolchimica, Fabbrica di Castellanza,
Castellanza, Italy

A study of the principal ABS plastics on the market was made in order to determine the relations between some main physical–mechanical properties and composition specifications. The statistical methodology of multiple regression was used in the investigation. The mathematical models obtained for most of the tested properties were well explained by second order and linear polynomials. The expansion of mathematical models enabled us to calculate the best estimates expected for the physical properties whenever the ABS compositions were known. The plotting of these polynomials obviously represents a qualitative picture of the physical properties–composition relations in the wide experimental range of the variables.

Some general information is available about the relationships between the properties and the compositions of acrylonitrile–butadiene–styrene (ABS) resins (*1–6*). It usually presents a qualitative picture of the dependence on the ABS composition of only particular properties (*7–15*). An early mathematical approach was by Dinpes and Schuster in West Germany (*16*).

In this work, data on 48 commercial ABS polymers from five different producers were elaborated by statistical methods (*17*) in order to determine in more detail the relationships between some physical properties and the composition of ABS resins. Because of data uniformity, ABS polymers containing α-methylstyrene were not included in this study.

Experimental

The physical properties of 48 commercial ABS samples from five different producers were related to the composition by regression analysis. The following were chosen as independent variables.

X_1—**Polybutadiene Content (PB).** X_1 was measured by IR spectrophotometry on hot-pressed film samples; it is expressed as wt % of the whole ABS (*i.e.* including all additives). The observed variation range was 7–28%.

X_2—**Acrylonitrile Content (AN) of the Free SAN Copolymer Fraction.** X_2 was determined from the nitrogen content of the acetone-soluble fractions; it is expressed as wt % of the whole ABS. The observed variation range was 13–25%.

X_3—Grafted SAN Copolymer Content (SANI). X_3 was determined by gravimetric analysis of the acetone-insoluble fractions; it is expressed as wt % of the whole ABS. The observed variation range was 0.5–20.5%.

X_4—Intrinsic Viscosity of the Free SAN Fraction $[\eta]$. X_4 was measured in dimethylformamide at 25°C on the acetone-soluble fraction; it is expressed in dl/g. The observed variation range was 0.45–1.10 dl/g.

X_5—Extractable Substances Content. X_5 was determined by extraction in a cyclohexane–methanol mixture; it is expressed as wt % of the whole ABS. The observed variation range was 1–5.5%.

X_6—Ash Content. X_6 was determined by burning a sample to ashes at 800°C; it is expressed as wt % of the whole ABS. The observed variation range was 1.0–7.0%.

The following physical properties were considered dependent variables.

Y_1, Y_2, and Y_3—Fluidity. These values were determined by measuring the length of injection-molded spirals at 190°, 210°, and 230°C respectively; they are expressed in cm.

Y_4—Tensile Strength. Y_4 is the yield strength measured according to ASTM D 638; it is expressed in kg/cm².

Y_5—Izod Impact. Y_5 was measured on notched samples according to ASTM D 256; it is expressed in kg cm/cm.

Y_6—Rockwell Hardness. Y_6 was measured according to ASTM D 785; it is expressed in units of the R scale.

Y_7—Deflection Temperature. Y_7 was measured according to ASTM D 648 under a load of 18.5 kg/cm²; it is expressed in °C.

Other characteristics that were mainly structural (*e.g.*, dimensions and structures of elastomer particles, and molecular weight distribution of the plastomers) and that also have considerable effect on the physical properties of the ABS polymers were not considered in this study.

Mathematical–Statistical Analysis

The data on the commercial ABS polymers were elaborated according to the statistical methodology of multiple regression (*17, 18, 19, 20, 21*). The relation between a given ABS physical property (Y) and the composition variables (X) can take the form of the following quadratic polynomial equation (*22, 23, 24*):

$$Y = \beta_0 + \sum_{i=1}^{K} \beta_i X_i + \sum_{\substack{i=1 \\ (i<j)}}^{K-1} \sum_{j=2}^{K} \beta_{ij} X_i X_j = \sum_{i=1}^{K} \beta_{ii} X^2_i + \varepsilon \qquad (1)$$

where Y is the observed value (measure) of the property; K is the number of independent variables; β_0, β_i, β_{ij}, and β_{ii}, are the coefficients of regression or parameters of the model; β_0 is the constant or fixed term; β_i are the coefficients referred to the effect of the first order factors; β_{ij} are the coefficients referred to the effect of interaction of the factors; and β_{ii} are the coefficients referred to the effect of the quadratic terms. ε is the difference between the observed value, Y, of the property and the true or expected value, $E(Y)$; it is assumed to be a casual variable, distributed normally and independently with a mean value of zero and a variance σ^2.

The coefficients (β_0. . . . β_{ii}) were unknown, and they could be calculated by the least squares method as suggested by the regression theory in order to obtain the best estimate. This calculation was performed on a UNIVAC 1106 computer by the stepwise method perfected by M. A. Efroymson (of Esso Research Engineering Co.) and reelaborated in Fortran V in the Newreg program of Montedison. The best estimated coefficients were then introduced into the polynomial equation, and the estimated value, Y, of the property was then calculated.

The significance of the estimated coefficients was tested by using the Student's t-distribution at the 0.05 level of significance. The multiple correlation coefficient (R) was derived from the total variability of measured properties and the residual variability of the calculated values. Variability was expressed as sums of squares. These calculations were performed for every property studied.

Results

In accordance with what has been specified, the polynomial can be usefully represented by graphs. It was appropriate to ascribe two independent variables, x_1 and x_2, to the orthogonal axis so that the property values (Y) could be represented by isolevel curves. The remaining independent variables then had to be fixed at constant values. In the graphs, the isolevel curves were plotted as solid lines within the experimental X_2 range and as broken lines beyond this range in order to allow for extrapolation.

Fluidity (Y_1, Y_2, and Y_3). The ABS fluidities, measured as spiral lengths at 190°, 210°, and 230°C (and expressed respectively by Y_1, Y_2, and Y_3), were related to the independent variables by the following polynomial equations.

$$\hat{Y}_1 = 33.708 - 0.387X_2 - 0.780X_1X_4 - 0.012X_2X_3 + 1.329X_4X_5 - 0.663X_4X_6 \quad (2)$$
$$R_1 = 0.867$$

$$\hat{Y}_2 = 41.894 - 0.446X_1X_4 - 0.010X_1{}^2 - 0.015X_2X_3 - 0.761X_2X_4 - 0.024X_2X_6 + 1.225X_4X_5 \quad (3)$$
$$R_2 = 0.904$$

$$\hat{Y}_3 = 40.156 - 0.992X_1X_4 - 0.017X_2X_3 - 0.996X_4X_6 + 0.273X_5{}^2 \quad (4)$$
$$R_3 = 0.866$$

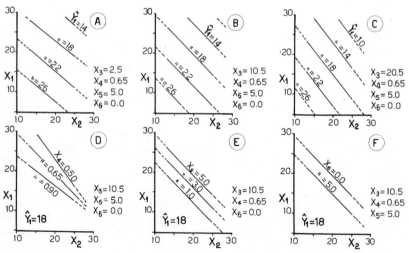

Figure 1.   Spiral length at 190°C, Y_1, (cm) as a function of ABS composition variables

$X_1 = PB$, %; $X_2 = AN$, %; $X_3 = SANI$, %; $X_4 = [\eta]$, dl/g; $X_5 = $ extractable substances, %; $X_6 = $ ash, %

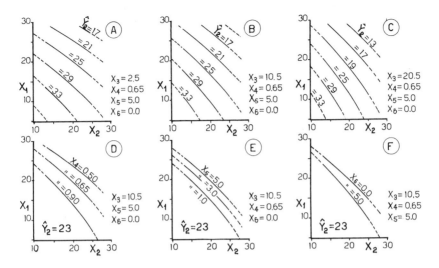

Figure 2.   Spiral length at 210°C, Y₂, (cm) as a function of ABS composition variables

Variables same as in Figure 1

All the correlation coefficients (R_1, R_2, and R_3) were highly significant. All the independent variables (X_1, . . . X_6) were involved in the equations, mainly as negative interaction ($X_i X_j$) terms.

The coefficients of the $X_2 X_3$ term were clearly unaffected by the temperature. Those of the $X_1 X_4$ and $X_4 X_6$ terms were weakly affected by the temperature increase, as is shown in the Y_1 and Y_3 equations (the $X_1 X_4$, X_1^2, and $X_2 X_6$ terms have nearly the same effects in the Y_2 equation). Only the X_5 terms had positive rather than negative effects on fluidity. The effects of X_2 (AN) on fluidity, either alone or in interaction, decreased as the temperature increased in accordance with well known aspects of the polarity of nitrile groups. The main effects on fluidity at the higher temperature were attributable to the SAN intrinsic viscosity (X_4) and the extractable substances content (X_5).

Figures 1A, 1B, and 1C depict the effect of grafted SAN content (X_3) on the fluidity at 190°C, with X_4, X_5, and X_6 being constant. The isolevel curves were shifted to the left by the increase in X_3, and the $X_2 X_3$ interaction was responsible for steepening the slope. In Figure 1D, the shift in the $Y_1 = 18$ cm isolevel with variation in the intrinsic viscosity of the plastomer (X_4) is plotted for fixed values of X_3, X_5, and X_6. The $X_1 X_4$ interaction was responsible for the change in slope. Figures 1E and 1F show the shift of the same isolevel ($Y_1 = 18$ cm) with variations in X_5 and X_6, respectively.

Figures 2A, 2B, and 2C demonstrate the effects of variation in X_3 on the fluidity at 210°C (Y_2). The previous considerations for Y_1 were applicable here too; the negative quadratic term (X_1^2) was responsible for the downward curvature of the isolevels. Figures 2D, 2E, and 2F depict the effects of the remaining variables, as well the shift in the $Y_2 = 23$ cm isolevel.

Figure 3 is a plot of the same effects on the fluidity at 230°C (Y_3). The isolevel curves were nearly horizontal for the decrease in the effect of AN (X_2); only the $X_2 X_3$ term was present. The greatest effects on the shift of the $Y_3 = 29$

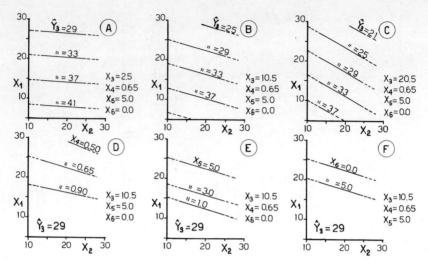

Figure 3.   Spiral length at 230°C, Y_3, (cm) as a function of ABS composition variables

Variables same as in Figure 1

cm isolevel were those of SAN intrinsic viscosity (X_4) and the extractable substances content (X_5) (*see* Figures 3D and 3E).

Tensile Strength (Y_4). The tensile strength of ABS resins was measured on suitable injection-molded specimens as yield strength. This property was related to composition by regression equation 5.

$$\hat{Y}_4 = 258.650 + 7.158X_2 - 1.279X_1X_5 + 4.869X_2X_4 + 1.108X_3X_5 \tag{5}$$
$$R_4 = 0.841$$

The main positive effects on the value of Y_4 were those expected from the X_2 and the X_2X_4 terms. The ash content (X_6) did not seem to have a significant effect on the yield strength, at least not in the examined composition range.

The extractable substances content (X_5) appeared in a positive (X_3X_5) as well as a negative (X_1X_5) interaction term. The resultant effect was negative because, in an ABS resin, X_1 (PB) is usually greater than X_3 (SANI). Moreover, there were some X_1 and X_3 values for which the yield strength was independent of X_5, at least in the examined variation range. This was not tested experimentally.

These interaction terms could be referred to some particular phenomena that could occur at the interfaces of the grafted elastomer and the SAN copolymer phases.

Figures 4A, 4B, and 4C illustrate the effect of the grafted copolymer content (X_3) on Y_4 when X_4 and X_5 were constant. The isolevel curves were parallel shifted as X_3 increased because there were no interaction terms with X_1 and X_2.

Figure 4D shows the shift in the $Y_4 = 440$ kg/cm² isolevel with variation in SAN intrinsic viscosity (X_4).

Figure 4E shows the previously discussed single point which was the intersection of the $Y_4 = 440$ kg/cm² isolevel as it was rotated by variation in X_5. Display of the intersection points parallel to the X_2 axis was described by the equations.

Since Y_4 was independent of X_6, the position of the $Y_4 = 440$ kg/cm^2 isolevel did not change with variation in the ash content (Figure 4F).

Izod Impact Strength (Y_5). Izod is very likely the most common impact test for ABS resins. It is usually performed on notched specimens ¼- and ⅛-inch thick, and both specimen types are simultaneously injection-molded.

Izod data for the ⅛-in. specimens correlated well with the composition variables, but the correlation was not satisfactory for the ¼-in. specimens. Thus, distribution of orientations should be more regular for the ⅛-in. specimens than for those measuring ¼ inch. Such greater regularity could increase the dependence of Izod on the ABS composition. This can be true only statistically because the intrinsic properties of the elastomer are always very important. As an example, the Izod value of an ABS resin containing very large elastomer particles can deviate significantly from the value calculated by the regression equation (Equation 6). The constant term is a negative, and there are no X_3 or X_5 terms.

$$\hat{Y}_5 = -51.278 + 4.476X_1 - 0.064X_1^2 + 1.396\ X_2X_4 + 0.144X_2X_6 - 0.388X_6^2 \quad (6)$$
$$R_5 = 0.818$$

As expected, PB (X_1) exerted the main positive effect on the Izod, but this was balanced by a relatively small quadratic term. The ash content (X_6) had a considerable negative quadratic term, balanced by a positive interaction term, X_2X_6. The latter became prominent at higher AN contents (X_2). The X_2X_4 (AN x intrinsic viscosity) term had a positive effect on the Izod, and it was also present in the yield strength equation (Equation 5). Since ABS failure is ductile, under both the tensile and the impact tests at room temperature, the X_2X_4 term assumes greater importance.

Figure 5A illustrates the dependence of Izod on X_1 and X_2 with the values of X_4 and X_6 being constant; the X_1 quadratic term was responsible for the weak upward curvature of the isolevel curves. Figure 5B depicts the $Y_5 = 25$

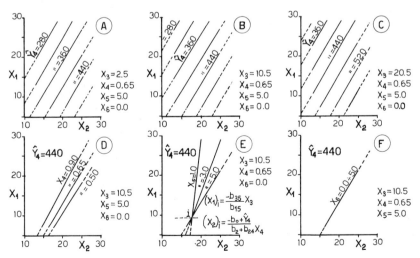

Figure 4. Tensile strength, Y_i, (kg/cm^2) as a function of ABS composition variables

Variables same as in Figure 1

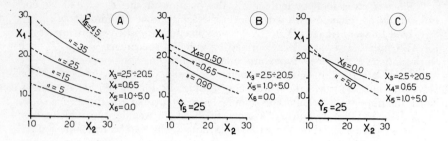

Figure 5. Izod impact strength, Y_5, (kg cm/cm) as a function of ABS composition variables

Variables same as in Figure 1

kg cm/cm isolevel shifted by variation in SAN intrinsic viscosity (X_4). Figure 5C shows that the ash content (mostly TiO_2 rutile) had a strengthening effect on the ABS resins when the AN content (X_2) was not too low.

Rockwell Hardness (Y_6). The hardness regression was the most simple (*see* Equation 7), and the correlation coefficient was the highest. The equation

$$\hat{Y}_6 = 97.678 + 0.858X_2 - 0.029X_1{}^2 \tag{7}$$

$$R_6 = 0.932$$

is composed of only three terms: the first is a positive constant, the second is a positive X_2 (AN) term, and the third is a negative quadratic X_1 (PB) term. Its meaning is obvious. Neither the extractable substances content (X_5) nor the ash content (X_6) seemed to affect Rockwell hardness over the observed composition range.

Figure 6 illustrates the dependence of hardness on variations in X_1 and X_2. The X_1 quadratic term was responsible for the downward curvature of the isolevels.

Deflection Temperature under Load (Y_7). The deflection temperature under load was the only thermal property related to ABS composition (*see* Equation 8). Because the correlation coefficient was not as high as for the other equations ($R^2 = 0.58$), the model must be accepted with some reserva-

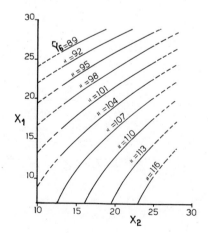

Figure 6.   Rockwell hardness, Y_6, (R) as a function of polybutadiene content, X_1, (%) and acrylonitrile content, X_2, (%)

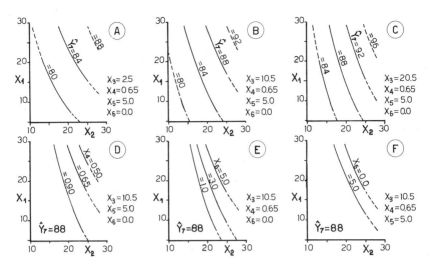

Figure 7. Deflection temperature, Y_7, (°C) as a function of ABS composition variables

Variables same as in Figure 1

$$\hat{Y}_7 = 73.662 + 0.013X_1X_2 + 0.018X_2X_3 + 0.637X_2X_4 - 0.050X_2X_5 + 0.078X_6^2 \quad (8)$$
$$R_7 = 0.761$$

tions. There were four X_2 (AN) interaction terms. The positive effect of the X_2X_4 term was remarkable; this term also appeared in the yield strength and Izod impact equations (Equations 5 and 6).

Figures 7A, 7B, and 7C illustrate the effect of variation in X_3 on the deflection temperature values. The isolevel curves shifted to the left as X_3 increased. The X_1X_2 term was responsible for the isolevel curvature. Figures 7D, 7E, and 7F show the effects on the same Y_7 isolevel curve of variations in the X_4X_5 and X_6 terms.

Table I. Calculated and Experimental Values for Properties of Urtal ABS Polymers

Property	Test	Data Source[a]	Urtal ABS Polymer A 12	M 122	B 32
Spiral length, cm	190°C	C	22	20	18
		E	24	21	18
	210°C	C	28.5	26.5	24
		E	29	27	25
	230°C	C	37	34	31
		E	35	33	30
Tensile strength, kg/cm²	ASTM D 638	C	444	430	388
		E	440	405	380
Notched Izod impact strength, kg cm/cm	ASTM D 256	C	12	21	30
		E	15	25	35
Rockwell hardness, R	ASTM D 785	C	113	109	102
		E	112	107	100
Deflection temperature under load, °C	ASTM D 648	C	83	85	86
		E	87	87	87

[a] C, calculated from mathematical model; and E, measured experimentally.

Application

Although the regression equations give only empirical relations between the properties and the ABS compositions, there is an important practical interest in their applications. As an example, there was good agreement between the measured and the calculated values of properties for some ABS resins (*see* Table I for these data for Urtal, an ABS resin from Montedison). It should be noted that Montedison data on ABS resins were not used in deriving the equations.

Conclusion

This work is the report of a regression analysis of the measured properties and the composition data of several commercial samples of ABS resins. The found empirical relations are very interesting from a practical point of view, as can be seen from their application. The correspondence of the equations was verified for a series of Urtal (an ABS resin produced by Montedison), and there was indication of the excellent reliability of the proposed interpretative models. When some theoretical considerations are made, the meaning of several terms in the empirical equations is not as trivial as it may seem at first.

Literature Cited

1. Frazer, W. J., *Chem. Ind.* (1966) 1399.
2. Hagermann, E. M., *J. Appl. Polym. Sci.* (1969) **13**, 1873.
3. "Encyclopedia of Polymer Science and Technology," Vol. 1, p. 436, Interscience, New York, 1965.
4. Merz, E. H., Claver, G. C., Baer, M., *J. Polym. Sci.* (1965) **22**, 325.
5. Thompson, M. S., "Gum Plastics," Reinhold, New York, 1958.
6. Davenport, N. E., Hubbard, L. W., Pottit, M. R., *Brit. Plast.* (1959) **32**, 549.
7. Keskkula, H., Turly, S. G., Boyer, R. F., *J. Appl. Polym. Sci.* (1971) **15**, 351.
8. Casale, A., Ducci, F., Rottondi, B., ANTEC XXX; (1972) Chicago.
9. Dasch, J., *Kunststoffe* (1967) 117.
10. *Ibid.* (1967) 328.
11. *Ibid.* (1968) 769.
12. *Ibid.* (1970) 113.
13. *Ibid.* (1970) 149.
14. Ronzoni, I., Binotto, G., Orsatti, E., *Mater. Plast. Elast.* (1963) **5**, 1.
15. Zitek, P., Musik, S., Zelinger, J., *Angew. Makromol. Chem.* (1969) **116**, 6.
16. Dinpes, K., Schuster, H., *Makromol. Chem.* (1967) **101**, 200.
17. Brownlee, K. A., "Statistical Theory and Methodology in Science and Engineering," p. 334, Wiley, New York, 1967.
18. Draper, N. R., Smith, H., "Applied Regression Analysis," Wiley, New York, 1966.
19. Williams, E. J., "Regression Analysis," Wiley, New York, 1967.
20. Tuker, J. W., Anscombe, F. J., *Thecnometrics* (1963) **5**, 141.
21. Bartlett, M. S., *Biometrics* (1947) **3**, 39.
22. Guest, P. G., "Numerical Methods of Curve Fitting," Cambridge University, Cambridge, 1961.
23. Ralston, A., Wilf, H. S., "Mathematical Methods for Digital Computers," Wiley, New York, 1960.
24. Catoni, M., Pizzigoni, G., Internat. Cong. Plast. Elast., Milano, Italy, 1972.

RECEIVED August 19, 1974

Graft Copolymers of EPDM Elastomers as Reinforcing Agents for Brittle Thermoplastic Materials

F. SEVERINI, A. PAGLIARI, C. TAVAZZANI, and G. VITTADINI

Montedison S.p.A., Centro Ricerche Bollate, Milano, Italy

Styrene and vinyl chloride were polymerized in the presence of EPDM elastomers with low ethylidene norbornene content. The viscosity of the reacting mass during polymerization was studied in the styrene–EPDM system. The diene monomer content in the elastomer markedly affected the grafting yield of vinyl chloride, and less so that of styrene. The physical-mechanical characteristics of the materials were determined. High impact strength, high resistance to aging, and considerable thermal stability were achieved by using EPDM elastomers instead of high diene rubbers in the radical polymerization of vinyl monomers.

The impact strength of brittle thermoplastic materials is generally improved by adding small amounts of rubber, either pure or modified by grafting with the monomer or monomers constituting the matrix to be reinforced (1, 2, 3, 4, 5). As a rule, modification is achieved by monomer polymerization in the presence of the reinforcing elastomer, which is usually a butadiene polymer or copolymer (6, 7).

The reaction proceeds *via* grafting and crosslinking of the elastomer; after the reaction is complete, the elastomer is dispersed as particles in the reaction mass. Because of the side chains, these particles adhere to the matrix, and, through a mechanism that is not yet completely clear, they increase the plastomer tenacity which is high even at temperatures below zero.

The physical–mechanical properties of materials reinforced with butadiene-based elastomers rapidly decline from action by atmospheric agents; some work indicated that degradation begins in the unsaturated rubber phase (8).

Recent findings indicated that this drawback can be overcome by using particular reinforcing agents, *i.e.* graft copolymers obtained by modifying saturated or low-unsaturated elastomers such as ethylene–propylene (EP) or ethylene–propylene–diene monomer (EPDM) rubbers and acrylic elastomers (9–16).

In previous papers (4, 13, 14), we described the preparation of high impact products of polystyrene (PS) and poly(vinyl chloride) (PVC) by radical polymerization of the respective monomers in the presence of saturated EP rubbers. Mechanical mixtures of such elastomers with PVC or PS having the same composition as that of the products modified by grafting produced materials in which significant variations of impact strength were not detected, and macroscopic incompatibility phenomena such as the surface migration of rubber were observed.

The presence of elastomers as grafted products increased the impact strength of the matrix, rendered the components of the system compatible, and hence allowed the procurement of technically valid materials.

In this paper we describe some findings from the study of the grafting reactions that occur during the polymerization of styrene and of vinyl chloride in the presence of different EPDM elastomers with various ethylidene norbornene (ENB) contents. The presence of unsaturated monomer units considerably improved the grafting yield of rubber and, in general, the radical reactivity of the elastomer without altering the resistance to aging.

Experimental

Materials. EPDM rubbers had the following properties:

Type	Propylene, wt %	ENB, wt %	η_{inh}, dl/g
A	59.5	0.0	1.8
B	50	3.7	1.8
C	40	4	2.1
D	40	7	1.7
E	39	9.2	2.1

The inherent viscosity was determined in Tetralin at 135°C.

The polybutadiene used was Intene from I. S. Rubber; $[\eta] = 2.1$ dl/g in cyclohexane at 30°C. Vinyl chloride and styrene monomers from Montedison were 99.99 and 99.5% pure respectively. Azobis(isobutyronitrile) (AIBN) from Fluka was 99% pure. Di-*tert*-butyl peroxide was 95% pure. *Tert*-butyl peracetate was used at a 50% dilution with dimethyl phthalate. The *n*-butyl ester of 3,3-(di-*tert*-butylperoxy)valerianic acid was 85% pure. *Tert*-dodecyl mercaptan, a commercial product of the Societé Nationale des Petroles d'Aquitaine, was used as a modifier. The solvents used were commercial products purified by fractionation.

The chromatographic separations of the products grafted with vinyl chloride were effected using Celite K-535 from Mascia and Brunelli S.p.A. as absorbent.

Procedures. GRAFT POLYMERIZATION OF VINYL CHLORIDE. Runs were conducted in a stainless steel reactor that was equipped with an anchor-shaped stirrer and was heated to the desired temperature by circulating fluid from a thermostated bath. Rubber was introduced into the reactor as granules; the monomer was then introduced after the air had been removed by repeated washings with nitrogen. After contact for several hours, the elastomer either swelled or was dissolved in the monomer. The initiator and an aqueous solution of the suspending agent were then introduced, and the reacting mass was heated to the desired temperature. At the end of the reaction, the polymer was separated from the suspending agent by centrifugation, and then it was washed with water and dried.

GRAFT POLYMERIZATION OF STYRENE (9). Polymerization was effected by the mass-suspension technique under nitrogen. The radical initiator was added to the rubber solution in styrene, with *tert*-dodecyl mercaptan being added either simultaneously or later during the polymerization. For prepolymerization, the reacting mass was heated and stirred until monomer conversion reached 25–30%. The syrupy product was poured into a reactor containing a suspending agent in aqueous solution, and the reaction was completed by heating the mass to 155°C under vigorous stirring. At the end of the reaction, the polymer was recovered as beads, which were washed and dried.

FRACTIONATION OF PVC–EPDM RAW POLYMERIZATES. *Determination of Ungrafted Rubber.* The amount of ungrafted rubber was determined by the Desreux chromatographic method under the described conditions (17) in the presence of n-heptane at 30°C.

Determination of the Components: Modified Rubber + Free Rubber. The dry product (10 g) was placed in a pear-shaped vessel with cyclohexanone in high excess. Stirring continued for 18 hrs at room temperature; then the material was filtered on Gooch. The residue was washed with acetone and dried to constant weight. Under these conditions, cyclohexanone dissolved all the PVC homopolymer, and it could dissolve small, insignificant amounts of non-crosslinked grafted copolymer with very high PVC content.

FRACTIONATION OF PS–EPDM RAW POLYMERIZATES. *Determination of the Components: Grafted Rubber + Free Rubber.* A few grams of material in the form of beads were placed in contact with a large excess of methyl ethyl ketone; the resultant dispersion was centrifuged. The solid, which consisted of graft copolymer and free rubber, was dried and weighed. The supernatant was concentrated and then precipitated with excess methanol; the precipitate consisted of practically pure PS.

Determination of Free Rubber. The residue from the methyl ethyl ketone extraction was dried and weighed; it was then placed in contact with excess n-heptane at room temperature for 24 hrs. The insoluble fraction was separated by filtration; after being washed and dried, it was weighed. The supernatant was evaporated in a calibrated flask, and a thin film of n-heptane-extracted rubber remained. From the data, the amount of grafted, crosslinked polymer was calculated, and its composition was determined.

DETERMINATION OF PROPERTIES. For all determinations, the PVC and the PS had molecular weights comparable to that of the homopolymer present in the crude reaction products. PVC was characterized under conditions described previously (15, 23). With PS, beads of product mixed with a specific antioxidant of the olefin elastomer (0.05%) and with a stearic acid derivative as lubricant (0.2%) were homogenized by extrusion in a double-screw extruder; thus the material was obtained as granules. The test specimens were machined to proper size and shape from compression-molded sheets (190°C, 40 kg/cm^2).

Results and Discussion

The formation of graft copolymers by radical polymerization of a monomer in the presence of a saturated polymer quite probably occurs through chain transfer reactions. The free radicals derived from the decomposition of the initiator or the growing chains of the monomer are generally supposed to transfer their activity to the preformed polymer chain, thereby originating radical active centers on which the monomer chains grow (18).

As is known (19), an index of the chain transfer processes is given by the values of the transfer constants of the monomer to the polymer and, in general, by the variations in the homopolymer average molecular weight with respect to the value obtained by operating in the absence of the transfer agent.

The use of unsaturated polymers with allyl hydrogen atoms favors the occurrence of transfer reactions and, therefore, the formation of graft copoly-

204 COPOLYMERS, POLYBLENDS, AND COMPOSITES

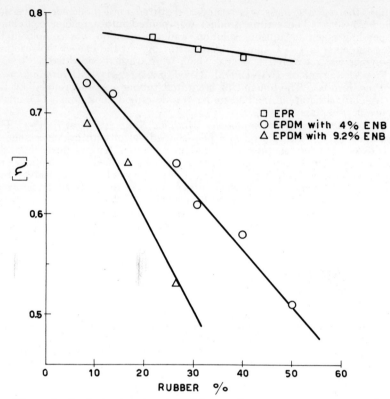

Figure 1. Intrinsic viscosity of PVC homopolymer vs. concentration of different EPDM elastomers in the initial elastomer + vinyl chloride mixture

T = 65°C; AIBN, 0.13%; suspending agent, variable amounts depending on amount of rubber present, range: 0.6–1.5%; monomer conversion = 70%; water/(rubber + monomer) = 1.4; and intrinsic viscosity determined in cyclohexanone at 30°C and expressed in dl/g

mers. Obviously it must be borne in mind that graft copolymers may also form through characteristic reactions of the double bond such as addition of free radicals and copolymerization with the monomer to be grafted.

With vinyl chloride, the decrease in intrinsic viscosity of the homopolymer that forms during the reaction with an increase in the initial amount of rubber was far more marked with EPDM elastomers than with saturated ethylene–propylene copolymer (Figure 1). This suggests that EPDM elastomers have a high radical activity, even with small amounts of ENB.

The increase in radical activity with unsaturation is also indicated by the data in Table I, which clearly indicate that products with a higher graft copolymer content can be obtained with EPDM elastomers than with EP rubbers; the composition of these products varies with the diene content (15).

Proof of the high reactivity of the unsaturated units present in the elastomers we used was obtained by polymerizing vinyl chloride in the presence of different amounts of two hydrocarbons, n-hexane and 2-methyl-2-butene; these may be taken as models of the saturated polyhydrocarbon and of the allyl system present in the EPDM terpolymer. AIBN (0.13%) was used as initiator.

Table I. Polymerization of Vinyl Chloride (VC) in the Presence of EPDM with Different ENB Contents[a]

ENB in EPDM, %	Conversion to PVC, %	EPDM in Crude Product, %	Grafted EPDM from 100 g Starting Rubber, %	Graft Copolymer from 100 g Starting Rubber, %	VC in Graft Copolymer, %
—	93.4	8.8	54.5	88.7	38.5
3.7	87.0	9.4	74.5	155.2	52.0
7.0	81.0	10.0	72.5	179.5	59.6
9.2	86.0	9.5	68.5	227.0	69.8

[a] Elastomer in starting mixture = 8.3%, T = 65°C, AIBN = 0.13%, suspending agent = 0.6%, and water/(rubber + monomer) = 1.4.

Calculation of the transfer constant C_s at 65°C by the Mayo equation (*20, 21, 22*) gave the following values:

$$\text{transfer to } n\text{-hexane } C_s = 2.9 \cdot 10^{-3}$$

$$\text{transfer to 2-methyl-2-butene } C_s = 4.4 \cdot 10^{-2}$$

It has been observed (*15, 23*) that the type of initiator used with vinyl chloride

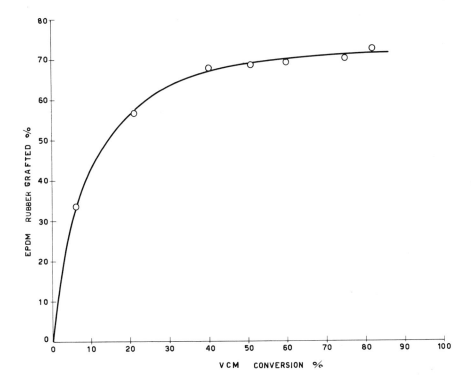

Figure 2. Grafted rubber vs. monomer conversion

Elastomer, EPDM type C; T = 65°C; AIBN, 0.13%; suspending agent, 0.6%; water/(rubber + monomer) = 1.4; and elastomer in the initial vinyl chloride + rubber mixture, 8.3%

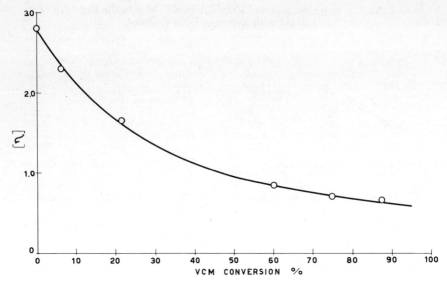

Figure 3.   Intrinsic viscosity of the n-heptane-soluble fraction of EPDM vs. monomer conversion

Conditions as indicated in Figure 2; intrinsic viscosities determined in cyclohexane at 30°C and expressed in dl/g

has practically no effect on the grafting yield of the reacting monomer and elastomer; this behavior suggests that the grafting reaction occurs predominantly on the radical centers that are generated on the elastomer by transfer with the growing monomer chains.

Figures 2 and 3 depict the variations in the amount of rubber participating in the grafting reaction and in the intrinsic viscosity of the still free elastomer

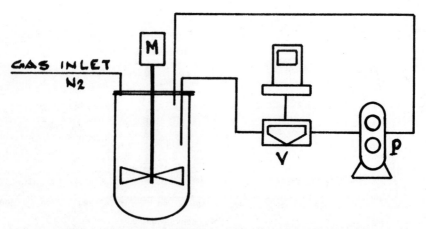

Figure 4.   System for measuring the viscosity of the reacting mass

M, stirrer motor; V, plate cone viscosimeter, Epprecht Rheomat 15 equipped with Viscosclav from Contraves (Zurich); and P, gear pump from Slack and Parr (England); capacity: 5 ml/rev, speed range: 3–35 rpm. The reacting mass and the measurement apparatus are dipped into the same thermostatted bath.

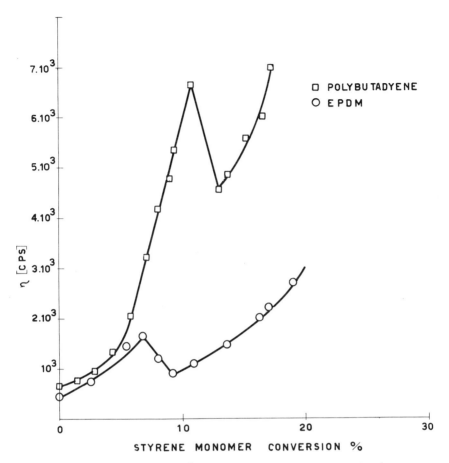

Figure 5. Viscosity of the reacting mass vs. monomer conversion

Measurements were made in apparatus depicted in Figure 4. T = 100°C; elastomer in the initial styrene solution, 10%. EPDM, type B; initiator, tert-butyl peracetate, 0.09%; modifier, tert-dodecyl mercaptan, 0.02%. Polybutadiene, Intene; initiator, tert-butyl peracetate, 0.02%; modifier, tert-dodecyl mercaptan, 0.1%.

that occurred with different degrees of monomer conversion. The data indicate that for conversions above *ca.* 60%, the modified rubber content did not increase, and the intrinsic viscosity of the extractable elastomer did not decrease. The limit value of the amount of grafted rubber may be explained (*15*) by assuming that, at conversions above 70%, the elastomer contained in the suspension particles cannot participate in the radical activity of the system. The available information is insufficient to interpret the decreasing values of the intrinsic viscosity of free rubber during the reaction. This behavior might be qualitatively explained by admitting that higher molecular-weight fractions preferentially participate in the grafting reaction, as was suggested for the polymerization of vinyl chloride in the presence of polyethylene (*24*).

With regard to styrene, we limited our investigations to the prepolymerization step with particular emphasis on phase inversion and on the behavior of grafting up to monomer conversions of *ca.* 30%. Phase inversion occurred at

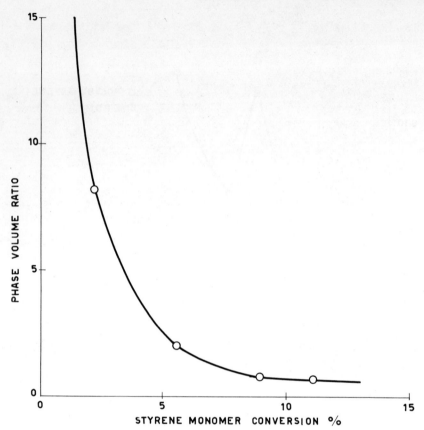

Figure 6.   Simulated variation of the ratio: rubber phase volume/styrene + poly-
styrene phase volume vs. styrene conversion
EPDM elastomer, type C; and polystyrene, [η] toluene/30°C = 0.8 dl/g

low degrees of monomer conversion; in fact, the transparent solution became opaque after polymerization began. This phenomenon was followed by measurements with a rotational viscosimeter at the reaction temperature. The viscosity of the reacting mass may be transferred in a closed cycle from the reactor to the measurement apparatus through a gear pump. The apparatus is diagrammed in Figure 4.

Variation in the viscosity of the reacting mass with monomer conversion was determined by using, under the described conditions, 10% solutions of EPDM type B elastomer and of Intene (*see* Materials Section above). The data are plotted in Figure 5. System viscosity increased up to a styrene conversion of 7%; then it decreased, reaching a minimum at 9% conversion; then it began increasing again. Phase inversion occurred within the indicated maximum–minimum range. With the butadiene elastomer, the viscosity pattern was different: inversion occurred at slightly higher conversions, and the phenomenon was more pronounced.

The range within which phase inversion occurred was determined by the Molau and Keskkula method (25). Measurements were made by using styrene solutions of EPDM type C rubber and polystyrene, simulating the empirical composition of reaction mixtures with different degrees of conversion. The data are plotted in Figure 6; they indicate that, in agreement with viscosimetric data, the phase inversion range occurred at conversions of from 5 to 10%.

The polymerization of styrene in the presence of an EPDM elastomer gives rise to the formation of a graft copolymer and of polystyrene homopolymer. In the prepolymerization step, the intrinsic viscosity of polystyrene homopolymer increased with monomer conversion (*see* Figure 7). This behavior may be explained by noting that while polystyrene formed, the elastomer separated from the initial homogeneous system, with a gradual decrease in the amount of rubber participating in the radical activity of the reacting mass (*26*). The amount of rubber participating in the grafting reaction as well as the composition of the pure graft copolymer separated from the reaction mixture are plotted in Figure 8 as functions of the degree of monomer conversion. At the same conversion levels, the elastomer reacted more readily with vinyl chloride than with styrene; this agrees with the fact that the radicals from styrene are expected to exhibit lower reactivity than those from vinyl chloride.

The low grafting yield agrees with the transfer constants determined by polymerizing solutions of EPDM types A, C, and E elastomers at 65°C in the presence of AIBN (0.13%) as an initiator; the reaction was interrupted at

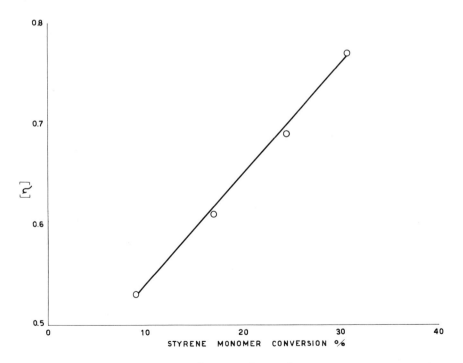

Figure 7. Intrinsic viscosity of polystyrene homopolymer vs. monomer conversion
T = 100°C; elastomer in the initial styrene solution, 10%; EPDM, type B; initiator, tert-butyl peracetate, 0.09%; and intrinsic viscosities determined in toluene at 30°C and expressed in dl/g

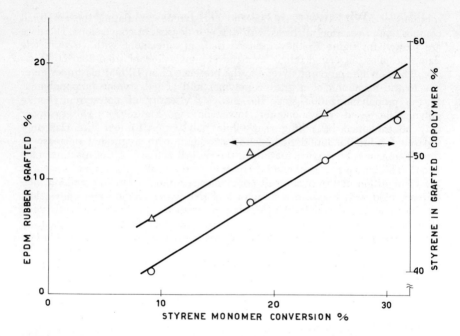

Figure 8. Grafted EPDM elastomer and composition of the grafted copolymer vs.
monomer conversion

T = *100°C; elastomer in initial styrene solution, 10%; EPDM, type B; initiator,* tert-*butyl per-acetate, 0.09%; and modifier,* tert-dodecyl mercaptan, 0.02%

conversions of between 5 and 10%. The findings (*21, 22*) were as follows:

Elastomer	C_s	*ENB in the Elastomer, %*
A	$2.4 \cdot 10^{-3}$	0
C	$3.6 \cdot 10^{-3}$	4.0
E	$3.0 \cdot 10^{-3}$	9.2

These data indicate that the reactivity of the radical from styrene was rather weak and that, unlike vinyl chloride, the ENB content affected the chain transfer processes of this radical only slightly. This agrees with the low values of the transfer constant as measured in the styrene polymerization in the presence of a highly unsaturated polymer such as natural rubber (*27*). The magnitude of the transfer constants was not high, and this explains the rather low grafting yields obtained when the reaction occurred in the presence of excess monomer.

Quite probably, under our reaction conditions grafting essentially occurred by chain transfer without directly involving the double bonds.

Preliminary findings indicate that in the graft copolymers of vinyl chloride and of styrene isolated from the reaction products, the unsaturation caused by

Table II.   Polymerization of Styrene in the Presence of EPDM Elastomers with Different ENB Contents[a]

Elastomer Type	ENB in Elastomer, %	EPDM in Crude Product, %	Grafted EPDM from 100 g Starting Rubber, %	Graft Copolymer in 100 g Crude Product, %
A	0	10	80	14.4
B	3.7	10	93	18.0
E	9.2	10	97	20.5

[a] Data obtained by fractionation of products (*see* Table IV).

the ENB in the starting elastomer remained practically unaltered. Unsaturation considerably affects the formation of insoluble gels consisting of grafted and crosslinked elastomer (*see* Table II).

Morphological examination of the reaction products also provided evidence that, with EPDM elastomers, the grafted and crosslinked rubber remained dispersed in the matrix as particles with an average diameter of 2–5μ. The structure remained unaltered, even after mechanical actions such as those required for transformation into manufactured articles. On the other hand, with EP rubbers, particle morphology was markedly altered under the same transformation conditions.

Crude products obtained by polymerizing styrene or vinyl chloride in the presence of ∼ 10% EPDM elastomer yield by the usual transformation tech-

Table III.   Physical–Mechanical Properties

Property	Test Method	PS/EPDM-10[a]	PVC/EPDM-9[b]	PS	PVC
Specific gravity, g/ml	ASTM D 792	1.04	1.32	1.05	1.39
Softening point, °C	ASTM D 1525[c]	93	77	98	80
HDT, °C[d]	ASTM D 648[e]	87.5	67	—	70
Notched Izod impact strength, kg cm/cm					
at 23°C	ASTM D 256	8	100	1.2 ÷ 1.5	6
at 0°C	ASTM D 758	5	89	—	—
at −10°C	ASTM D 758	—	20	—	—
Tensile strength, kg/cm²					
at yield	ASTM D 638	301	420	—	550
at break	(v = 20 mm/min)	291	320	—	350
ultimate elongation		30.7	35	1	28
Flelastic modulus	(v = 5 mm/min)	16.000	—	30.000	31.000
exural elastic modulus, kg/cm²	ASTM D 790	16.450	23.000	30.000	30.000
Rockwell hardness	ASTM D 785	58 (L)	50 (L)	80 (M)	90 (L)

[a] EPDM elastomer, type C; and elastomer in starting solution, 10%. Initial polymerization: T = 100°C; styrene conversion, 30%; and initiator, *tert*-butyl peracetate, 0.09%. Suspension polymerization: suspending agent, 0.6%; *tert*-butyl peroxide, 0.4%; and water/(rubber + monomer) = 1.2. Suspension cycle: 2 hrs at 120°C, 1 hr at 140°C, and 2 hrs at 155°C.
[b] EPDM elastomer, type C; elastomer in starting mixture, 8.3%; T = 65°C; AIBN = 0.13%; suspending agent = 0.6%; and water/(rubber + monomer) = 1.4.
[c] Vicat, 55 kg, oil.
[d] HDT: heat distortion temperature.
[e] At 264 psi.

niques manufactured articles endowed with high impact strength. In addition, they are characterized by a high resistance to atmospheric agents (*11, 15, 23*), which makes them convenient for outdoor applications.

Table III summarizes the most important physical–mechanical properties, as determined by ASTM methods, of high impact PS and PVC with EPDM type C elastomer contents of 10% and 9.2% respectively (*see* Experimental section above for preparative techniques). These products are characterized in Table IV.

Table IV. Main Properties of the Characterized High Impact Products[a]

Property	PVC/EPDM	PS/EPDM
Type of elastomer	C	C
Total EPDM, %	9.2	10.0
Average diameter of elastomer particles, μ	0.1–0.5	2.8
Free EPDM, % total EPDM	29.0	7.0
Grafted copolymer per 100 g product, g	20.6	21.0
EPDM in the grafted copolymer, %	31.5	45.0
$[\eta]$ of homopolymer, ml/g $\times$ 10^2	0.72[b]	0.7[c]

[a] *See* Table III.
[b] Intrinsic viscosity measured in cyclohexanone at 30°C.
[c] Intrinsic viscosity measured in toluene at 30°C.

Conclusions

Grafting of vinyl chloride and of styrene in the presence of EPDM elastomers occurred under the described conditions with formation of graft copolymers of rubber and of poly(vinyl chloride) and polystyrene homopolymers.

The grafting reaction is determined primarily by chain transfer processes; these decrease until they become negligible as the polymerization of excess vinyl monomer proceeds. Formation of the vinyl polymer causes separation of the elastomer from the initial system and a gradual decrease in the amount that participates in the radical activity of the reacting mass.

The transfer constants of the EPDM elastomer are relatively low, but the rubber modifications that occur with grafting can make rubber an effective reinforcing agent for the brittle matrixes under consideration.

With vinyl chloride, the content of diene units in the elastomer considerably affects the yield of graft copolymer. With styrene, the values of the transfer constants of elastomers with different ENB contents suggest that this effect is much less.

In comparison with poly(vinyl chloride) and with polystyrene, crude polymerizates with a rubber content of *ca.* 10% are characterized by higher impact strength and by lower values of stiffness, which are nevertheless acceptable for the expected applications of such products.

Literature Cited

1. Frey, H. H., *Kunststoffe* (1959) **49**, 50.
2. Göbel, W., *Kaut. Gummi* (1969) **22**, 116.
3. Martin, J. R., *Rubber Plast. Age* (1966) **47**, 1321.
4. Natta, G., Pegoraro, M., Severini, F., Dabhade, S., *Rubber Chem. Technol.* (1966) **39**, 1667.
5. Sharp, J. J., *Brit. Plast.* (1959) **32**, 431.
6. Augier, D. J., Fettes, E. M., *Rubber Chem. Technol.* (1965) **38**, 1164.

7. Keskkula, H., *Appl. Polym. Symp.* (1970) **15**, 51.
8. Gesner, B. D., *J. Appl. Polym. Sci.* (1965) **9**, 3701.
9. Baer, M. J., *J. Appl. Polym. Sci.* (1972) **16**, 1125.
10. Barkhuff, R. A., U.S. Patent **3,408,424** (1963).
11. *Chem. Eng. News* (1970) Dec. 7, 55.
12. Meredith, C. L., *Rubber Chem. Technol.* (1971) **44**, 1130.
13. Natta, G., Beati, E., Severini, F., Toffano, S., Italian Patent **698,014** (1963).
14. Natta, G., Severini, F., Pegoraro, M., Beati, E., Aurello, G., Toffano, S., *Chim. Ind. Milan* (1965) **47**, 960.
15. Severini, F., Mariani, E., Cerri, E., Pagliari, A., *Chim. Ind. Milan* (1973) **55**, 270.
16. Zahn, E., *Appl. Polym. Symp.* (1969) **11**, 209.
17. Desreux, V., *Rec. Trav. Chim. Pays Bas* (1969) **68**, 769.
18. Smets, G., Hart, R., *Fortschr. Hochpolym. Forsch.* (1960) **2**, 173.
19. Flory, J. P., "Principles of Polymer Chemistry," Chap. 4, Cornell University, Ithaca, 1953.
20. Mayo, F. R., *J. Amer. Chem. Soc.* (1943) **65**, 2334.
21. Severini, F., Pegoraro, M., Pagliari, A., Tavazzani, C., *Chem. Zvesti* (1972) **26**, 241.
22. Severini, F., Pagliari, A., Tavazzani, C., unpublished data.
23. Severini, F., Mariani, E., Pagliari, A., Cerri, E., ADVAN. CHEM. SER. (1971) **99**, 260.
24. Wollrab, F., Declerk, F., Dusmoulin, J., Obsomer, M., Georlette, P., *Amer. Chem. Soc., Div. Polym. Chem., Prepr.* **13**, 499 (Boston, April, 1972).
25. Molau, E. G., Keskkula, H., *J. Polym. Sci. Part A-1* (1966) **14**, 1595.
26. Rosen, S. L., *J. Appl. Polym. Sci.* (1973) **17**, 180.
27. Minoura, Y., Mori, Y., Imoto, M., *Makromol. Chem.* (1957) **24**, 205.

RECEIVED March 5, 1974.

20

Grafting of Styrene and Acrylonitrile onto Ethylene Polymers

HEINRICH ALBERTS, HERBERT BARTL, and RAINER KUHN

Zentrale Forschung, Wissenschaftliches Hauptlaboratorium der Bayer AG, Leverkusen, West Germany

Styrene was grafted onto low-density polyethylene under swelling conditions. The homogeneity of the graft copolymers depended not only on the temperature of the styrene diffusion, but also on the comonomer content and the crystallinity of the grafting backbone. The grafting efficiencies were affected primarily by the activity of the initiator radical and, to a smaller extent, by the mutual ratio of monomer to initiator to substrate. The supramolecular structure of the graft products could be elucidated by differential thermal analysis and by electron microscopy. The presumed molecular structure of the graft copolymer could be deduced from light scattering measurements.

In 1968 we reported on the effect of the monomer radical activities on the grafting of various vinyl monomers onto ethylene–vinyl acetate copolymers (EVA) (1). We found that the grafting activity of vinyl monomers corresponds to the monomer radical activity published by Mayo and Walling (2) and others (3, 4). This feature of vinyl monomers is depicted in Figure 1.

Vinyl chloride is one of the most active monomers in grafting reactions. When EVA was used as grafting substrate, the resultant graft copolymer had improved compatibility with poly(vinyl chloride) (PVC). Blending of these grafts with PVC produced high-impact PVC. When styrene was grafted onto EVA, the graft product had a very low graft copolymer content. The grafting of acrylonitrile onto EVA was also studied by Bartl and Hardt (1). Unlike the vinyl chloride graft products, the corresponding acrylonitrile grafts were incompatible despite comparable grafting efficiencies. When styrene–acrylonitrile combinations were grafted onto EVA, the graft products had an average graft copolymer content of about 50 wt %; however, the reaction products were incompatible in all cases.

The objective of these studies was to obtain more information about the factors that affect the grafting efficiencies and the compatibility of graft polymers. Since incompatible grafts often have very poor mechanical properties, it was of interest to explore grafts of improved compatibility which should have improved mechanical properties. One approach to this problem was to obtain information as exact as possible about the molecular structure of the graft

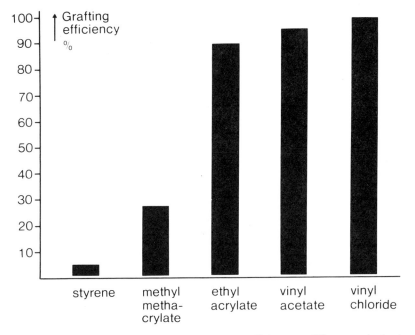

Figure 1. Grafting of various monomers onto Levapren 450, a product of Bayer AG

polymers and the number of the graft chains per molecule as well as their molecular weight, molecular weight distribution, and compatibility with the grafting substrate. For this study, the grafting of styrene onto low-density polyethylene (LDPE) was chosen as the model reaction. The theoretical assumptions underlying the evaluation of the measurements of light scattering have been discussed elsewhere (5).

Materials and Experimental Techniques

Materials. LDPE of grade A or B was used as the grafting substrate; it was in the form of pellets. Grade A is a homopolyethylene with a weight average molecular weight $M_w = 5.7 \times 10^5$, a molecular nonuniformity $U = (M_w/M_n - 1) \approx 18$, a molecular weight of the unbranched chain sections (long-chain branching) of $M_c = 3.5 \times 10^3$, and a density $\rho = 0.916$ g/cm³. Grade B is a chemically uniform ethylene–vinyl acetate random copolymer (EVA) with highly branched short and long chains, vinyl acetate content = 8.5 wt %, $M_w = 5.2 \times 10^5$, $U = 14$, $M_c = 5 \times 10^3$, and $\rho = 0.925$ g/cm³.

The styrene and acrylonitrile monomers used were freshly distilled. Azodi-(isobutyronitrile) (Porofor N) (PN), benzoyl peroxide (BPO), and *tert*-butyl peroctoate (*t*-BO) with a purity higher than 99% were used as initiators.

Grafting under Swelling Conditions. Grafting under swelling conditions was carried out in aqueous dispersion. This grafting method increased the problem of diffusional control during the reaction. On the other hand, this method avoided transfer reactions by solvents and permitted the use of lower temperatures than did bulk grafting in the melt.

At constant temperature, the individual components diffused into the pellets. Subsequently, the polymerization process was started by increasing

Table I.　Grafting Efficiencies[a]

Experiment	PE Grade	Initiator		Swelling		Polymerization		Styrene wt %	Grafting Efficiency[b] %	Preparation Procedure
		Type	Conc., mole %	T, °C	Time, hrs	T, °C	Time, hrs			
M	A	PN	1.0	65	5	80	8	38.6	6.2	I
R		BPO	0.70	50	24			35.9	12.9	I
ZB				60	3			44.3	10.9	I
ZA				70	3			41.5	9.1	I
U				70/50	3/3			43.9	10.0	II
A				85/50	3/1			43.7	11.0	II
N				85/50	3/5			42.7	14.9	II
E		t-BO	0.784	50	3			42.4	56.4	I
L			0.784	50	24			36.5	59.7	I
ZG				70	3			44.0	51.8	I
S				70/50	3/5			42.9	52.2	II
G				85/50	3/1			43.7	51.6	II
OA	B	BPO	0.091	50	5	85	5	20.2	17.0	III
O		BPO	0.70	85/50	3/5	85	8	43.2	23.2	II
OC		t-BO	0.23	75	3	90		20.4	76.0[c]	III
OE		t-BO	0.72	80/50	2/3	90		18.5	69.0	III
Q		t-BO	0.784	85/50	3/5	85		44.3	68.8	II

[a] Referred to the styrene polymerized in the pellets.
[b] *See* text.
[c] Chain regulator, isobutene.

the temperature, and it was completed at constant temperature. In several cases, the initiator was not added to the system until after the pellets had been swollen with styrene in order to allow it to diffuse into the pellets at a lower temperature thereby preventing premature, unwanted decomposition of the peroxide. The swelling period was generally 3–5 hrs, but it was also extended up to 24 hrs.

Preparation Procedures. PREPARATION PROCEDURE I. In a 6-l stirring vessel, 2000 g deionized water, 600 g LDPE, 80 ml 10% dispersant solution (1:1 copolymer of methacrylic acid and methyl methacrylate), 490 g styrene, and 8 g radical former were stirred under nitrogen with conditions as indicated in Table I. Subsequently the pellets were collected on a suction filter, washed with plenty of water, and dried for 48–72 hrs *in vacuo* at 50°C. The styrene content of the pellets was determined from the weight increase of the pellets:

$$styrene\ (wt\ \%) = \frac{yield - 600}{yield}$$

PREPARATION PROCEDURE II. In a 6-l stirring vessel, 2000 g deionized water, 600 g LDPE, 400 g styrene, and 80 ml 10% dispersant solution were stirred under nitrogen with conditions as indicated in Table I. The mixture was then cooled to 50°C, and a solution of 8 g radical former in 92 g styrene was added. The batch was stirred at 50°C for the time indicated. Polymerization was started by increasing the temperature. At the end of the reaction, the pellets were suctioned off, washed, and then dried for 48–72 hrs *in vacuo* at 50°C. The styrene content of the beads was calculated from the weight increase as in Procedure I.

PREPARATION PROCEDURE III. The procedure was the same as in I and II except that 250 g styrene and 1000 g LDPE were used. The aliphatic mono-olefins used as chain regulators were passed through the batch *via* an ascending pipe.

Analysis of the Polymer. Grafting efficiency and degree of grafting were determined by fractional precipitation. After the graft product was dissolved in toluene or toluene–dimethylformamide mixtures at 100°C, ungrafted poly-styrene (PS) or polystyrene–acrylonitrile remained in solution after cooling and was separated. After the precipitated product was isolated and dried, its styrene content was determined, and from this the grafting efficiency was calculated:

$$\text{grafting efficiency } (\%) = \frac{\text{PS grafted}}{\text{PS grafted} + \text{homo-PS}} \times 100$$

The light scattering of this material could be measured even though it still contained ungrafted polyethylene. In the new method of light scattering, ungrafted portions of polyethylene do not interfere with the determination of the molecular weight of the graft chains, nor, after refractionation according to the molecular weight of the graft chains, with the determination of the molecu-lar weight distribution of the graft chains (5). In order to determine the degree of grafting of the substrate, however, it was necessary to determine the amount of ungrafted polyethylene. By careful precipitation of the graft product after separation of ungrafted polystyrene from a 1:1 toluene–*n*-heptane solution at 94°C with 1-butanol, it was possible to separate the free polyethylene almost quantitatively since it remained in solution under these conditions.

The polystyrene content of the individual fractions was determined by NMR studies in tetrachloroethylene at 80°C from the ratio of the peak areas of the aromatic and aliphatic protons. When styrene content was low, determination was made by IR spectroscopy of films, using a calibration curve. Acrylonitrile was determined by nitrogen analysis. Light scattering was measured in toluene at 90°C at the wavelength $\lambda = 5461$ A. Under these conditions, solvents and grafting bases were almost isorefractive, *i.e.* only the graft chains were visible. The solutions used in all measurements were purified by centrifugation at 90°C and approximately 10,000 g.

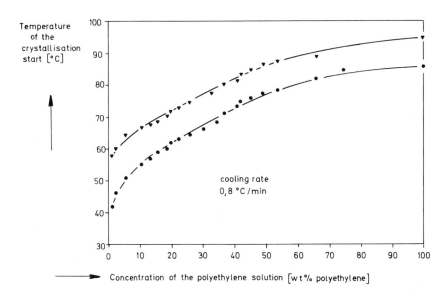

Figure 2.   Temperature of the start of crystallization as a function of polyethylene concentration in xylene solution
▼, *LDPE A; and* ●, *LDPE B*

Results

Styrene Grafting. In all phases of grafting under swelling conditions, the spherical polyethylene pellets retained their shape; the only changes were an increase in diameter as a function of the styrene content, and an increase in transparency as a function of the degree of grafting and the styrene content. From the increase in the transparency of the grafted pellets, one could conclude that there was an increase in compatibility with increasing grafting efficiency.

DIFFUSION PROCESSES. The rate of diffusion of styrene into LDPE was a complex function of the temperature (6, 7) and also of the crystallinity of the LDPE (8). As the temperature and content of solvents like xylene and styrene increased, the degree of crystallinity of polyethylene decreased (Figure 2).

In other words, when a homogeneous polyethylene solution was cooled at a constant rate, the point of beginning crystallization was reached. This point was indicated by the first visual turbidity. At all concentrations, homopolymer A had a point of beginning crystallization about 10°C higher than that of copolymer B. This finding seemed reasonable since the crystallization temperature depends on the comonomer content and on the degree of short-chain branching of the polyethylene. From the curve we concluded that the temperature at which styrene diffused into LDPE A was about 10°C above that for diffusion into LDPE B at a comparable rate of diffusion.

The homogeneity of styrene distribution in the swollen pellets after incipient styrene polymerization was another determinant of proper process conditions (Figure 3). With homopolyethylene A, for example, 24 hrs of swelling

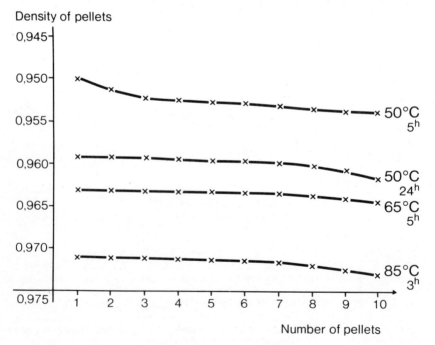

Figure 3.   Effect of swelling conditions on the styrene content of the pellets

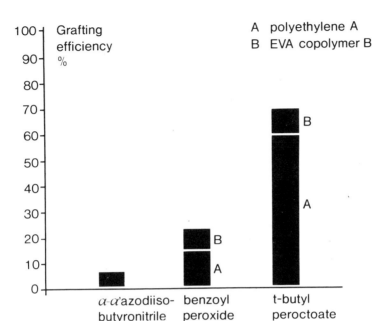

Figure 4.   Initiator and substrate activity in grafting styrene

at 50°C resulted in a much less uniform styrene distribution in the pellets than swelling at 85°C for 3 hrs. When swelling was accomplished at ambient temperature with subsequent polymerization at 60°–90°C, homogeneous styrene distribution in the pellets could not be obtained. The polystyrene was concentrated at the surface of the pellets forming an outer skin; the grafting efficiency was very poor. However, a homogeneous styrene distribution could be obtained by swelling at elevated temperature (65°–85°C, *see* Figure 3). This condition was essential for obtaining high grafting efficiencies.

In 1959, Hoffman *et al.* (9) reported the importance of the diffusion temperature in radiation-induced grafting of styrene under swelling conditions. They used the term "diffusion-controlled grafting," and they noted that when styrene was grafted stepwise (first swelling, then grafting), only the first step was diffusion-controlled. Of course, since the medium inside the pellets is highly viscous, the termination reaction should be diffusion-controlled. This is discussed below.

GRAFTING EFFICIENCY. In all the batches mentioned above, polymerization yields exceeded 90%. The graft products contained 10–45 wt % polystyrene. When the grafting efficiencies attained with the various peroxides under otherwise identical conditions were compared, the following activity scale was established: azodi(isobutyronitrile) (*1*) < benzoyl peroxide (*10, 11, 12*) < *tert*-butyl peroctoate (Figure 4).

Under identical reaction conditions, the vinyl acetate content of the backbone polymer had a slight effect on the grafting efficiency. The grafting efficiency with EVA was about 30 wt % greater than that with homopolyethylene A (Figure 4). Saponification of the EVA backbone did not change the polystyrene content of the graft copolymer; this meant that styrene was grafted mainly in the C-chain of the substrate molecule.

The factors that decisively affected the grafting efficiency in grafting styrene were: (*a*) the activity of the initiator radical (the predominant factor), and (*b*) the activity of the grafting substrate, *i.e.* the chemical composition, structure, and crystallinity of the substrate.

In order to investigate the mutual effects of the polyethylene, styrene, and initiator ratios on the grafting efficiency, the following experiments were performed.

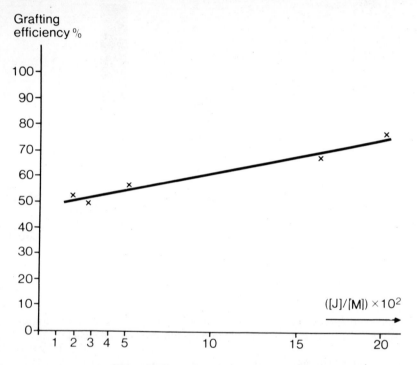

Figure 5. Grafting efficiency as a function of monomer concentration
[LDPE] and [I], constant; and [M], decreasing

(*a*) The styrene concentration was varied while the quantities of initiator and polyethylene A were kept constant. The grafting efficiency increased linearly as styrene concentration decreased (*see* Figure 5). This finding could be interpreted as follows: with decreasing styrene concentration, the probability of a transfer reaction between initiator radical and polymer chain increased, and hence the possibility of chain-growing started by substrate radicals also increased.

(*b*) The quantity of polyethylene was kept constant while the styrene and initiator concentrations were decreased in such a way that the ratio of their concentrations was kept constant. The grafting efficiency increased up to a practically constant value (Figure 6). This series of experiments suggested that, although the transfer reaction towards the substrate was promoted at first,

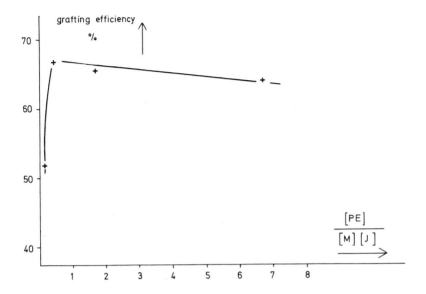

Figure 6.   Grafting efficiency as a function of initiator concentration
[LDPE], constant; [M] and [I], decreasing; and [M]/[I], constant

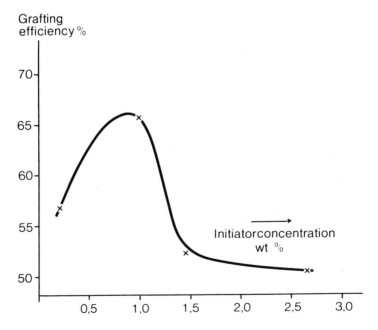

Figure 7.   Grafting efficiency as a function of initiator concentration
[LDPE] and [M], constant; and [I], increasing

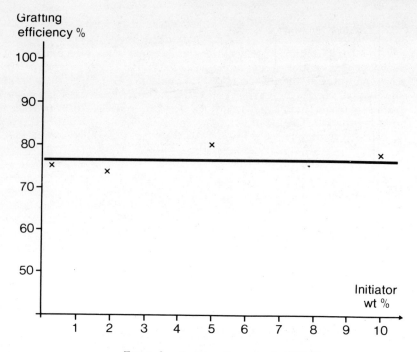

Figure 8.   Grafting styrene onto EVA B
Chain transfer agent. isobutene; [LDPE] and [M], constant; and [I], increasing

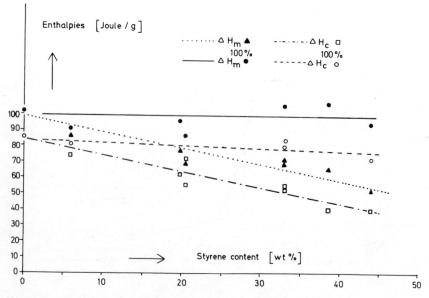

Figure 9.   Heat of melting (ΔH_m) and heat of crystallization (ΔH_c) vs. styrene content of LDPE A-g-styrene

its statistical frequency decreased as the result of the constant reduction in amount of initiator.

(*c*) The quantities of polyethylene and styrene were kept constant while the initiator concentration was increased. First an increase in grafting efficiency was observed and then a decrease (Figure 7). The decrease in grafting efficiency at very high initiator concentrations suggested an increase in the number of chain termination reactions between EVA polymer radicals and primary radicals. On the other hand, with increasing quantity of primary radicals, the starting reaction of styrene polymerization will be promoted.

(*d*) The grafting reaction was accomplished as in *c* but in the presence of a chain transfer agent such as isobutene. Grafting efficiency remained constant when initiator concentration was increased (Figure 8). This striking effect could be interpreted easily. Because of the mechanism of radical polymerization in the presence of chain transfer agents, the initiator first effected the formation of a regulator radical. With monoolefins as chain transfer agents, the regulator radical did not initiate chain-growing, but only the transfer reaction to the EVA substrate. When initiator concentration increased, the content of ungrafted substrate molecules decreased in accordance with the postulated transfer mechanism.

STRUCTURAL INVESTIGATIONS. The melting and crystallization points and the melting and crystallization enthalpies of styrene graft polymers with different styrene contents and different grafting efficiencies were determined by differential scanning calorimetry. The melting and crystallization points were

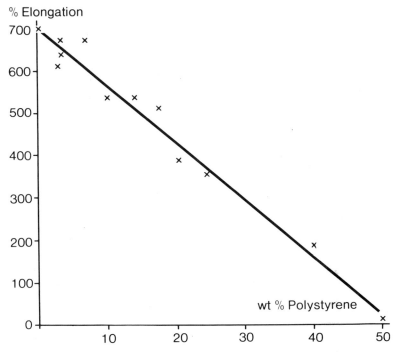

Figure 10. Grafting styrene onto EVA B—elongation at break vs. polystyrene content

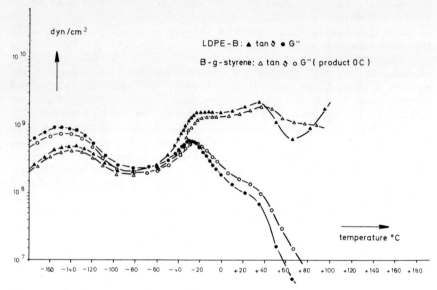

Figure 11. Dynamic mechanical behavior of LDPE B and LDPE B-g-styrene (Experiment OC in Table I)

The transitions were obtained from the maximum in G″ at about −25°C. The simultaneous transition at about −140°C is the γ-transition of polyethylene.

not affected by the grafting process (Figure 9). The levels of the crystallization and melting enthalpies of the graft polymers decreased with increasing polystyrene content. However, when the measured data were standardized to 100% polyethylene, no deviation from the melting enthalpy of the original polyethylene was detectable within the limit of error of this method. The grafting process had no appreciable effect on the crystallinity of the grafted substrate, and only a slight effect on post-crystallization. From the fact that the crystallization and melting enthalpies decreased linearly as styrene content increased, it was concluded that polystyrene was present as a virtually inert filler in a continuous polyethylene phase.

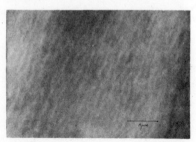

Figure 12. Electron photomicrograph of styrene grafted onto LDPE B (Experiment OC in Table I)

Styrene content, 20.4 wt %; and grafting efficieny, 76%

Figure 13. Electron photomicrograph of styrene grafted onto LDPE B (Experiment OA in Table I)

Styrene content, 20.2 wt %; grafting efficiency, 17%; and ungrafted polystyrene removed with THF

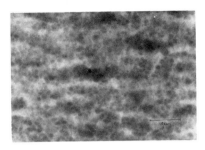

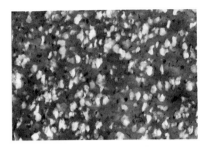

Figure 14.   Electron photomicrograph of styrene grafted onto LDPE A (Experiment A in Table I)
Styrene content, 43.7 wt %; and grafting efficiency, 11.0%

Figure 15.   Electron photomicrograph of graft product A at higher magnification after removal of ungrafted polystyrene with THF (cf. Figure 14)

   This conception was supported by the elongation measurements of the graft polymers (Figure 10). As styrene content increased, the maximum elongation decreased linearly, regardless of the grafting efficiency. This behavior is well known for ethylene polymers loaded with fillers (*13*). The data suggested a decrease in the entanglement between the polyethylene crystallites as a result of the surrounding, incompatible polystyrene molecules.

   In Figure 11, the dynamic mechanical behaviors of EVA B and of the graft product from Experiment OC (*see* Table I), which had a styrene content of 20.4 wt % and a grafting efficiency of 76.0%, are compared. The glass transition temperature T_g of the graft polymer was $-26°C$ whereas the T_g of EVA B was $-27°C$; therefore, the difference was not significant. The modulus of the graft was slightly changed from that of LDPE B, so a slight mutual effect of the polystyrene and polyethylene regions was reasonable.

   The morphology of the graft products was elucidated from electron photomicrographs of microtomed, thin sections of the grafts. In Figures 12 and 13, the grafting products had a styrene content of about 20 wt %, and EVA B was the substrate in the grafting reaction. Despite the widely different grafting efficiencies, only slight structural differences were apparent. When LDPE A was the substrate, it was reasonable to assume styrene occlusions in a polyethylene matrix (Figure 14). The styrene content of this graft polymer was about 40 wt %. After the ungrafted polystyrene units were removed with tetrahydrofuran (THF), the styrene occlusions were clearly visible (Figure 15). Since in this case the substrate was LDPE A which has a considerably greater tendency toward crystallization (Figure 2) and a crystallization temperature about 10°C higher than that of EVA B, the following could be assumed. Cooling to 50°C during the peroxide diffusion period resulted in the formation of crystalline regions which caused the styrene to accumulate in occlusions. When polymerization was started by an increase in temperature, the rate of styrene polymerization was faster than that of styrene diffusion through the slowly disappearing crystalline regions, and, hence, the styrene occlusions were retained. When EVA was the grafting substrate, the styrene occlusions did not form to the same extent as in product A because of the clearly lower crystallization degree, tendency, and temperature of EVA.

   Compatibility of Graft Copolymers. From the increase in the transparency of the grafted pellets with increasing degree of grafting, an increase in com-

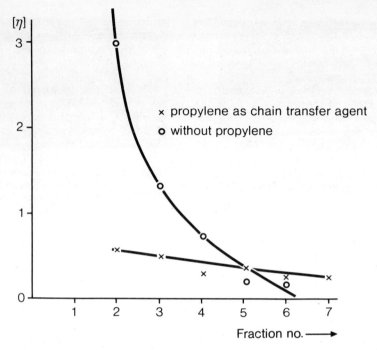

Figure 16. Intrinsic viscosities of different fractions of EVA-g-styrene-co-acrylonitrile

patibility with increasing grafting efficiency could also be concluded. A quantitative measure of polymer compatibility could be derived from measurements of light scattering from the determination of the second virial coefficient and the radii of gyration (5, 14, 15) as well as from the determination of the first visual turbidity point (5). Hence, the shorter the graft chains, the more compatible they were with the grafting base. When the ratio of monomer-to-grafting substrate and the grafting efficiency were kept constant, a decrease in the molecular weight of the graft chain increased the number of grafting sites. The resultant consequences for the property pattern of graft polymers could be exemplified by the grafting of styrene–acrylonitrile combinations onto an ethylene–vinyl acetate copolymer.

   Styrene–Acrylonitrile Grafting. Thus, the key to improving the compatibility of graft copolymers was not only an improvement in the grafting efficiency, but it was also control of the number and molecular weight of the graft chains. Although the molecular weight of the graft chains could be reduced by regulators such as mercaptans, a higher mercaptan concentration also resulted in a decreased grafting efficiency (16). An optimum system of regulators was found; it not only reduced the molecular weights of the graft chains but also increased their number through chain transfer to the polymeric substrate. These regulators were monoolefins such as ethylene, propylene, and isobutene. When a combination of styrene and acrylonitrile was grafted onto ethylene polymers in the presence of these monoolefins, the graft products were extremely homogeneous, as was proved by analysis of the graft products after preparative precipitation fractionation (Figure 16). Because these graft products have drastically reduced proportions of free styrene–acrylonitrile copolymer and because they have more graft chains with reduced molecular weight than do the graft products obtained without regulators, their compatibility was considerably improved. As a result of their homogeneity, their processing has been improved.

In order to obtain a better understanding of the relations between molecular weight and the compatibility of the grafting substrate, it was necessary to determine the molecular weight of the grafted chains of the graft copolymer. Such investigations could be facilitated by oxidative degradation of the backbone polymer and determination of the molecular weights and molecular weight distributions of the isolated graft chains when polybutadiene was used as the grafting substrate (*17, 18, 19*). Ethylene polymers, however, could not be subjected to such degradation reactions since this would lead unavoidably to degradation of the grafted chains too.

Molecular Weights of the Graft Chains. The molecular weights of the graft chains were determined by measuring the light scattering under those conditions when the grafting substrate was isorefractive with the solvent (*5*). This new method of determination was tried on EVA–polystyrene graft copolymers which were obtained by grafting approximately 20 wt % styrene onto LDPE B. When the polyethylene/styrene ratio was kept constant, the initiator concentration could be varied to obtain graft chains differing in length (Figure 17).

Evaluation of the anomalous Zimm diagrams for graft polymer solutions (*5*) revealed only a negligible deviation of the molecular weights of the graft

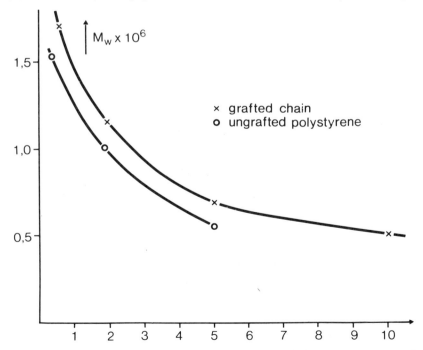

Figure 17. Molecular weight of grafted and ungrafted chains as a function of initiator concentration (wt %)

chains from those of the ungrafted polystyrene. In addition to the expected decrease in molecular weights as initiator concentration increased, there was a decrease in the amount of ungrafted polyethylene and an increase in the compatibility of the graft chains with the grafting substrate (5). Figure 18 represents the integral molecular weight distributions of the graft chains as determined by measuring the light scattering after preparative fractionation. The weight averages of the molecular weights calculated from these distributions agreed well with the values obtained for the ungrafted polystyrene as well as for that fraction from which only the ungrafted polystyrene was removed by fractionation. In addition, the molecular non-uniformities of the graft chains calculated from Figure 18 were the same as those determined for the ungrafted polystyrene by gel permeation chromatography (5). From the correspondence of the weight averages of the molecular weights and of the non-uniformities of the grafted and ungrafted polystyrenes, it was concluded that, at best, one graft chain was linked to one substrate molecule.

Conformation of the Graft Chains in Solution. The findings from differential thermal analysis (DTA) could be interpreted to mean that graft chains were practically not incorporated into the polyethylene crystalline regions. Information about the grafting site, *i.e.*, whether the grafting process occurred in fact within the polyethylene molecule, could be derived from the measured radii of gyration of the graft chains. As the molecular weight of the graft chains decreased, the values for the radii of gyration of the grafted polystyrene increased relative to those for the homopolystyrene (Figure 19). This increase could be interpreted as follows. The LDPE B used as the grafting substrate was highly long-chain-branched. Even in solution, this polyethylene molecule was rather compact. The graft copolymer (product OC in Table I) contained 20 wt % styrene. After removal of the ungrafted polystyrene and fractional precipitation of the graft copolymer, the radii of gyration of the fractions were measured. By the new method of light scattering measurements, only the polystyrene molecules were visible. The radii of gyration of the polystyrene molecules depend on their conformation in solution. Molecules with coiled conformation in solution have smaller radii of gyration than do stretched molecules with the same molecular weight.

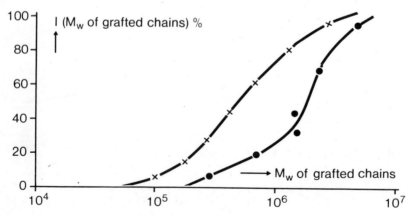

Figure 18. Molecular weight distributions of grafted chains
Initiator concentration: ×, 0.48 wt % and ●, 5.0 wt %

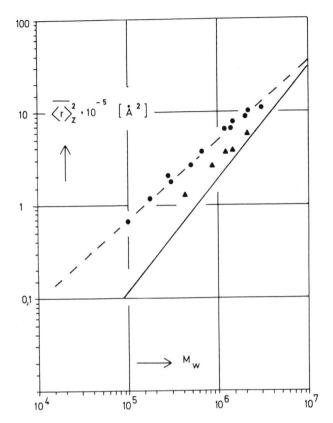

Figure 19.   Radii of gyration of polystyrene grafted onto different substrates
●, *LDPE B;* ▲, *EVA (vinyl acetate content, 45 wt %); and* ———, *homopolystyrene with a molecular non-uniformity of zero*

When the polystyrene graft chain was anchored to the substrate within the compact polyethylene molecule, the graft chain should have one section within and one section outside the polyethylene molecule. The conformations of these sections would be different. The inner section was forced to enter into as few interactions with the surrounding polyethylene molecule as possible, which should lead to stiffening of the inner polystyrene chain. The section of graft chain outside the polyethylene molecule would be free and movable, and it would have an ordinary, coiled conformation in solution. The measured value of gyration should be considered the combined value for an extended chain component and a coiled chain component. In order to explain the increasing deviation of the radii of gyration of the graft chains from those of homopolystyrene as the molecular weight decreased, we have to assume that the average length of the inner graft chain was constant. Then, as the molecular weight of the graft chain decreased, the coiled chain component of the radius of gyration decreased, too, and the extended chain component increased.

If this interpretation is correct, the effect of inner graft-chain stiffening should be reduced when the grafted polymer becomes more compatible with the grafted substrate and/or when the substrate molecule becomes less compact. Therefore styrene was grafted onto an ethylene–vinyl acetate copolymer (vinyl acetate content, 45 wt %) with a lower long-chain branching frequency (20) than LDPE and improved compatibility with polystyrene (5) by the same preparation procedure (III) as for grafting styrene onto LDPE B. After removal of ungrafted polystyrene and fractional precipitation of the EVA-g-styrene copolymer, the radii of gyration of the fractions were measured (see Figure 19). In fact, the radii of gyration of EVA-g-styrene, when compared with those of LDPE B-g-styrene, showed the expected smaller deviation from the radii of gyration of ungrafted polystyrene. Perhaps, this remaining deviation could be reduced if the radii of gyration of the graft chains would be calculated with corrections for the small non-uniformity. Furthermore, the little effect of molecules with two or more graft chains was neglected (5) in the calculations.

Conclusion

The objective of this study, namely improvement of the compatibility of graft polymers, as mentioned at the beginning of this paper, was reached. Contrary to the findings that the grafting of styrene–acrylonitrile onto EVA copolymers results in incompatible graft products—as was reported in 1968 (1) —an improved grafting technique now permits the production of graft thermoplastics with improved compatibility.

Acknowledgment

The authors are grateful to M. Hoffmann for interesting discussions on structural problems, to G. Kaempf and L. Morbitzer for investigations on the morphology of the graft copolymers, and to H. Kroemer for differential thermal analysis.

Literature Cited

1. Bartl, H., Hardt, D., ADVAN. CHEM. SER. (1969) **91**, 477.
2. Mayo, F. R., Walling, C., Chem. Rev. (1950) **46**, 191.
3. Bamford, C. H., Jenkins, A. D., Johnston, R., Trans. Faraday Soc. (1959) **55**, 41.
4. Burnett, G. M., Wright, W. W., Proc. Roy. Soc. London (1954) **A 221**, 41.
5. Kuhn, R., Alberts, H., Bartl, H., Makromol. Chem. (1974) **175**, 1471.
6. Bent, H. A., J. Polym. Sci. (1957) **24**, 387.
7. Pinsky, J., Mod. Plast. (1957) **April**, 145.
8. Fuhrmann, J., Diremeyer, M., Rehage, G., Ber. Bunsenges. Phys. Chem. (1970) **74**, 842.
9. Hoffman, A. S., Gilliland, E. R., Merrill, E. W., Stockmayer, W. H., J. Polym. Sci. (1959) **34**, 461.
10. Fischer, J. P., Angew. Makromol. Chem. (1973) **33**, 35.
11. Allen, P. W., Ayrey, G., Moore, C. G., J. Polym. Sci. (1959) **36**, 55.
12. Czvikovszky, T., Dobo, J., J. Polym. Sci. Part C (1967) **16**, 2973.
13. Miller, S. A., "Ethylene and Its Industrial Derivatives," p. 473, Ernest Benn, London, 1969.
14. Kuhn, R., Cantow, H.-J., Liang, S. B., Angew. Makromol. Chem. (1971) **18**, 93.
15. Kuhn, R., Bugdahl, V., Cantow, H.-J., Angew. Makromol. Chem. (1971) **18**, 109.
16. Hayes, R. A., J. Polym. Sci. (1953) **11**, 531.
17. Rieke, J. K., Hart, G. M., Saunders, F. L., J. Polym. Sci. Part C (1964) **4**, 589.
18. Locatelli, J. L., Riess, G., Angew. Makromol. Chem. (1973) **28**, 161.
19. Hoffmann, M., Pampus, G., Marwede, G., Kaut. Gummi Kunstst. (1969) **22**, 691.
20. Bartl, H., Kaut. Gummi Kunstst. (1972) **25**, 452.

RECEIVED April 3, 1974.

Transparent Acrylic/PVC–Graft/Blend Polymers

R. G. BAUER and M. S. GUILLOD[1]

Research Division, The Goodyear Tire & Rubber Co., Akron, Ohio 44316

Transparent graft/blend polymers were formed when acrylonitrile and 2-ethylhexyl acrylate were polymerized in homogeneous solution in the presence of poly(vinyl chloride). Physical properties of films cast from a solution blend of poly(vinyl chloride) and a copolymer of acrylonitrile and 2-ethylhexyl acrylate were generally inferior to those of the aforementioned graft/blends. A study of the morphology showed why the blend films were opaque while the graft/blends were transparent. Accelerated UV aging revealed that the graft/blend films remained serviceable approximately four times as long as the parent poly(vinyl chloride).

This study was initiated to find good aging transparent films for outdoor applications. Poly(vinyl chloride) (PVC) has several excellent physical attributes, and, no doubt for this reason, annual commercial production of PVC and copolymers has grown to 4.29 billion pounds (ranking 26th in volume in the chemical industry) (1). As Grassie (2) has stated, however, "of all the high tonnage polymers . . . PVC is by far the least stable in its pure state." Only the use of various stabilizers has made possible outdoor applications of PVC.

On the other hand, acrylic polymers have significantly better stability to UV degradation, and their addition to PVC to improve outdoor weathering has been suggested (3). Furthermore, copolymer films of acrylic esters and acrylonitrile were shown to have outstanding weathering properties, and their physical behavior is similar to that of plasticized PVC (4).

The prospects of combining the advantages of both polymer systems in a single polymer led to solution and melt blend studies. Although there was no gross evidence of incompatibility with the blend polymers, opaque films were obtained. This was not unexpected since refractive indexes [acrylic copolymer, $n_D^{20} = 1.49$; poly(vinyl chloride), $n_D^{20} = 1.54$] were not matched, and solubility parameters [acrylic copolymer, δ (calc.) $= 9.5$; poly(vinyl chloride), $\delta = 9.7$] differ slightly.

Both thermodynamic considerations (5) and microscopic evidence (6) suggest that when two different high-molecular-weight polymers are blended, a heterogeneous mixture will be formed. It has been suggested that the ultimate

[1] Present address: Alza Corp., 950 Page Mill Road, Palo Alto, Calif. 94304.

state of molecular mixing, characteristic of liquid mixtures, is only approached by polyblends as a limit (7).

Several investigators have looked at PVC blends with nitrile–butadiene rubber (6, 8), and, although compatibility increased with acrylonitrile content (up to about 40 wt %), micro-heterogeneity was still evident in electron photomicrographs.

Would copolymerization of the acrylic monomers in a homogeneous solution of PVC lead to graft reactions that might enhance the compatibility of the two polymer systems? Evidence has been presented that substantial grafting occurs on PVC in emulsion (9) or homogeneous solution (10), but not under heterogeneous conditions in methyl methacrylate which is not a suitable solvent for PVC (11). The preparation of graft copolymers on PVC by mechano-chemical reactions has been described by Guyot et al. (12, 13, 14). These same investigators have studied further the thermal degradation of these graft copolymers in some detail (15, 16). More recently, cationically polymerizable olefins were grafted onto PVC and the thermal behavior of the product has been the subject of some controversy (17, 18).

The present investigation focuses on polymers prepared by the reaction of the monomers 2-ethylhexyl acrylate and acrylonitrile on PVC in a homogeneous solution using free radical initiation.

Experimental

Monomer Purification. All polymers were obtained from monomers purified by passing them through Rohm and Haas Amberlyst exchange resins (salt forms).

Polymer Preparation. POLY(VINYL CHLORIDE). The PVC polymers were all commercial materials. Representative suspension grade PVC polymers are Goodyear's Pliovic K656 and K906 and Dow's 144. K656 (IV = 0.77) was used for the melt blends, and K906 (IV = 1.1) and 144 (IV = 1.0) were used for the solution-cast films.

ACRYLIC ESTER/ACRYLONITRILE COPOLYMERS. The preparation of these copolymers is described in Ref. 4.

GRAFT/BLEND POLYMERS. These polymers were prepared by dissolving PVC in either methyl ethyl ketone or tetrahydrofuran, followed by the appropriate 2-ethylhexyl acrylate/acrylonitrile monomer charge and the free radial initiator, e.g., Lupersol 11 (tert-butyl peroxypivalate). This solution was then heated to 50°–70°C in 4-oz bottles, 2-l. glass reactors, or larger glass-lined autoclaves. Polymerization times were normally 12–16 hrs; a conversion check was then made, and, if conversion was complete, films were cast from the polymer solution.

Physical Measurements. IMPACT STRENGTH. Impact strength was measured by the notched Izod technique at 23°C. Samples were conditioned at these temperatures after notching, prior to testing.

TRANSMISSION AND HAZE MEASUREMENTS. Light transmission was measured with a Gardner color difference meter, and haze was measured with a Hazemeter attachment on the Gardner colorimeter. Test pieces were films approximately 2 mils thick. All measurements were made at 23°C.

DIFFERENTIAL THERMAL ANALYSIS (DTA). These thermal profiles were obtained using a calorimeter cell with a DuPont 900 DTA instrument. Samples were cut from molded or cast films of these polymers, and they weighed 10–20 mg. A programmed heating rate of 10°C/min and a sensitivity of 0.2°/inch were used in this study. All thermal profiles were obtained with the sample flushed with an N_2 stream.

GEL PERMEATION CHROMATOGRAPHY (GPC). The GPC data were obtained with a Waters Associates Model 100 which was operated on tetrahydrofuran at room temperature.

OPTICAL AND ELECTRON MICROSCOPY. Samples of film for optical and electron microscopy were prepared by microtoming. The samples for optical phase contrast microscopy were approximately 15–20μ thick whereas those for electron microscopy were ultra microtomed with a diamond knife to about 0.05–0.1μ thickness. A Leitz Ortholux microscope was used for the phase contrast microscopy and an RCA-EMU-36 electron microscope at 40,000 × magnification was used for the electron microscopy.

Results and Discussion

Purely physical polymer blends are most commonly prepared by either mechanical mixing (melt) or dissolution in a common solvent followed by casting and solvent removal. In this study, both techniques were used; the latter method was more readily applicable for film formation in small-scale laboratory batches. It was recognized that certain morphological differences between melt- and solution-fabricated polymers are often observed; these include phase inversions and distortions, especially with graft and block polymers. However, casual observation by optical and electron microscopy revealed no dramatic differences between the melt- and solution-cast films, and this cannot be readily explained.

One noteworthy aspect of the blend study of polymers mixed in the melt pertained to optimization of impact strength relative to acrylonitrile content in the copolymer. For this study, the polymer blends were compression-molded and tested with a Tinius Olsen Izod Tester. The composition of the acrylic copolymers and the blend composition are given in Table I. The change in

Table I.   Blend Composition of Acrylic Ester/Acrylonitrile Copolymers with PVC

Constituent	Content, %
Pliovic K656 PVC	83.7
Copolymer[a]	12.5
PVC stabilizer	1.84
Lubricant	1.96
Total	100.00%

[a] Acrylonitrile/2-ethylhexyl acrylate copolymers were of the following compositions, by wt ratio: 0/100, 10/90, 12/88, 15/85, 17/83, 20/80, 22/78, 25/75, and 30/70.

impact strength with acrylonitrile content is shown in Figure 1; it can be seen that impact strength is at a maximum at about 20–30 wt % acrylonitrile in the copolymer.

With this background information, it was decided to see how these blends would behave as films. In Table II are tabulated some physical properties of extruded films of PVC/acrylic copolymer blends (9/1, 1/1, and 1/9). In Table III are tabulated physical properties of solution-cast blend films, cast from tetrahydrofuran solution. The cast films contained no lubricants or stabilizers (which were present in the extruded films); nevertheless, their optical property deficiencies were similar to those of the extruded films. The additives used were: PVC stabilizer (*e.g.*, organotin compounds), lubricant (*e.g.*, metallic stearates), and plasticizer (phthalate type).

The question then arose: what would be the nature of the polyblend which could be prepared if the acrylic monomers were polymerized in the

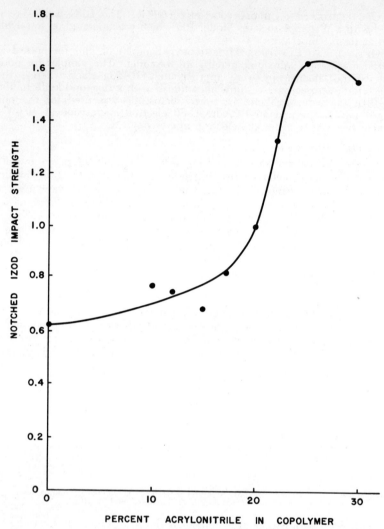

Figure 1.   Notched Izod of PVC/acrylic blends vs. *concentration of acrylonitrile in the acrylic*

presence of PVC? It was necessary to find a suitable solvent medium for PVC
and acrylic copolymer. Both methyl ethyl ketone and tetrahydrofuran were
found to be suitable for preparing 10–15% solutions of PVC/acrylic graft/
blend after more than 10 hrs reaction at 50°–70°C, using 1–12 parts of an
organic free radical initiator (based on 100 parts monomer charge). Typically,
the weight ratio of acrylic monomers to PVC was varied from 1/1 to 1/9 since
insolubilization invariably occurred when the ratio of acrylic monomer exceeded 1/1. Although the weight ratio of acrylic ester (2-ethylhexyl acrylate)
to acrylonitrile was varied from 0.43/1.0 to 2.34/1.0, the optimum for film
strength properties appeared to be *ca.* 1.5/1.0.

Table II. Properties of Extruded Film of PVC/Acrylic Copolymer Melt Blends

Property	Blend		
	1	*2*	*3*
Composition: PVC/(AN/2-EHA)[a]	9/1	1/1	1/9
Lubricant	0.1	0.5	0.5
Stabilizer	2.05	1.7	1.1
Plasticizer	10.0	5.0	5.0
Light transmission, %	88.7	83.2	83.0
Haze, %	19.9	52.9	44.9
Tensile strength, psi	2850	3044	2254
Ultimate elongation, %	5.0	115	120
Drop height at $-20°F$, in.	3	6	20
Crescent tear, lbs/in.	420	485	183

[a] AN/2-EHA: acrylonitrile/2-ethylhexyl acrylate, 1.0/1.5 wt %.

Unexpectedly, when films were cast from these poly(vinyl chloride-g-2-ethylhexyl acrylate/acrylonitrile) solutions, they were crystal clear with very low haze values. Table IV lists some of the physical properties of these poly(vinyl chloride-g-2-ethylhexyl acrylate/acrylonitrile) cast films; values for a solution blend of the acrylic copolymer and PVC are also included. While tensile strengths of the 1.0/1.5 and 1.0/4.0 acrylic copolymer/PVC–graft/blend films were about 80–90% that of the homopolymer PVC film, they were significantly higher than that of the blend polymer. Furthermore, the crescent tear strengths were higher than those of the blend and the PVC film. Most importantly, the haze values of the graft/blend films were much improved over that of the blend polymer, and they more nearly approached that of the homopolymer PVC.

Similar polymer compatibilization effects were observed by Wellons and co-workers (*19*) on radiation graft copolymers of cellulose acetate and polystyrene and by Riess and his colleagues (*20*) on various block copolymers. Hughes and Brown (*21*) also reported some evidence of compatibilization in a

Table III. Properties of Solution-Cast PVC/Acrylic Copolymer Blends

Property	Blend		
	1	*2*	*3*
Composition: PVC/(AN/2-EHA)[a]	9/1	1/1	1/9
Tensile strength, psi	3400	3300	1200
Elongation, %	100	100	200
Crescent tear, lbs/in.	150	425	800
Light transmission, %	93	93	93
Haze, %	23	40	20

[a] AN/2-EHA: acrylonitrile/2-ethylhexyl acrylate, 1.0/1.5 wt %.

polymer solution study in which a single hazy phase existed from a graft polymer whereas the physical mixture of homopolymers separated into two phases in a common solvent.

What is the evidence for the occurrence of graft reactions of these acrylic monomers onto PVC? How would this grafting, if it occurred, affect the compatibility of the two polymer system? These and other perplexing questions led to several experiments which were intended to clear up the uncertainty.

A DTA thermogram between $-100°$ and $+140°C$ (Figure 2) of the 1.5/1.0-PVC/acrylic copolymer (1.5/1.0-2EHA/AN) graft/blend showed a

Table IV.   Physical Properties of Cast Films

Description[a]	Tensile Properties				Tear Strength		Optical Properties	
	Yield,		Break,		Cres-cent, lbs/in.	Elmen-dorf, g/mil	Trans., %	Haze, %
	psi	%	psi	%				
1.5/1.0 PVC/acrylic copolymer solution blend	4625	9	3640	60	159	10	93	49
1.5/1.0 PVC/acrylic solution graft/blend	—	—	5190	5.5	192	11	93	4.0
4.0/1.0 PVC/acrylic solution graft/blend	—	—	5760	6.0	228	20	93	1.0
PVC hompolymer	—	—	6250	6.0	104	18	93	1.5
Acrylic copolymer	—	—	2530	300	105	137	93	2.1

[a] Acrylic copolymers were 1.5/1.0 (wt ratio) 2-ethylhexyl acrylate/acrylonitrile charge. All films were 2.0 mils thick, cast on glass plates with a precision draw-down blade.

single inflection at about 85°C which presumably was that of the graft PVC. There was no indication of any inflection, characteristic of a glass transition for the acrylic copolymer, which would have been expected at about −40°C if this were a blend.

Furthermore, a GPC curve (Figure 3), obtained on this 1.5/1.0-PVC/ acrylic copolymer (1.5/1.0-2EHA/AN) graft/blend showed only a single peak, although it was considerably broader than that of the parent homopolymer PVC which had an MWD of 2.1.

Several of these poly(vinyl chloride-g-2-ethylhexyl acrylate/acrylonitrile) films were examined by phase contrast light microscopy, and none showed evidence of two phases, i.e., there were no suspended phases larger than 0.25–0.50μ.

Next, several of these poly(vinyl chloride-g-2-ethylhexyl acrylate/acrylonitrile) polymers and a solution blend copolymer were examined by electron microscopy (see Figures 4, 5, and 6). As would be expected, the blend polymer (Figure 4) showed gross irregular shaped domains much larger than 0.25μ. Figures 5 and 6 are electron photomicrographs of poly(vinyl chloride-g-2-ethylhexyl acrylate/acrylonitrile) with essentially all of the domain sizes smaller than 0.25μ. Consequently, the graft/blend polymers would be expected to be transparent whereas the solution blends would scatter light and appear turbid.

Recognizing the difficulties which one could expect to have with a fractionation or extraction procedure for analyzing these polymers, we searched for suitable solvent–nonsolvents for the PVC and acrylic copolymers. Applica-

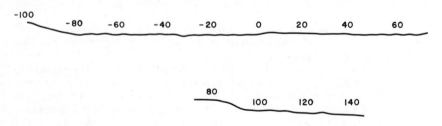

Figure 2.   DTA thermogram of 1.5/1.0–PVC/acrylic copolymer graft/blend

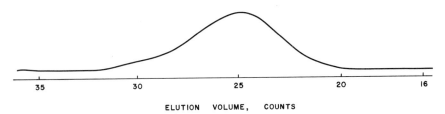

ELUTION   VOLUME,   COUNTS

Figure 3.   Molecular weight distribution curve for poly(vinyl chloride-g-2-ethylhexyl acrylate/acrylonitrile)

tion of the fractional separation procedures of Smets and Claesen (*11*) was inconclusive as the addition of nonsolvent to the graft/blend polymer solution led to the formation of a gelatinous mixture which was too well emulsified to isolate. Molau (*22*) pointed out that isolation of graft copolymers by means of a fractional precipitation technique from a reaction mixture containing the

Figure 4.   Electron photomicrograph of 1.5/1.0 blend of PVC and acrylic copolymer (1.5/1.0 2-EHA/AN)

Figure 5. Electron photomicrograph of 1.0/4.0–acrylic copolymer (1.5/1.0 2-EHA/AN)/PVC–graft/blend

two parent homopolymers is extremely laborious. Ikada and co-workers (23) recently confirmed this with graft copolymers of poly(vinyl alcohol-g-methyl methacrylate) and poly(vinyl acetate-g-methyl methacrylate); they observed that a stable colloidal suspension was formed which resisted aggregation during isolation.

A Soxhlet-type extraction was partially successful in separating the ungrafted polymer from the grafted PVC. Examination of all conventional solvents for the acrylic copolymer which were complete nonsolvents for PVC was not especially fruitful. Chloroform completely dissolved the acrylic copolymer, but it had only limited solubilization of the PVC (8.5% at 23°C, 21.2% soluble at reflux temperature). A Soxhlet extraction, under N_2, on a 1.5/1.0-PVC/acrylic copolymer (1.5/1.0-2EHA/AN) graft/blend yielded an insoluble fraction (about 67% of the graft/blend polymer) which had IR absorption bands in the ester carbonyl region (1750–1700 cm^{-1}); this might indicate grafted 2-ethylhexyl acrylate. This band was absent from a similarly treated homopolymer PVC sample. The presence of ester oxygen in the graft/blend polymer was confirmed by oxygen analysis. There was no evidence for the presence of a —C≡N nitrile group in IR spectrum of the graft/blend sample. Since an appreciable fraction of the PVC homopolymer was soluble in refluxing chloroform, there can be no explanation for the absence of the nitrile group in the insoluble fraction, nor can there be a quantitative measure of graft efficiency.

In an effort to exclude the possibility that conditions prevailing during the grafting reaction (Reaction B) would produce ester oxygen sites on the PVC backbone, control reactions were run in a non-oxygen-containing solvent (1,2-dichloroethane) instead of methyl ethyl ketone (Reaction C) and in the absence of the free radical initiator, *tert*-butyl peroxypivalate (Reaction A) (*see* Table V). Each of these treated PVC homopolymers was then Soxhlet extracted with chloroform, and the insoluble fraction was examined by IR. There was no absorption in the carbonyl ester region or evidence of other foreign compounds; hence, the solvent and initiator were ruled out as sources of carbonyl oxygen in the PVC graft. One other explanation for the presence

Figure 6.   Electron photomicrograph of 1.0/1.5–acrylic copolymer (1.5/1.0 2-EHA/AN)/PVC–graft/blend

Table V.   Controls for Oxygen Content of Graft/Blend Polymers[a]

	Reaction		
Material	A	B	C
Methyl ethyl ketone, reagent grade	110	110	—
1,2-Dichloroethane, reagent grade	—	—	130
Pliovic K906 PVC	7.5	7.5	7.5
Lupersol 11 (tert-butyl peroxypivalate, 75% act.)	—	0.95	0.95

[a] All three experimentals were prepared by first dissolving PVC in solvent at room temperature, then treating the solution under N₂ at 60°C for 24 hrs. Finally, the treated PVC was isolated and Soxhlet extracted with chloroform under N₂ for 36 hrs.

of carbonyl ester groups in the Soxhlet insolubles which cannot be precluded would be entrainment of acrylic ester copolymer in the PVC matrix because of polymer entanglements.

It was observed that these transparent 1.5/1.0-PVC/acrylic copolymer (1.5/1.0-2EHA/AN)-graft/blend films were somewhat more resistant to degradation from UV exposure than unstabilized homopolymer PVC (see Table VI). The graft/blend films retained clarity and were noticeably less prone to shrinkage in the Fadeometer where the service life was extended from 500 hrs for the control to 2000 hrs for the graft/blend films.

A DTA profile to 400°C to the transparent 1.5/1.0-PVC/acrylic copolymer (1.5/1.0-2EHA/AN) indicated that while the decomposition mechanism may be somewhat altered by this graft procedure (there was no strong exotherm), the onset of degradation as indicated by an endotherm occurred at about the same temperature as for the homopolymer (see Figure 7). This appears to be contrary to the observations of Gaylord and Takahashi (17) and Thame et al. (18) on cationic graft polymers since they observed an increase in thermal stability with graft polymers. Presumably, a free radical graft reaction would occur to a lesser extent at the labile allylic and tertiary chlorine atoms than would a cationic alkylation or graft polymerization reaction. It is more likely that the α-hydrogen on the PVC backbone is the preferred site for hydrogen atom abstraction occurring with the free radical graft reaction. Hence, in the free radical graft product, graft sites would be more prevalent at the α-hydrogen position and less prevalent at the tertiary or allylic carbon–chlorine site than in the ionic graft products. This could account for differences in thermal stability of the various graft products.

Table VI.   Effect of Carbon Arc Fadeometer Exposure on Appearance of Films

	Time of Exposure, hrs			
Polymer	0	500	1500	2000
Dow 144 PVC	transparent	white, opaque	opaque, failed[a]	—
Dow 144 PVC (1 phr Cyasorb UV531 added)	transparent	transparent	transparent	transparent, failed[a]
Dow 144 PVC/acrylic copolymer graft/blend[b]	transparent	transparent	transparent	transparent, failed[a]
Dow 144 PVC/acrylic copolymer graft/blend[b] (1 phr Cyasorb UV531 added)	transparent	transparent	transparent	transparent, failed[a]

[a] Failure indicates that the film could not withstand a simple 90° flex test without fracture.
[b] The acrylic copolymer was a 1.5/1.0 2-ethylhexyl acrylate/acrylonitrile charged ratio, where PVC/acrylic copolymer was in the ratio of 1.5/1.0.

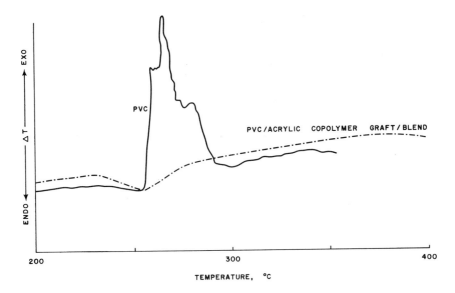

Figure 7.   Differential thermal analysis (10°C/min in N₂) of PVC and poly(vinyl chloride-g-2-ethylhexyl acrylate/acrylonitrile)

A more thorough examination of the thermal behavior of these poly(vinyl chloride-g-2-ethylhexyl acrylate/acrylonitrile) polymers would be appropriate to elucidate further the mechanism of PVC degradation.

Conclusions

A transparent PVC–acrylic copolymer product can be formed by polymerizing 2-ethylhexyl acrylate and acrylonitrile in the presence of a homogeneous solution of PVC using a free radical initiator. In contrast, products formed by the solution blending of PVC and a copolymer of 2-ethylhexyl acrylate and acrylonitrile were translucent to opaque.

The transparent PVC–acrylic graft/blend products are somewhat more resistant to UV degradation than the parent PVC polymer and have physical properties which are of interest in film applications. These properties include improved tear strength over the PVC parent polymer and somewhat higher tensiles than are obtainable with the acrylic copolymer.

Acknowledgments

The authors thank Goodyear Tire and Rubber Co. for permission to publish this paper, and Goodyear Research Microscopy Section for assistance in the light and electron microscopy studies.

Literature Cited

1. Anderson, E. V., "Top 50 Chemicals Reflect Industry Recovery," *Chem. Eng. News* (1973) May 7, 8–10.
2. Grassie, N., "Polymer Science," A. D. Jenkins, Ed., Vol. 2, p. 1495, North Holland, Amsterdam and London, 1972.

3. Deanin, R. O., Orroth, S. A., Eliasen, R. W., Greer, T. N., *Polym. Eng. Sci.* (1970) **10** (4), 228.
4. Bauer, R. G., Wathen, T. M., Mast, W. C., *Amer. Chem. Soc., Div. Polym. Chem., Prepr.* **14** (1), 606 (Detroit, May, 1973).
5. Hughes, L. J., Britt, G. E., *J. Appl. Polym. Sci.* (1961) **5**, 337.
6. Matsuo, M., Nozaki, C., Jyo, Y., *Polym. Eng. Sci.* (1969) **9**, 197.
7. Yu, A. J., Advan. Chem. Ser. (1970) **99**, 2.
8. Horvath, J. W., Wilson, W. A., Lundstrom, H. S., Purdon, J. R., *Appl. Polym. Symp.* (1968) **7**, 95–126.
9. Hayes, R. A., *J. Polym. Sci.* (1953) **11** (6), 531.
10. Prabhakara Rao, S., Santappa, M., *J. Polym. Sci. Part A-1* (1967) **5**, 2681.
11. Smets, G., Claesen, M., *J. Polym. Sci.* (1952) **8**, 289.
12. Guyot, A., Michel, A., *J. Appl. Polym. Sci.* (1969) **13**, 911.
13. Michel, A., Galin, M., Guyot, A., *J. Appl. Polym. Sci.* (1969) **13**, 929.
14. Michel, A., Bert, M., Guyot, A., *J. Appl. Polym. Sci.* (1969) **13**, 945.
15. McNeill, I. C., Neil, D., Guyot, A., Bert, M., Michel, A., *Eur. Polym. J.* (1971) **7**, 453.
16. Guyot, A., Bert, M., Michel, A., McNeill, I. C., *Eur. Polym. J.* (1971) **7**, 471.
17. Gaylord, N. G., Takahashi, A., Advan. Chem. Ser. (1971) **99**, 302.
18. Thame, N. G., Lundberg, R. D., Kennedy, J. P., *J. Polym. Sci. Part A-1* (1972) **10**, 2507-2525.
19. Wellons, J. D., Williams, J. L., Stannett, V., *J. Polym. Sci. Part A-1* (1967) **5**, 1341.
20. Riess, G., Kohler, J., Tournut, C., Banderet, A., *Makromol. Chem.* (1967) **101**, 58.
21. Hughes, L. J., Brown, G. L., *J. Appl. Polym. Sci.* (1963) **7**, 59.
22. Molau, G. E., "Characterization of Macromolecular Structure," D. McIntire, Ed., p. 245, National Research Council, National Academy of Sciences, Washington, D.C., 1968.
23. Ikada, Y., Horii, F., Sakurada, I., *J. Polym. Sci.* (1971) **11**, 27.

Received February 20, 1974. Work supported by Contribution #524 from the Goodyear Tire and Rubber Co. Research Laboratory.

22

Rubber-Modified Polymers. Location of Block Copolymers in Two-Phase Materials

G. RIESS and Y. JOLIVET

Ecole Supérieure de Chimie, 3 rue A. Werner, 68093 Mulhouse Cédex, France

In heterophase polymeric materials, outstanding mechanical properties are obtainable only by regulating dispersed phase particle size and adhesion between phases, usually by adding block or graft copolymers. The emulsifying effect of block and graft copolymers was demonstrated for the PS–PI system. Since characteristics of the blend polymers (e.g., molecular weight and composition) determine whether the copolymer is situated in the continuous phase, in the dispersed phase, or at the interface, we tried to establish the conditions for the last. Different techniques for locating a PS–PI block copolymer were studied: use of a copolymer containing a fluorescent group, x-ray scanning microanalysis, and quantitative analysis of the gel formed after crosslinking the elastomeric phase by γ-irradiation. For these PS–PI blends, the amount of copolymer at the interface correlated with impact resistance.

In heterophase polymeric materials such as rubber modified polystyrene or acrylonitrile–butadiene–styrene (ABS) resins, outstanding mechanical properties can be obtained only by regulating the dispersed rubber particle size and by achieving adhesion between the rubber and the resin phase. This can usually be achieved by adding block or graft copolymers, or by their formation *in situ*, as in industry.

In this paper we report on interfacial adhesion and emulsifying properties in relation to locating block copolymers in two-phase materials. The polystyrene (PS)–polyisoprene (PI) system was studied in order to locate PS–PI block copolymers and to define the conditions under which these copolymers are situated at the resin–rubber interface.

Emulsifying Effect

Oil-in-Oil Emulsions. We demonstrated previously that block and graft copolymers act as oil-in-oil emulsifiers for the corresponding incompatible homopolymers, *e.g.*, PS and PI (*1, 2, 3*). To prove that this emulsifying effect of AB block and graft copolymers is general, we studied the simpler case of two nonmiscible liquids a and b which do not have the inconvenient high viscosity typical of polymeric systems. Furthermore, a is a good solvent for component A of the copolymer and a nonsolvent for component B; conversely, b is a solvent for component B and a nonsolvent for component A. For PS–PI block copolymers, these requirements are met by the system dimethylforma-mide (DMF)–hexane as nonmiscible solvents (4, 5).

The emulsifying effect of a copolymer can be characterized by determining the type of emulsion (DMF in hexane or hexane in DMF), its stability, its viscosity, and the particle size of the dispersed phase. These characteristics of oil-in-oil emulsions obtained with PS–PI block copolymers were studied as functions of solvent volume ratio, molecular weight, composition, and structure of the copolymer (5). Although Bancroft's rule was established for conven-tional oil–water emulsions, it appears to apply also to oil-in-oil emulsions—the continuous phase of the emulsion is preferentially formed by the solvent having the best solubility for the emulsifier (6, 7). Thus, block or graft copolymers can be prepared giving hexane/DMF, DMF/hexane, or both types of emulsions.

For high molecular weight block copolymers, the emulsifying efficiency seems to be greater for two block copolymers than for three, and efficiency is maximum when the two blocks, PS and PI, are of similar molecular weight. Under these conditions, copolymer solubility is reduced, and most of it is located at the interface. For other compositions, and especially for low molecu-lar weights, the copolymer can form micelles in either solvent phase, and its emulsifying efficiency is thereby greatly decreased. A block or graft copolymer is efficient as emulsifier if it is situated mainly at the interface, i.e., there is no preferential solubility in either phase. Regulating the molecular characteristics of the copolymer (composition, molecular weight, and structure) can alter its solubility and therefore its efficiency. PS–PI block copolymers with the lowest association degree (micellization), in hexane as well as in DMF, are most efficient as oil-in-oil emulsifiers for the DMF–hexane system (8).

Emulsifying Effect of Copolymers in the Solid State. The next step was to ascertain if the findings for two nonmiscible liquids are applicable to two incompatible polymers such as PS and PI, and if a block copolymer also functions as an emulsifier.

A relatively simple method of studying the dispersion degree for polymers in the solid state is to define the limits where films made by solvent evaporation of polymer mixtures pass from hazy to clear. Films of a two-phase system may appear hazy if the average particle size of the dispersed phase is greater than 800–1000 A, or roughly 1/5 the wavelength of visible light ($\lambda/5$), and also if the refractive indexes of the two polymers are different. Therefore, if the polyblend film is clear, the dimensions of the dispersed phase are less than 800–1000 A, and only by electron microscopy can a two-phase system be distinguished from a compatible blend (9). So the emulsifying effect of a copolymer for polymers in the solid state is apparent if the copolymer can transform a hazy blend into a clear one.

The film preparation technique has been described elsewhere (3). It should be noted that films are observed directly and by phase contrast micros-copy. Thus, we can compare the emulsifying effect of different styrene–isoprene copolymers (random, block, and graft copolymers) of the same overall com-position (40 wt % PS, 60 wt % PI) and practically the same molecular weight ($\overline{M}_n \simeq 50,000$) in a given PS–PI blend, where $\overline{M}_n$ of PS is 45,000 and $\overline{M}_n$ of PI is 25,000. The appearance of the films obtained with different blend compo-sitions is depicted in Figure 1. Hazy or opaque areas are striped. From Figure 1, it appears that random copolymers always cause hazy films which means

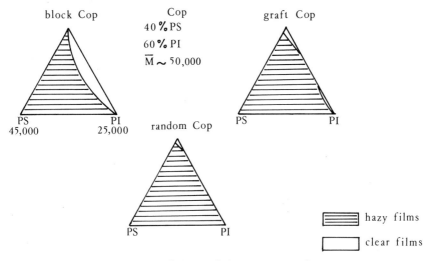

Figure 1. Emulsifying effect of PS–PI copolymers

that this type copolymer has no emulsifying effect. This agrees with the find-ings for oil–oil emulsions. On the other hand, the fact that block and graft copolymers at some polyblend compositions do give transparent films proves that these copolymers act as oil-in-oil emulsifiers in the solid state, and that block copolymers are generally more efficient than graft copolymers.

On the basis of a systematic study of the emulsifying effect of block copolymers in PS–PI and in polystyrene–poly(methyl methacrylate) (PS–PMM) polyblends (*3*), it was possible to represent schematically the appear-ance of the films for different blend compositions as functions of molecular weight and composition of the block copolymer (Cop), as well as of molecular weight of the homopolymers (*see* Figure 2). Thus in a polyblend containing PS and PI of practically the same molecular weight ($\overline{M}_1 \simeq \overline{M}_2$), the best emulsifying properties are obtained with a two block copolymer whose com-position is about 50:50 and whose molecular weight is higher than those of the homopolymers ($\overline{M}_3 > \overline{M}_1 \simeq \overline{M}_2$). In fact the area where transparent films

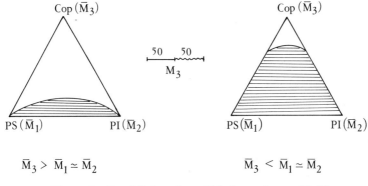

Figure 2. Emulsifying effect of block copolymers PS–PI

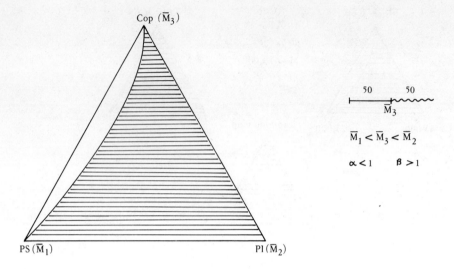

Figure 3.   Emulsifying effect of block copolymers PS–PI

are obtained is small when $\overline{M}_3 < \overline{M}_1 \simeq \overline{M}_2$. Both PI/PS and PS/PI emulsions are obtained, which means that the copolymer has no preferential solubility in either phase.

A given Cop is therefore a better emulsifier in a polyblend if smaller amounts are required to obtain transparent films, *i.e.*, small particle size of the dispersed phase. In order to compare the emulsifying efficiencies of different polyblends, it is convenient to define the ratio of the molecular weights of the homopolymer and the corresponding component of the block copolymer as follows:

$$\alpha = \frac{\overline{M} \text{ of PS homopolymer}}{\overline{M} \text{ of PS component of copolymer}} \qquad (1)$$

$$\beta = \frac{\overline{M} \text{ of PI homopolymer}}{\overline{M} \text{ of PI component of copolymer}} \qquad (2)$$

when $\overline{M}_3 > \overline{M}_1 \simeq \overline{M}_2$, $\alpha < 1$ and $\beta < 1$. When $\overline{M}_3 < \overline{M}_1 \simeq \overline{M}_2$, $\alpha > 1$ and $\beta > 1$. Thus, Cop is less efficient as an emulsifier if $\alpha > 1$ and $\beta > 1$. It should be noted, however, that this distinction is only semiquantitative because the limits between hazy and clear areas correspond to a given but arbitrary particle size of the dispersed phase. Nevertheless, it is important to note that the emulsifying effect of Cop depends on the molecular weight of the different species in the polyblend. Furthermore, the symmetrical evolution of the emulsifying effect if $\overline{M}_1 \simeq \overline{M}_2$ in the presence of a balanced copolymer (copolymer whose composition is $\sim$ 50:50), which means that $\alpha \simeq \beta$, indicates that there is no preferential solubility in either phase.

On the other hand, only a small amount of PI can be emulsified by changing the molecular weight of one of the homopolymers, *e.g.*, $\overline{M}_1 < \overline{M}_3 < \overline{M}_2$. In this case, $\alpha < 1$ and $\beta > 1$. With such polyblends, the emulsifying effect is best when PS forms the continuous phase (*see* Figure 3). According to Ban-

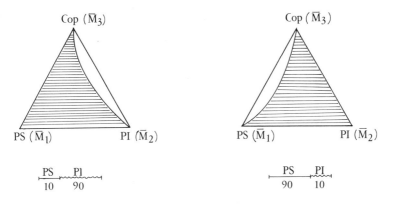

Figure 4.   Emulsifying effect of block copolymers PS–PI

croft's rule $(6, 7)$, this may indicate that Cop has a given solubility, certainly as a micellar dispersion, in the PS phase. The reverse situation, solubility in the PI phase, is achieved if $\alpha > 1$ and $\beta < 1$.

The effect is similar with block copolymers whose compositions are quite different from 50:50. Figure 4 demonstrates that with copolymers containing a large amount of PI (90%) and if $\alpha > 1$ and $\beta < 1$, only blends containing

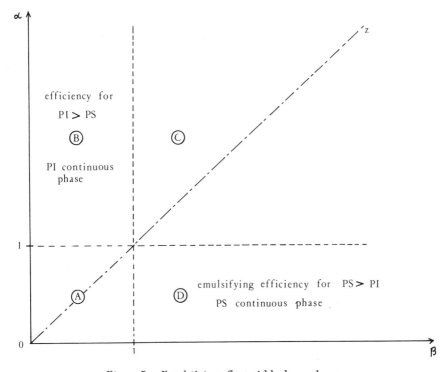

Figure 5.   Emulsifying effect of block copolymer

mainly PI are well emulsified. Since PI is the continuous phase of the blend, this would mean, again according to Bancroft's rule, that Cop is more soluble in PI than in PS. The reverse occurs with copolymers containing mainly PS (90%) if $\alpha < 1$ and $\beta > 1$.

The emulsifying effect of two block copolymers, based on the observation of films made by solvent evaporation, can be summarized in the plot of α vs. β (Figure 5). In this diagram we can distinguish four areas. The emulsifying properties of Cop seem to be best in area A and especially along the Oz axis where Cop may not have preferential solubility in either phase. Therefore, it may be concluded that for $\alpha = \beta < 1$, a certain amount of Cop is at the interface of the two phases. Although the emulsifying effect of Cop is less in area C ($\alpha > 1$, $\beta > 1$), on the Oz axis a certain amount of Cop is also at the interface. In area B ($\alpha > 1$, $\beta < 1$), however, only a small amount of PS is emulsified when PI forms the continuous phase of the blend; therefore, according to Bancroft's rule, Cop may have a certain solubility, probably as a micellar dispersion in the PI phase. The reverse is observed in area D where $\alpha < 1$ and $\beta > 1$; emulsification of PS is better than that of PI when PS forms the continuous phase of the blend.

Inasmuch as the emulsifying effect was studied semiquantitatively by a given film technique, the boundaries between areas may not be as sharp as in Figure 5. Area size may depend on polyblend type and on the technique used for blending the different polymers. Although study of the emulsifying effect of block copolymers in the PS–PI and PS–PMM systems has led qualitatively to similar conclusions, the areas defined in Figure 5 cannot be superimposed exactly for these two systems. With the film technique, which seems to be the most reproducible, a shift of the different boundaries can also be observed when various solvents are used for film preparation, especially preferential solvents for either polymer (10).

It can be concluded that the best emulsifying properties are obtained with two-block copolymers, and that these are superior to three-block copolymers and to graft copolymers. Furthermore, a two-block copolymer is most efficient if its composition is about 50:50 and if its molecular weight is higher than those of the corresponding homopolymers. These results agree with those obtained recently by Skoulios and co-workers (11). An explanation of these phenomena by Meier is based on the different precipitation rates of the various polymers during solvent evaporation (12).

Mechanical Properties of Two-Phase Polymer Blends

Study of impact-resistant polymer blends has elucidated the fundamental role of block and graft copolymers (Cop) in a typical system like PS–PI–Cop (1, 2) and also in the case of two-phase materials formed by two resins

Table I.   Impact Resistance in Binary PS–PI Blends

PS		PI		
$\overline{\mathrm{M}}_w$	In Blend, %	$\overline{\mathrm{M}}_w$	In Blend, %	R[a]
91,000	87.5	50,000	12.5	0.6
420,000	87.5	50,000	12.5	2.0

[a] Impact resistance, R, was determined by Charpy method on compression-molded, unnotched, 60 ×10 ×1.7 mm samples of the polyblend. Blending was achieved by precipitating with methanol dilute solutions of the polymers in benzene or toluene. Values of R are only relative.

Table II. Impact Resistance, R, in Ternary PS–PI–Cop Blends

PS[a]		PI[a]			
$\overline{M}_w$	In Blend, %	$\overline{M}_w$	In Blend, %	Cop[b], %	R
91,000	67.5	50,000	16.25	16.25	1.2
123,000	67.5	270,000	16.25	16.25	2.0
420,000	67.5	50,000	16.25	16.25	2.6

[a] Homopolymers were prepared by anionic polymerization; $\overline{M}_w/\overline{M}_n \simeq 1.1$.
[b] Two-block copolymer PS–PI was prepared by anionic polymerization; $\overline{M}_n$ = 146,000; 41 wt % PI.

[*e.g.*, polystyrene–poly(methyl methacrylate)–Cop or polystyrene–poly(vinyl chloride)–Cop] in which one of the phases can be plasticized selectively (*13, 14*). The degree of dispersion of one phase in the other, even in the solid state, can be regulated by Cop. So Cop acts for the corresponding homopolymers as an oil-in-oil emulsifier, and the dispersion degree depends on Cop concentration, structure, composition, and molecular weight (*1, 2*).

Concerning the mechanical properties and especially the impact resistance of such blends, we have to consider the nature of the continuous phase, the nature and amount of the dispersed rubber phase, and the nature of the interface.

Nature of the Continuous Phase. Impact resistance is a function of the intrinsic properties of the polymer forming the continuous phase, especially its molecular weight (*1, 2*). This is summarized in Table I for binary PS–PI blends and in Table II for ternary PS–PI–Cop blends. Table I shows that, for a given blend, increase in molecular weight of PS forming the continuous phase of the material increases impact resistance. The same observation can be made from Table II for a ternary PS–PI–Cop blend.

Nature and Amount of the Dispersed Rubber Phase. The effect of the nature of the dispersed rubber phase became apparent during our work on selective plasticization of systems containing two resins A and B, a corresponding AB Cop, and a selective plasticizer of polymer A or B (*13, 14*) where A was polystyrene (PS) and B was poly(methyl methacrylate) (PMM) or poly(vinyl chloride) (PVC). Selective plasticization is a new method of obtaining resin elastomeric systems which have the advantage that the physical properties (*e.g.*, mechanical properties and refractive index) of the rubbery phase can be varied by the nature and amount of the plasticizer. For such systems, impact resistance is maximum when the energy absorption capacity of the rubbery phase is maximum (*e.g.*, for a given amount of plasticizer with respect to the dispersed phase).

Impact resistance for a system containing 50% PMM, 25% PS, and 25% block copolymer PS–PMM (42% PS, $\overline{M}_n$ = 120,000) is plotted in Figure 6. PS, the dispersed phase of the system, was plasticized selectively with increasing amounts of diisobutylazelate. Impact resistance is maximum when the amount of plasticizer is 50% with respect to the dispersed PS phase.

Nature of the Interface. The most important aspect is the nature of the matrix–dispersed phase interface which determines the adhesion between these two phases. The anchoring effect between two phases with a block copolymer located at the interface depends essentially on the amount of Cop in the blend, the amount of elastomer, and the particle size of the dispersed phase. Particle size, in turn, depends on the molecular characteristics of homopolymer and Cop, as well as on the composition of the blend. In order to take into account these

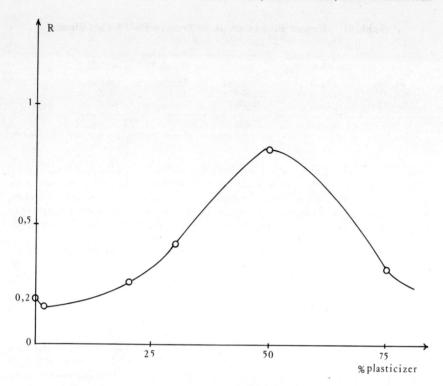

Figure 6. Impact resistance as a function of the amount of plasticizer in a ternary blend PS–PMM–Cop(PS–PMM) with PS selectively plasticized

50% PMM, M_n = 40,000; 25% PS, M_n = 38,000; and 25% Cop (42 wt % PS, M_n = 120,000)

different factors, the occupation density, D, is given by:

$$D = \frac{\text{number of AB links of the Cop in 1 cm}^3 \text{ polyblend}}{\text{total surface of rubber particles in 1 cm}^3 \text{ polyblend}} \tag{3}$$

D can be calculated when the polyblend composition is known and the number and the radius of the rubber particles are determined from microtomed sections.

A study of various PS–PMM polyblends in which the dispersed PS phase was selectively plasticized demonstrated that the impact resistance, R, varies in the same way as the occupation density, D (*see* Table III). Since the dispersed rubber particles are practically spherical, a simple relationship for R and D is:

$$R \sim D = K \frac{x}{ax + y} r \tag{4}$$

where x is the amount of Cop in the blend, y is the amount of elastomer in the blend (plasticized PS), a is the proportion of rubber sequence in the Cop, K is a constant, and r is the mean radius of the particles.

In the special case when the amounts of Cop and PS are the same ($x = y$), this relationship becomes:

$$R \sim D = K'r \tag{5}$$

Table III.   Effect of Blend Composition on Impact Resistance and
Occupation Density of PS–PMM Polyblends[a]

Variable	Blend					
	1	2	3	4	5	6
Blend composition, %						
PS	0	7	14	20	25	34
PMM	100	86	72	60	50	32
Cop	0	7	14	20	25	34
Radius of dispersed particles, μ	0	1.1	1.7	3.3	2.4	1.7
Occupation density, $D \times 10^{-14}$	0	0.29	0.43	0.84	0.60	0.45
Impact resistance	1	0.8	1	2.6	1.5	0.7

[a] PS $\overline{M}_w$ = 172,000; PMM $\overline{M}_w$ = 590,000; Cop $\overline{M}_n$ = 445,000, 48 wt% PS; amount of plasticizer (butoxyethyl stearate) with respect to the PS phase, 50%.

Indeed as is seen in Table III, for this special case the impact resistance varies in the same way as the particle size of the dispersed phase.

Furthermore, since r is a function of x, Equation 4 can be written as:

$$R \sim D = K \frac{x}{ax + y} f(x) \qquad (6)$$

where K is maximum if all the Cop is situated at the interface.

Location of the Copolymer in Polyblends

In heterophase polymeric materials, adhesion between the two phases is achieved only if Cop is situated at the interface. There are different techniques by which the location of Cop in a two-phase material can be determined.

One method is based on determining the refractive indexes of the two phases by interference microscopy. However this technique, which gives semi-quantitative information, can be applied only if the particles of the dispersed phase are not too small (*1, 2*). We therefore prepared by anionic polymerization different copolymers of PS–PI containing a fluorescent group like styrene–9-phenyl-10-anthracene (*see* Structure 7 for structure of fluorescent copolymers). In such a PS–PI blend, the PI phase can be detected first by phase

Table IV.   Location of Block Copolymer (Cop) in PS–PI
Polyblends by Fluorescence[a]

Blend 1		Blend 2		
PS Continuous Phase	PI Dispersed Phase	PI Continuous Phase	PS Dispersed Phase	Location of Cop
+	−	−	+	PS phase
+	−	+	−	not determined
−	+	+	−	PI phase
−	+	−	+	interface

[a] +, blue fluorescence of the Cop in this phase; −, no fluorescence.

contrast microscopy. On the same sample, UV fluorescence microscopy will then reveal the copolymer in the continuous phase, in the dispersed phase, or at the interface.

The location of the copolymer was studied as a function of the characteristics of the polymers in the blend (*e.g.*, homopolymer molecular weights and copolymer molecular weight and composition) which are related to the α and β of Equations 1 and 2. For a given PS–PI–Cop system, it was found that the location of Cop is practically independent of the amounts of PS, PI, and Cop in the polyblend. Therefore for such a system, two types of blends were generally prepared: one rich in PS with PS forming the continuous phase (Blend 1), and one rich in PI with a continuous PI phase (Blend 2). The different fluorescence possibilities are listed in Table IV. If blue fluorescence is observed in the continuous PS phase of Blend 1, it can be inferred that Cop is soluble in PS; and in fact the dispersed PS phase of Blend 2 (PI continuous phase) is also fluorescent.

However, with a polyblend film obtained by solvent evaporation or a slice of molded sample, the diameter of the dispersed particles is usually smaller than the thickness of film or sample. Consequently, it is difficult to say if Cop is soluble in the PI particles or if it is situated at the interface when the dis-

Table V.   Location of Block Copolymer in Polyblends

Sample	Technique[a]	α	β	Location[b]
film	fluor.	7.8	5.2	I
		7.8	3	I
		7.8	1.6	I
		3.2	5.2	I
		3.2	3	I
		3.2	1.6	I
		0.6	5.2	PS
		0.6	3	PS
		0.6	1.6	PS
		0.5	5	PS
		0.1	5	PS
		3.2	0.7	PI
molded[c]		5.4	5.2	I or PI
		3.2	3	I or PI
		0.9	3	PS
	x-ray	23	23	I or PI
		0.6	8	PS

[a] Fluorescence observed by UV microscopy, or x-ray scanning.
[b] I, at interface; PS, in PS phase; and PI, in PI phase.
[c] Compression-molded.

persed PI particles have a blue fluorescence. Only by observing a second blend, where PI is the continuous phase, is it possible to distinguish between the two. If the continuous PI phase is fluorescent, Cop is soluble in PI; but if the dispersed PS phase is fluorescent, it can be concluded that Cop is at the interface.

The data obtained by this technique for films and molded samples are presented in Table V and in Figure 7 where α *vs.* β is plotted on a logarithmic scale. In Figure 7 area C ($\alpha > 1$ and $\beta > 1$), Cop is mostly at the interface. In area B ($\alpha > 1$ and $\beta < 1$), Cop has a certain degree of solubility in PI, whereas in area D ($\alpha < 1$, $\beta > 1$) Cop is soluble in PS. It is more difficult to draw a conclusion about area A ($\alpha < 1$, $\beta < 1$) where some solubility in PS and in PI is indicated by blue fluorescence of both phases; however, because

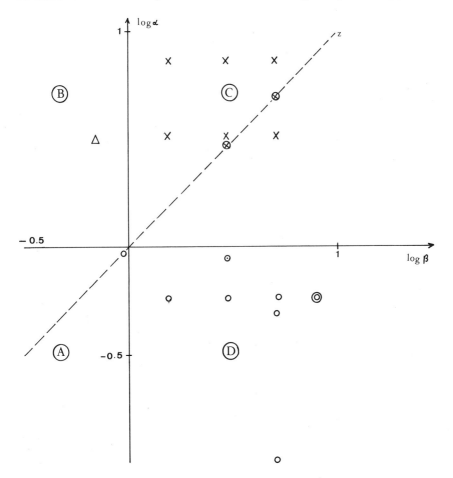

Figure 7.   Location of block copolymers in polyblends by fluorescent Cop and x-ray scanning techniques

×—*film, interface, fluorescence;* ○—*film, PS phase, fluorescence;* △—*film, PI phase, fluorescence;* ◉—*compression-molded, PS phase, x-ray scanning;* ⊙—*compression-molded, PS phase, fluorescence; and* ⊗—*compression-molded, PI phase or interface, fluorescence*

Table VI. Composition of Polyblends and Characterization of Polymers[a]

| | Homopolymers | | | | Copolymer | | |
| | PS | | PI | | | PS | |
Blend	$\overline{M}_n$	a, %	$\overline{M}_n$	b, %	$\overline{M}_n$	Fraction	c, %
II	300,000	60.5	41,500	25	44,000	0.72	14.5
III	300,000	60	22,000	20	26,000	0.51	20
IV	10,500	58.4	22,000	21.6	94,000	0.61	20
V	10,500	60	53,000	20	26,000	0.51	20
VI	300,000	58.4	53,000	21.6	94,000	0.61	20

[a] $a + b + c = 100\%$; PS + PI + Cop = 100%.

of the emulsifying effect demonstrated previously, some Cop is also at the interface, especially if $\alpha = \beta$ (axis $0z$).

The data obtained by this technique were confirmed by another technique using a PS–PI copolymer whose PI component was slightly chlorinated (1–2% total chlorine). Such a chlorinated copolymer is located in a PS–PI blend by an x-ray scanning microanalyzer (Castaing's Microsound). However this technique seems to have the same limitations as interference microscopy, especially for area A of Figure 7.

A more quantitative technique is based on the fact that in a blend containing polyisoprene (or polybutadiene) the rubber phase can be selectively crosslinked by γ-radiation (15). The blend was irradiated at 40–68 Mrad under nitrogen so that PI was quantitatively crosslinked. The insoluble fraction was then isolated by benzene extraction. Tests of binary blends of PS and PI homopolymers demonstrated that PI and PS can be well separated after irradiation, and that there is practically no solubility of one homopolymer in the other. This technique can be applied only if PS is the continuous phase.

If Cop is soluble in the PS phase, it is found in the benzene-soluble fraction, mostly as a colloidal dispersion. On the other hand, if Cop is soluble in the PI phase or if it is at the interface, it remains in the insoluble fraction. By analyzing the PS content of this fraction and from its weight increase with respect to PI homopolymer, the amount of Cop can be determined. However it is impossible to distinguish between Cop in the PI phase and Cop at the interface by this technique. Thus, by analyzing a given PS–PI–Cop blend for gel content and amount of PS in the gel, it is possible to locate the copolymer in either the continuous PS phase or at the interface (or in the rubber phase). For example, if all copolymer is in the continuous PS phase, the amount of gel equals the PI content and its PS content is zero. When Cop is at the interface or in the rubber phase, the amount of gel exceeds the PI content and its PS content is different from zero.

Table VII. Analysis of Polyblends by Irradiation Technique

Blend	α	β	Blend Type	Cop at Interface or in Rubber Phase, %
II	9.5	3.5	1 or 2	93 ± 4
III	23	1.7	1 or 2	91 ± 3
IV	0.18	0.61	3	26 ± 8
V	0.81	4.1	3	29 ± 3
VI	5.3	1.5	1 or 2	88 ± 3

The compositions of the different blends which have been analyzed and the characteristics of the polymers are given in Table VI.

Data on these blends obtained by the irradiation technique are given in Table VII, and it is apparent that the amount of copolymer at the interface or in the rubber phase varies. Blends can also be classified as follows: *type 1* or *2*—copolymer at the interface or in the rubber phase (however the two cannot be distinguished by this technique), and *type 3*—copolymer in the continuous PS phase.

Thus for $\alpha > 1$ and $\beta > 1$, Cop is at the interface or in the PI phase, and, from Figure 7, we may infer that it is at the interface. For blend V, only a small amount of Cop is at the interface, and the remainder is mostly soluble in the PS phase; this agrees with Figure 7 (area D). For blend IV (corresponding to area A with $\alpha < 1$ and $\beta < 1$), some Cop is probably soluble in PS, and the remainder which was insolubilized by irradiation may be distributed at the interface and in the PI phase.

It is interesting to note in Tables VI and VII that the same copolymer ($\overline{M}_n = 94,000$) with homopolymers of different molecular weights (blends IV and VI) has different values of α and β, and therefore 88% or only 26% of Cop is situated at the interface (or in the rubber phase).

Conclusion

From the data on the emulsifying properties and the location of Cop in polyblends, it appears that the emulsifying effect is best when $\alpha \simeq \beta < 1$, where the particle size of the dispersed phase is smallest, although some Cop may be soluble as micellar dispersion in the PS or PI phase. With increasing values of $\alpha \simeq \beta$, this solubility decreases, as do the Cop emulsifying properties for polyblends prepared by solvent evaporation or simultaneous precipitation. The optimum conditions therefore must be around $\alpha = \beta = 1$. This is depicted in Figure 8 for a ternary blend containing PS and PI of similar molecular weight in the presence of a two-block copolymer whose composition is about 50:50.

To correlate these findings with impact resistance, we studied three polyblends of same composition (67.5% Cop, 16.5% PI, and 16.5% PS) but with different homopolymers (*1, 2*). Since we had previously demonstrated the

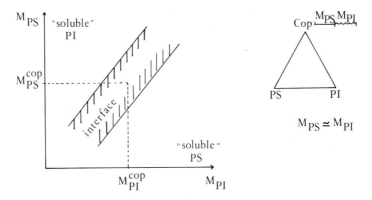

Figure 8.   Location of block copolymers in ternary blends

Table VIII. Effect of Homopolymers and of Block Copolymer (Cop) on Impact Resistance of Polyblends

| Blend | Homopolymer[a] | | Cop | | | | |
	PS, M_w	PI, M_w	PS, %	M_n	α	β	R
VII	123,000	270,000	59	145,000	1.45	4.3	2.0
VIII	165,000	160,000	39	475,000	0.9	0.5	2.9
IX	165,000	270,000	39	475,000	0.9	0.9	> 5

[a] Prepared by anionic polymerization; $\overline{M}_w / \overline{M}_n \simeq 1.1$.

effect of PS molecular weight on impact resistance, we selected blends containing PS with a molecular weight average of 140,000 (Table VIII). Since blends VII and IX have the same PI and practically the same PS, the tremendous increase in impact resistance can be attributed to the Cop. Furthermore, this increase in R is not caused by the small increase in total PI content (from 16.5% to 19.8%) resulting from the change in Cop type. Increasing PI molecular weight from 160,000 to 270,000 in a given PS–Cop system also increases impact resistance (blends VIII and IX).

It is interesting to note that this maximum is for $\alpha = \beta = 0.9$ which corresponds to the optimum area defined previously. This confirms that in impact-resistant material, the particle size of the dispersed rubber phase must be controlled, as well as the occupation density which characterizes somewhat the adhesion between the two phases.

Literature Cited

1. Kohler, J., Riess, G., Banderet, A., *Eur. Polym. J.* (1968) **4**, 173.
2. Kohler, J., Riess, G., Banderet, A., *Eur. Polym. J.* (1968) **4**, 187.
3. Riess, G., Kohler, J., Tournut, C., Banderet, A., *Makromol. Chem.* (1967) **101**, 58.
4. Periard, J., Banderet, A., Riess, G., *J. Polym. Sci. Part B* (1970) **8**, 109.
5. Periard, J., Riess, G., *Kolloid Z. Z. Polym.* (1973) **251**, 97.
6. Bancroft, W. D., *J. Phys. Chem.* (1913) **17**, 501.
7. Bancroft, W. D., *J. Phys. Chem.* (1915) **19**, 275.
8. Periard, J., Riess, G., *Eur. Polym. J.* (1973) **9**, 687.
9. Kohler, J., Job, C., Banderet, A., Riess, G., *Rev. Gén. Caout. Plast.* (1969) **46**, 1317.
10. Kohler, J., Thesis, Mulhouse, 1967.
11. Skoulios, A., Helfer, P., Gallot, Y., Selb, J., *Makromol. Chem.* (1971) **148**, 305.
12. Meier, D., private communication.
13. Periard, J., Banderet, A., Riess, G., *Angew. Makromol. Chem.* (1971) **15**, 37.
14. Periard, J., Banderet, A., Riess, G., *Angew. Makromol. Chem.* (1971) **15**, 55.
15. Jolivet, Y., Thesis, Mulhouse, 1971.

RECEIVED May 7, 1974. Work supported by the Compagnie Française de Raffinage.

Creep as Related to the Structure of Block Polymers

LAWRENCE E. NIELSEN

Monsanto Co., St. Louis, Mo. 63166

The creep of five commercial block copolymers, both amorphous and crystalline, was measured at 23° and 105°C with several loads. For a given load, the creep of block polymers was greater than that of the homopolymer making up the hard block. Specimens quench-cooled from the melt had more creep and a faster creep rate than slow-cooled specimens. The effect of molding conditions was most noticeable when creep was tested at 23°C, but annealing at 105°C for many hours did not eliminate the differences caused by initial molding conditions. Thermoplastic rubbers (block polymers with a high percent of soft block) had more creep than conventional crosslinked rubbers. For crystallizable block polymers, increasing the amount of crystalline phase reduced the creep.

Although the dynamic mechanical properties and the stress–strain behavior of block copolymers have been studied extensively, very little creep data are available on these materials (*1–17*). A number of block copolymers are now commercially available as thermoplastic elastomers to replace crosslinked rubber formulations and other plastics (*16*). For applications in which the finished object must bear loads for extended periods of time, it is important to know how these new materials compare with conventional crosslinked rubbers and more rigid plastics in dimensional stability or creep behavior. The creep of five commercial block polymers was measured as a function of temperature and molding conditions. Four of the polymers had crystalline hard blocks, and one had a glassy polystyrene hard block. The soft blocks were various kinds of elastomeric materials. The creep of the block polymers was also compared with that of a normal, crosslinked natural rubber and crystalline poly(tetramethylene terephthalate) (PTMT).

Materials and Techniques

Three of the block polymers were DuPont Hytrels 4055, 5555, and 6355 (*17, 18, 19*). These materials have a crystallizable PTMT hard block and a poly(tetramethylene glycol) (PTMG) soft block with a molecular weight of about 1000. Hytrels 4055, 5555, and 6355 contained about 42.6, 35, and 27% PTMG, respectively. The fourth polymer was Uniroyal TPR-19 which has a

257

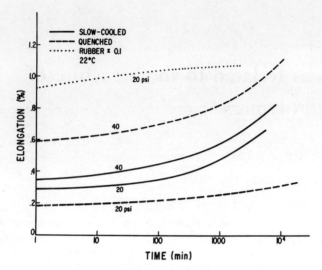

Figure 1. Creep of Kraton 1101 and a vulcanized natural rubber at 22°C (load in lbs/in.² is noted on curves)

crystalline polypropylene hard block and a soft block of ethylene–propylene elastomer with roughly 30% total ethylene. The fifth polymer was Shell Kraton 1101 triblock polymer with polystyrene ends and a polybutadiene central block; this material is about 24.5% styrene. Each polystyrene block has a molecular weight of about 12,500 while the polybutadiene block has a molecular weight of 75,000.

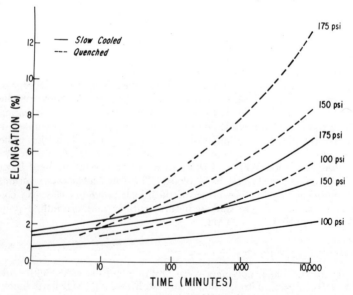

Figure 2. Creep of TPR-19 with crystalline polypropylene hard blocks at 23°C

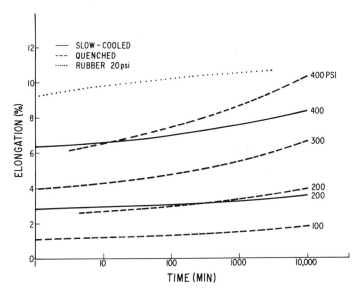

Figure 3.   Creep of Hytrel 4055 and a vulcanized natural rubber at 23°C

Specimens of each material were prepared under two very different molding conditions: slow-cooled and quench-cooled. Hytrels 6355 and 5555 were compression-molded at 270°C; quench-cooled specimens were then plunged into ice water whereas slow-cooled specimens were cooled to 200°C over 10 min and then cooled from 200°C to room temperature in about another 10 min.

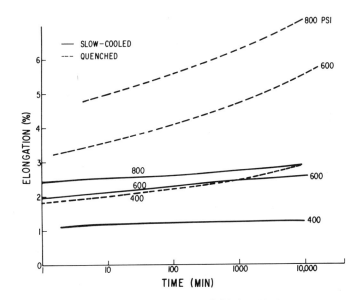

Figure 4.   Creep of Hytrel 5555 at 23°C

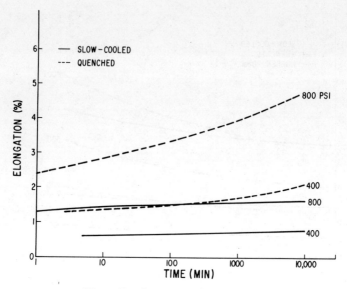

Figure 5.   Creep of Hytrel 6355 at 23°C

Hytrel 4055 was molded under the same conditions except that the molding temperature was 260°C, and the slow-cooled specimens were cooled to 180°C in 10 min. The initial molding temperature for PTMT was 255°C. TPR-19 was molded similarly except that the initial molding temperature was 200°C. Kraton 1101 was molded at 135°C. Most of the specimens measured 4 × 3/8 × 0.030 in. except for the crosslinked natural rubber specimens which were cut from large rubber bands.

The creep of the lowest modulus materials was the change in spacing of two ink dots on the specimens as measured with a 30-power traveling micro-

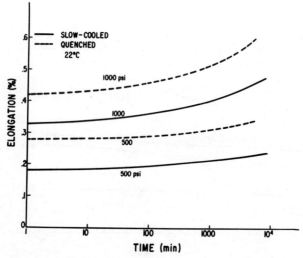

Figure 6.   Creep of poly(tetramethylene terephthalate) at
22°C

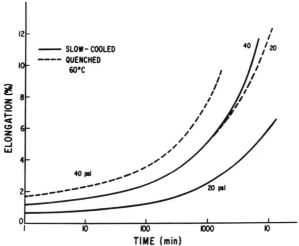

Figure 7.   Creep of Kraton 1101 at 60°C

scope. The creep of the more rigid materials, such as Hytrel 6355, was measured by a recording LVDT extensometer (*20*). Tests were conducted at
23° ± 1°C at a relative humidity of about 50%. Creep tests were also made
at 105°C on the Hytrels and TPR-19, and at 60°C on Kraton.

Results and Discussion

Typical creep curves are presented in Figures 1–12. The elastic moduli
of the block copolymers were between those of conventional crosslinked rubbers
and those of rigid crystalline polymers such as PTMT and polyethylene. Thus,

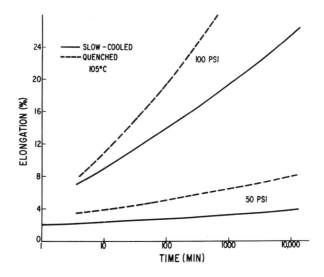

*Figure 8.   Creep of TPR-19 propylene block polymer at
105°C*

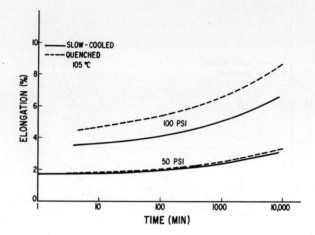

Figure 9.   Creep of Hytrel 4055 at 105°C

it is not surprising that the initial creep deformations of block polymers at a given load were greater than those of the rigid crystalline polymers but less than that of a crosslinked rubber. However, the creep rate, or the slope of the creep curve, was generally greater for block polymers than for either cross-linked rubbers or conventional rigid polymers.

Creep of block polymers is determined primarily by the dimensional stability of the hard blocks which can be either glassy or crystalline. The implication of this work is that the small size of the hard blocks leads to morphological structures with poorer dimensional stability than conventional polymers. The hard blocks must either break up or deform relatively easily when a load is applied to them. When the hard block is a crystalline material, as in the Hytrels and TPR-19, the creep depends on the degree of crystallinity of the

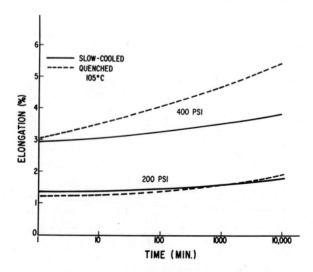

Figure 10.   Creep of Hytrel 5555 at 105°C

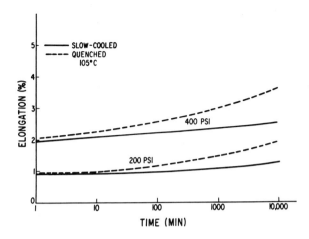

Figure 11.   Creep of Hytrel 6355 at 105°C

polymer. The higher the total crystallinity, the smaller the creep and the creep rate.

The morphology of the materials must also be important. Large changes in creep were found for very small changes in the degree of crystallinity as measured by x-ray techniques or by differential scanning calorimetry (DSC). This is apparent from a comparison of the creep data with data on crystallinity index and apparent heat of fusion (Table I). The crystallinity index is the area above the amorphous curve divided by the total area obtained from an

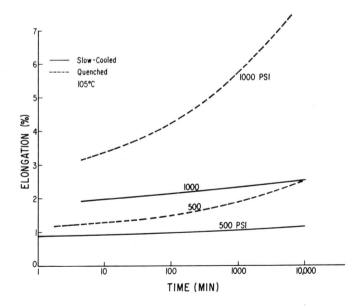

Figure 12.   Creep of poly(tetramethylene terephthalate) at 105°C

Table I. X-ray Crystallinity Index and DSC Apparent Heats of Fusion

Polymer	Heat Treatment	Crystallinity Index	Apparent Heat of Fusion, cal/g
TPR-19	slow-cooled	.24	9.8
	quenched	.21	8.0
Hytrel 4055	slow-cooled	0.9	4.8
	quenched	.07	4.5
Hytrel 5555	slow-cooled	.34	7.6
	quenched	.19	7.5
Hytrel 6355	slow-cooled	.24	7.6
	quenched	.20	8.6

x-ray scan. Both the crystallinity index and the apparent heat of fusion are relative indicators of the degree of crystallinity. Both the x-ray and the DSC curves showed some unexpected changes in shape in the Hytrel series which sometimes reversed the expected trends in crystallinity. DSC curves sometimes indicated higher crystallinities for the quench-cooled than for the slow-cooled Hytrels as measured by the area under the DSC melting peak. Differences in morphology are implied by the broad width of the melting peak of the quenched polymer compared with the peak for the slow-cooled polymer. Detailed studies of the morphology have not been made on these materials, but Cella (17) found a lamellar structure of the crystalline phase in similar Hytrels. Examples

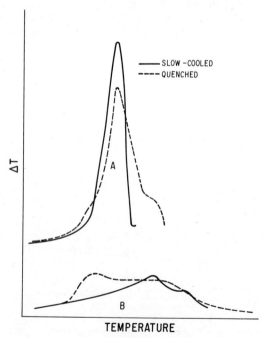

Figure 13.   DCS curves for A, Hytrel 6355 and B, Hytrel 4055 (polymers were annealed for 8 days at 105°C after molding)

of DSC curves for two polymers annealed for 8 days at 105°C are presented in Figure 13.

Molding conditions greatly affected the creep of block polymers, especially crystalline ones measured at 23°C. Both initial creep deformation and creep rate were generally higher for quenched specimens than for slow-cooled ones. The same effects, although less dramatic, were found at 105°C (specimens were held at 105°C at least 15 hrs before the creep tests were started). Even with this long annealing time, the differences produced by the molding conditions were not eliminated, but only reduced in magnitude.

Another example of the large effect of morphology on creep behavior may have been found with the Kraton polymer. It was impossible to obtain good reproducible creep values with this polymer, and supposedly identical specimens often differed greatly in creep behavior. The cause of this erratic behavior was not determined, but it might be related to cylinder-to-lamellae morphological changes. The composition of the Kraton is about that at which this transformation occurs. As an example of this erratic behavior, the creep of the slow-cooled polymer at both 20 and 40 psi fell between the corresponding values of the quench-cooled material (Figure 1).

Summary

Block polymers have a higher creep rate than either crosslinked rubbers or crystalline homopolymers. The hard blocks are apparently not large enough to be stable to high loads, but they apparently break up or allow chain slippage to occur so that very high creep rates are found at high loads. There are great differences between slow-cooled and quenched specimens. Quenched specimens had more creep and a higher creep rate than slow-cooled ones. High temperature annealing below the melting point of the hard block did not eliminate the differences between quenched and slow-cooled specimens. This indicates that the morphology is permanently locked-in at the time of initial molding.

Acknowledgment

Most of the data were obtained by Jerry Sugarman.

Literature Cited

1. Baer, M., *J. Polym. Sci. Part A* (1964) **2**, 417.
2. Kenney, J. F., *Polym. Eng. Sci.* (1968) **8**, 216.
3. Zelinske, R., Childers, C. W., *Rubber Chem. Technol.* (1968) **41**, 161.
4. Beecher, J. F., Marker, L., Bradford, R. D., Aggarwal, S. L., *J. Polym. Sci. Part C* (1969) **26**, 117.
5. Holden, G., Bishop, E. T., Legge, N. R., *J. Polym. Sci. Part C* (1969) **26**, 37.
6. Harrell, Jr., L. L., *Macromolecules* (1969) **2**, 607.
7. Morton, M., McGrath, J. E., Juliano, P. C., *J. Polym. Sci. Part C* (1969) **26**, 99.
8. Estes, G. M., Cooper, S. L., Tobolsky, A. V., *J. Macromol. Sci. Rev. Macromol. Chem.* (1970) **C4**, 313.
9. Miyamoto, T., Kodama, K., Shibayama, K., *J. Polym. Sci. Part A-2* (1970) **8**, 2095.
10. Shen, M., Kaelble, D. H., *J. Polym. Sci. Part B* (1970) **8**, 149.
11. Charrier, J.-M., Ranchoux, R. J. P., *Polym. Eng. Sci.* (1971) **11**, 318.
12. Lim, C. K., Cohen, R. E., Tschoegl, N. W., ADVAN. CHEM. SER. (1971) **99**, 397.
13. Cohen, R. E., Tschoegl, N. W., *Int. J. Polym. Mater.* (1972) **2**, 49.
14. Fielding-Russell, G. S., *Rubber Chem. Technol.* (1972) **45**, 252.
15. Kraus, G., Rollmann, K. W., Gardner, J. O., *J. Polym. Sci. Part A-2* (1972) **10**, 2061.

16. Scheiner, L. L., *Plastics Technol.* (1973) **19** (5), 36.
17. Cella, R. J., *J. Polym. Sci. Part C* (1973) **42**, 727.
18. Brown, M., Witsiepe, W. K., *Rubber Age* (1972) **104** (3), 35.
19. Hoeschele, G. K., Witsiepe, W. K., *Angew. Makromol. Chem.* (1973) **29/30**, 267.
20. Nielsen, L. E., *Trans. Soc. Rheol.* (1969) **13**, 141.

RECEIVED May 7, 1974.

Morphology, Crystallization, and Surface Properties of Styrene–Tetrahydrofuran Block Polymers

AKIRA TAKAHASHI

Department of Industrial Chemistry, Faculty of Engineering,
Mie University, Tsu, Mie-Ken, Japan

YUYA YAMASHITA

Department of Synthetic Chemistry, Faculty of Engineering,
Nagoya University, Nagoya, Japan

AB and ABA styrene–tetrahydrofuran block copolymers were synthesized by ion-coupling reactions. Morphologies in thin films cast from cyclohexane solutions were governed by crystallization and microphase separation; however, the prevalence of either factor controlled the final morphology. Melting and glass transition temperatures of block copolymers changed only slightly with variations in styrene content. Crystallization curves could be superimposed on the rate curve of homopolytetrahydrofuran, indicating that nucleation has occurred in polytetrahydrofuran domains. When a small amount of styrene–tetrahydrofuran block copolymer is added to a homopolystyrene, the block copolymer is surface active; this indicates rapid surface tension lowering of the polystyrene melt.

Microphase separation and domain formation in block copolymers, which are the result of incompatibility of block chains, have been studied extensively (*1, 2*). In addition to being incompatible, block chains in a copolymer generally have different thermal transition temperatures. The surface tensions of molten block chains also differ. When a crystalline block chain is incorporated into a block copolymer, it is expected that crystallization of the crystalline block chain causes considerable change in resultant morphology. Surface properties of a block copolymer and of its blend with a homopolymer should also be modified by the surface tension difference between block chains and the homopolymer. Since these factors determine the morphological features of a block copolymer both in bulk and at surface, a unified study of morphology, crystallization, and surface activity of any block copolymer is important to our understanding of its physical properties.

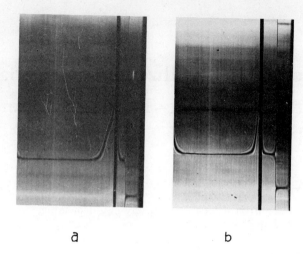

a   b

Figure 1. Comparison of sedimentation patterns of ST-5 block copolymer (a) before and (b) after fractionation

The melting temperature of polytetrahydrofuran is *ca.* 43°C, and its glass transition temperature is *ca.* −86°C whereas polystyrene is usually an amorphous polymer at its glass transition temperature of 90°C. Furthermore, the surface tension of polytetrahydrofuran is slightly lower than that of polystyrene. Thus, styrene–tetrahydrofuran is an interesting block copolymer. This paper presents the results of morphological, crystallization, and some surface chemical studies of this block copolymer.

Experimental

Materials. An AB block copolymer of styrene and tetrahydrofuran (ST) and an ABA block copolymer of styrene–tetrahydrofuran–styrene (STS) were synthesized by the ion-coupling reaction between the living ends of polystyryl anions and polytetrahydrofuran cations as reported previously (*3, 4*). To re-

Table I.   Observed Molecular Weights and Compositions of ST and STS Polymers

Sample	PS Block	PTHF Block	Block Polymer	PS, mole %
	$M_n \times 10^{-4}$			
S–T1	0.90	6.05	7.0	12.1
−2	3.01	6.05	10.3	26.9
−3	8.46	6.05	13.4	44.2
−4	13.7	6.05	22.4	57.8
−5	34.2	6.05	(45.6)	91.1
−6	3.4	13.4	21.0	11.4
−7	6.0	13.4	19.2	19.4
−8	10.3	13.4	23.0	40.5
STS–1	0.90	7.20	10.3	14.3
−2	3.01	7.20	(6.8)	34.7
−3	8.46	7.20	28.7	42.2
−4	13.7	7.20	26	57.6
−5	34.2	7.20	50	87.6

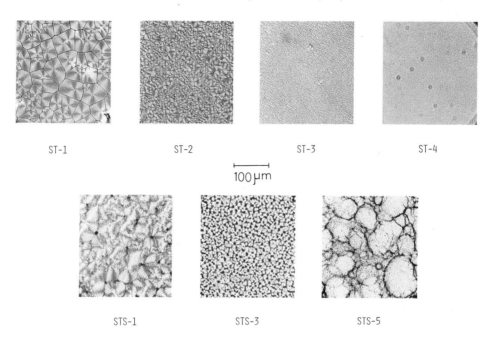

ST-1   ST-2   ST-3   ST-4

|———————|
100 μm

STS-1   STS-3   STS-5

Figure 2. Optical micrographs (crossed polars) of ST and STS block copolymer films cast from cyclohexane at 20°C

move homopolymers, they were fractionated by tetrahydrofuran–isopropyl alcohol mixed solvents. Polytetrahydrofuran (PTHF) ($M_n = 5400$) was synthesized by cationic polymerization, using BF_3–THF as catalyst. The Pressure Chemical polystyrene (PS) ($M_n = 1 \times 10^4$) was purified further.

Characterization. Number average molecular weights of the block copolymers in toluene were determined at 25°C by a Mechrolab high speed membrane osmometer fitted with an 0–8 membrane. Styrene mole fractions in the block copolymers were determined by H′–NMR in CCl_4 at 60°C. To determine the effect of fractionation procedure, sedimentation patterns of tetrahydrofuran solutions of fractionated samples were examined using a Spinco model E ultracentrifuge at 59,780 rpm.

Methods. Thin films of block copolymers were cast from their cyclohexane solutions at 20°C, and the films were observed under a polarizing microscope. Ultrathin films (50–500 nm) were cast from the same solutions onto sheet meshes at 20°C and stained by osmium tetroxide vapor; they were then observed under a Hitachi HU-11 electron microscope.

Melting temperatures and crystallization rates of the cast films were determined by dilatometry. The glass transition temperature of the styrene block was determined by dilatometry; the glass transition temperature of the tetrahydrofuran block was estimated using a Vibron EDV-3 viscoelastometer. A Perkin-Elmer differential scanning calorimeter was used to measure the degree of crystallinity of the films. A Rigaku Denki x-ray small angle scattering camera was used to measure the long period in the cast films.

Surface tensions of both the ST copolymers in homopolymers and the PS–PTHF homopolymer blend were measured in argon by the pendant drop method described by Roe (5).

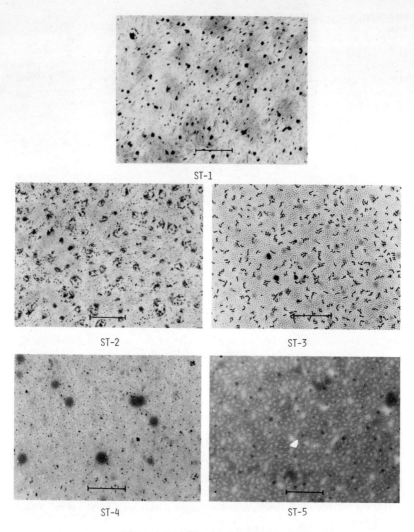

Figure 3. Electron micrographs of ST-1–5 block copolymer films cast from cyclohexane at 20°C (stain, OsO₄; scale = 1μm)

Results and Discussion

Characterization of the Block Copolymer. Figure 1 shows sedimentation patterns of the ST-5 polymer before and after fractionation. Note that the fractionation is nearly satisfactory.

Polymer characteristics are summarized in Table I. For ST copolymers, the agreement between the observed molecular weights and those calculated from the molecular weights of PS and PTHF blocks is good. However, for STS copolymers, the agreement is not as good, perhaps because of the rather inefficient fractionation of ABA block copolymers. The data in Table I indicate the purity of our samples.

Morphology. Ultrathin cast films of polystyrene and PTHF homopolymers stained by osmium tetroxide vapor were examined by electron microscopy. Only the PTHF was stained.

Photographs of both ST and STS block copolymers taken by the polarizing microscope and the electron microscope (Figures 2, 3, 4, and 5) reveal that when large spherulites are observed, white fibrils are seen in the electron micrographs. The electron micrograph of ST-1 (Figure 6a) shows a spherulitic primary growth stage of fibrils and their branching. For further comparison, the electron micrograph of the Pt–Pd-shadowed surface of the cast film of STS-1 (Figure 6b) also reveals fibrils and their branching. Therefore, the fibrils have the feature of crystalline lamellae. Fibril width is about 12–20 nm in ST-1 film and 20–40 nm in STS-1 film. These values are almost identical to the crystalline lamellar thickness observed in a fractured surface of bulk crystallized PTHF of $M_n = 6 \times 10^4$ (6).

The long periods (L) were determined by applying the Bragg's condition to the peak positions of small angle x-ray scattering curves (Figure 7). The L values for ST-1, ST-6, STS-1, as well as the bulk crystallized PTHF, range from 30 to 40 nm, and they are almost identical to the fibril width (*see* Table II). The correspondence of fibril width and long period together with the

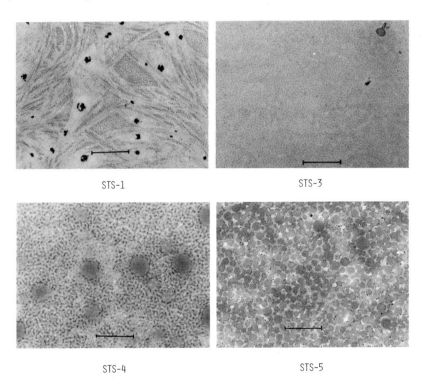

STS-1   STS-3

STS-4   STS-5

Figure 4.   Electron micrographs of STS-1, -3, -4, and -5 copolymer films cast from cyclohexane at 20°C (stain, OsO₄; scale = 1 μm)

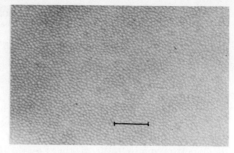

ST-6

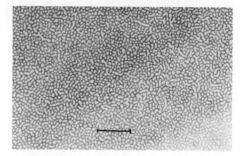

ST-7

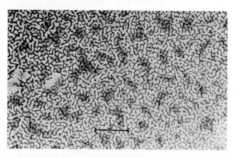

Figure 5.   Electron micrographs of ST-6, -7, and -8 copolymer films cast from cyclohexane at 20°C (stain, OsO₄; scale = 1µm)   ST-8

spherulitic character of fibrils lead to the conclusions that the fibrils are crystalline lamellae oriented perpendicularly to the film surface, and that fibril width corresponds to lamellar thickness even though they are coated with PS block chains.

Closer inspection reveals that besides fibrils, spherical domains are also observed in STS-1 and to a lesser extent in ST-1 and ST-2. Thus, fibrils and spherical domains co-exist. Increase in PS content in the block copolymers results in progressively smaller spherulites, and a very rough surface for the STS-5 film is observed in the optical micrograph.

In the electron micrographs of ST-3 and ST-4, lamella or layer type morphology appears and fibrils disappear. Lamella type morphology also appears in STS-3 and STS-4. Moreover, in ST-4, the continuous phase is PS. Therefore, the type of block copolymer—i.e., ST or STS, does not make much difference in the morphology. At still higher mole % PS, the surfaces of cast films of ST-5 and STS-5 show considerable accumulation of PTHF blocks. For ST-5, the continuous phase is PTHF blocks, and for STS-5, many PTHF block discs cover the surface. Accumulation of the PTHF block in ST and STS block copolymers may be attributed to the surface active nature of the PTHF block in ST and STS block copolymers (*see* the section on Surface Activity). Average domain sizes determined from the electron micrographs are also presented in Table II.

ST-1, -2, and -3 polymers and ST-6, -7, and -8 polymers nearly correspond in their PS contents, but the M_n of PTHF block in the latter series is almost twice as large as that in the former. In electron micrographs of cast films of ST-6, -7, and -8 (Figure 5), the following structures are observed: for ST-6, PS domain, which looks like paving stones, in PTHF matrix; for ST-7, bending rod-like domain of PS in PTHF matrix; and for ST-8, PS–PTHF lamellae. This structure change with increasing PS content is quite analogous to that observed for non-crystalline block copolymers (2, 7). Average domain sizes determined

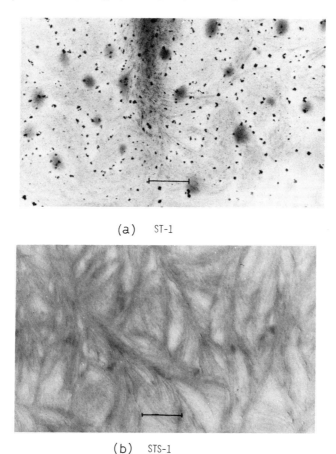

(a)   ST-1

(b)   STS-1

Figure 6.   Electron micrograph of (a) the cast film of ST-1 (stain, OsO₄; scale = 1μm) and (b) the surface of cast film of STS-1 (shadow, Pt–Pd)

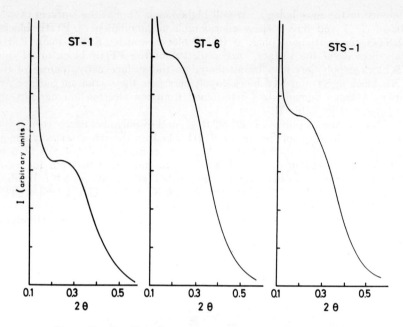

*Figure 7. Small angle x-ray scattering curves for ST-1, ST-6,
and STS-1 block copolymer films*

from the electron micrographs are given in Table II. Although spherulites
could not be observed in ST-6, -7, and -8 films, they are definitely crystalline
as evidenced by their DSC measurements. A comparison of Figures 3 and 5
clearly indicates that morphology is not determined entirely by PS content.
The molecular weights of the blocks also contribute to morphological develop-

**Table II.   Morphology Parameters Determined from Electron Micrographs
and Long Periods (*L*) for PTHF, ST, and STS Polymers**

Sample	Domain Shape	Size[a], nm		L, nm
PTHF	—		—	33
ST-1	PTHF fibril	T	= 12–22	35
-2	PTHF fibril	T	= 12–20	48
-3	{ PS rod	D	= 40	32
	PTHF lamella	T	= 12	
-4	{ PTHF layer-lamella	T	= 42	—
	PS rod	D	= 42	
-5	PTHF	T	= 50~60	—
-6	PS sphere	D	= 65	40
-7	PS sphere-rod	D	= 59~65	40
-8	PTHF rod-lamella	T	= 37~45	40
STS-1	PTHF fibril	T	= 21~42	40
-3	{ PS lamella	T	= 50	—
	PTHF lamella	T	= 12–25	
-4	{ PS rod	D	= 50	—
	PTHF lamella	T	= 12~17	
-5	PTHF disc	D	= 200	—

[a] T = thickness, D = diameter.

Table III.   Comparison of Volume Fractions of PS Blocks Determined from PS Contents and Electron Micrographs

Sample	PS, mol %	Structure	Domain	Structure Parameter[a], nm	ϕ[b] I	ϕ[b] II
ST–6	11.4	sphere	PS	D = 65 A = 110	0.137	0.15
ST–7	19.4	sphere-rod	PS	D = 59~65 A = 110	0.234	0.25
ST–8	40.5	rod-lamella	PTHF	PTHF 37~45 PS 45~65	0.469	0.57

[a] For ST–6 and ST–7, D = domain diameter and A = distance between domain centers for ST–8, thickness.
[b] Calculated (I) from the PS mol % using ρ_{PS} = 1.05 and ρ_{PTHF} = 0.95, and (II) from electron micrographs using Equations 1, 2, and 3.

ment. Despite the fact that the casting temperature was much lower than the theta temperature for PS blocks, in ST-1 and ST-2 crystallization prevailed over microphase separation during casting; however, in ST-6 and -7 domain formation occurred first, and crystallization of the PTHF domain was followed after evaporation of the casting solvent. The paving stone morphology in the ST-6 film also seems to support this conclusion since the morphology is considered to result from the deformation of PS spheres caused by crystallization of PTHF blocks.

By considering the paving-stone-like PS domains as spheres (ST-6), the bending rods as cylindrical rods (ST-7), and the lamellae as plane parallel layers (ST-8), the volume fraction, ϕ, of each domain was calculated from the diameter, D, and the distance, A, measured on the electron micrographs using the following equations (8):

for spheres,
$$\phi_s = 0.74 \left(\frac{D_s}{A_s}\right)^3 \tag{1}$$

for rods,
$$\phi_r = 0.91 \left(\frac{D_r}{A_r}\right)^2 \tag{2}$$

and for layers or lamellae,
$$\phi_l = \frac{D_l}{A_l} \tag{3}$$

The results are presented in Table III. These values for the volume fractions agree well with those calculated from the mole % PS. The same conclusion was reached for non-crystalline block copolymers. Therefore, the agreement also supports the theory that the domain is formed first (8).

The dependence of morphological features on the composition of block copolymers is summarized in Table IV.

Degree of Crystallinity. The degree of crystallinity $(1 - \lambda)$ of the cast films was calculated from the ratio of the observed heat of fusion by DSC to the heat of fusion (ΔH_u) of 100% crystalline PTHF. A ΔH_u of 3800 cal/mole was obtained by analyzing the melting point depression data of PTHF–diluents systems (6) and was used in the calculation. As shown in Figure 8, the degree of crystallinity decreases with increased PS content in the block copolymers. Moreover, the data for ST and STS copolymers show approximately the same $(1 - \lambda)$ vs. PS content relationship. The degree of crystallinity was also determined for isothermally crystallized samples. In dilatometers, the cast films were melted at 80°C for 1 hr and then crystallized at 20°C for 10^4 min. The

Table IV.   Dependence of Morphology on Composition of ST and STS Block Copolymers for Solvent-Cast Films

(I) The Case of Microphase Separation Predominance[a]

A spheres in B matrix	A rods in B matrix	alternate lamellae	B rods in A matrix	B spheres in A matrix
		decreasing B content →		
(ST-6)	(ST-7)	(ST-8)		

(II) The Case of Crystallization Predominance

	fibrils (spherulites)	──────────→ (small spherulites)		fibrils disappear (rough surface)
		decreasing B content →		
(ST-1) (STS-1)	(ST-2)	(ST-3) (STS-3)		(ST-5) (STS-5)

[a] A = PS block, B = PTHF block.

$(1 - \lambda)$ was determined from the dilatometer reading and is also plotted in Figure 8. The degree of crystallinity for these samples is somewhat higher than the value for cast films.

Crystallization Kinetics. Typical isothermal crystallization curves, which were measured using cast films of both types of copolymers, are presented in Figures 9 and 10. At the early stage of crystallization, the effect of the non-crystalline PS block on the rate curves was only shifts of the degree of crystallinity $[1 - \lambda(t)]$ *vs.* time curve along the time axis. However, the extent of shift does not correspond to PS content in the block copolymers. The change in crystallization temperature also causes the crystallization curves to shift. At the initial stage of crystallization, these rate curves could be superimposed on the rate curve of the homo-PTHF ($M_n = 6 \times 10^4$), and an Avrami exponent

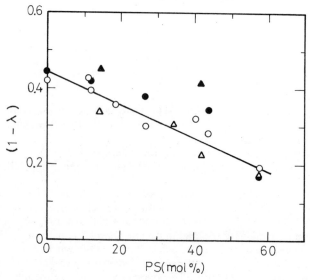

Figure 8.   Degree of crystallinity $(1 - \lambda)$ vs. PS content (mole %)

O, *ST and* △, *STS copolymers determined by DSC; and* ●, *ST and* ▲, *STS copolymers by dilatometry*

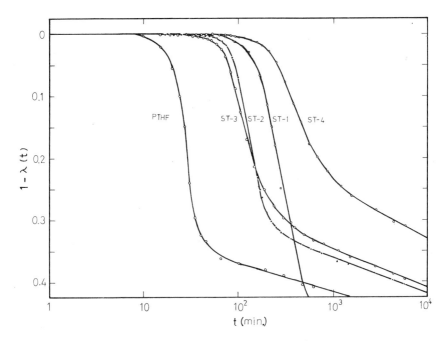

Figure 9. Kinetics of isothermal crystallization at 20°C of the PTHF homopolymer and the ST-1, -2, -3, and -4 copolymers after quenching from 80°C

of 4 was obtained. This value and the insensitivity of crystallization rates to the copolymer composition can be understood by remembering that in the cast films nucleation and subsequent crystallization take place in PTHF domains since these domains have the same role as that of isolated droplets in the nucleation experiments.

The effects of melt temperatures of the films before quenching on the crystallization curves were compared at 80° and 100°C. The latter is higher than the glass transition temperature (T_g) of PS block, and the former is lower than the T_g. However, there were no substantial differences in the observed rate curves. It appears that melting above or below T_g of the PS block does not cause substantial change in the size of the PTHF domains. Therefore, the crystallization rate (G) of the block copolymers may be approximated by the equation for homopolymer crystallization (9):

$$\ln G = \ln G_0 - \frac{E_D}{RT_c} - \frac{\Delta F^*}{RT_c} \tag{4}$$

where G_0 is a constant, E_D is the activation energy for transport of molecules across the crystal–liquid interface, ΔF^* is the free energy required to form a crystal nucleus of critical size, T_c is the crystallization temperature, and R is the gas constant.

In the limit of high molecular weight, $\Delta F^*/RT_c$ is expressed for a three-dimensional nucleation process

$$\frac{\Delta F_3^*}{RT_c} = \left[\frac{8\pi \, \sigma_{en} \, \sigma_{un}^2}{R \, \Delta H_u^2} \right] \left[\frac{T_m^{\circ 2}}{T_c \, (T_m^\circ - T_c)^2} \right] \tag{5}$$

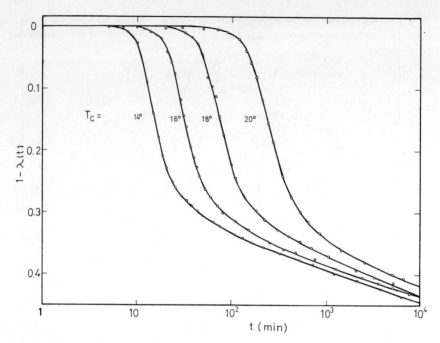

Figure 10. Kinetics of isothermal crystallization at T_c *of copolymer STS-3 after quenching from 80°C*

and in the case of a two-dimensional nucleation process

$$\frac{\Delta F_2{}^*}{RT_c} = \left[\frac{4\,\sigma_{en}\,\sigma_{un}{}^2}{R\,\Delta H_u}\right]\left[\frac{T_m{}^\circ}{T_c\,(T_m{}^\circ - T_c)}\right] \tag{6}$$

where σ_{en} and σ_{un} are the end and lateral interfacial free energies of nucleus, ΔH_u is the enthalpy of fusion per repeating unit, and $T_m{}^\circ$ is the equilibrium melting temperature. For PTHF, $T_m{}^\circ = 57°C$ (6).

To avoid the complexity arising from the secondary crystallization stage, it is convenient to put $G = (1/\tau_{0.15})$ where $\tau_{0.15}$ is the time elapsed to attain 15% crystallinity. Then, $\ln(1/\tau_{0.15})$ *vs.* $(T_m{}^\circ{}^m/T_c)(1/\Delta T)^m$ should be a straight line according to the dimension of nucleus formation—*i.e.*, $m = 1$ or 2. Figure 11 shows the two-dimensional case. The lines are almost parallel including the homo-PTHF, thus satisfying the demand of linearity. Parallel and linear plots between $\ln(1/\tau_{0.15})$ and $(T_m{}^\circ{}^2/T_c)(1/\Delta T)^2$ were obtained for the three-dimensional case also. The products of interfacial free energies of crystalline nucleus had the following values: $\sigma_{en}\sigma_{un} = 2.2 \times 10^5$ cal²/mole² and $\sigma_{en}\sigma^2{}_{un} = 185 \times 10^5$ cal³/mole³. Thus, both the homo-PTHF and the block copolymers gave almost the same values for the products; furthermore, dimensional differentiation of the nucleation process could not be attained. The fact that the interfacial free energy products are independent of PS content of block copolymers and are the same as that of the homo-PTHF clearly indicates that the nucleation process has proceeded in the PTHF domains. The overall crystallization rates decrease in the order: homo-PTHF > STS > ST. However, it cannot be explained adequately why the crystallization rate of the triblock copolymers is faster than that of the diblocks.

Thermal Transition Temperatures. To determine the melting temperatures (T_m) of block copolymers, the cast films were isothermally crystallized at 20°C for 10^4 min after quenching from 80°C and then heated at 0.1°C/min. The dilatometer readings are presented in Figure 12. A somewhat broad melting range is observed, which is characteristic of copolymers but to a lesser extent than for random copolymers. The data are summarized in Table V, and melting temperatures are plotted against mole % PS in Figure 13. The melting temperatures change only gradually with increasing PS content in the block copolymers. No block type effect on T_m was observed. Since the molecular weights of the PTHF block are sufficiently high, the melting point depression with increasing PS content is not the chain length effect of the PTHF block.

Flory (*10, 11*) has extended his treatment of homopolymer melting to block copolymers. According to his theory, melting point depression is still a function of the sequence length of crystallizable units A, but it is not as great as for a random copolymer which has the same copolymer composition. The lowering of the equilibrium melting point is given by

$$\frac{1}{T_m{}^e} - \frac{1}{T_m{}^\circ} = - \frac{R}{\Delta H_u} \ln p \tag{7}$$

where $T_m{}^\circ$ is the melting point of an infinitely long crystal of homopolymer A, and p is the sequence propagation probability defined as the probability of an A unit being succeeded by another A. However, the true equilibrium value

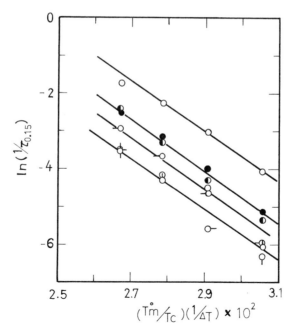

Figure 11. The two dimensional case: plots of ln(1/τ₀.₁₅) vs. (Tₘ°/Tᶜ)(1/ΔT)
O, PTHF; ◐, ST-1; -O, ST-2; Q, ST-3; O-, ST-4; ◑, STS-1; and ●, STS-3

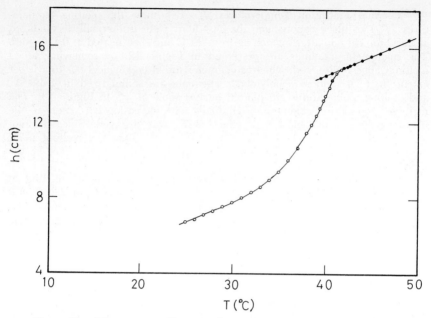

Figure 12. Dilatometer readings for the melting process of ST-2 copolymer

of $T_m{}^e$ cannot be determined experimentally since equilibrium is never achieved because of kinetic factors. Moreover, the small domain size of the PTHF block restricts considerably both longitudinal and lateral growth of the PTHF blocks. The melting point depression observed in Figure 13 indicates the presence of various types of defects which have been imposed on the PTHF crystallites, including interfacial (or interzonal) disorders.

It is expected that the electron density difference between the PS and PTHF domains will result in a small angle x-ray scattering peak. The observed long periods are about 40 nm, which is smaller than the domain center distance. It is concluded that the peak position corresponding to the domain center distance is located at a much smaller angle and therefore could not be observed since it is beyond the resolution of our instrument. The observed long periods (L) are closer to the PTHF lamellar thickness seen in the electron micrographs. Moreover, the $L = 33$ nm value for the bulk crystallized homo-PTHF ($M_n = 6 \times 10^4$) is slightly smaller than the L for the block copolymers. PTHF lamellar (or fibril) thickness seen in the electron micro-

Table V.   Thermal Transition Temperatures of PTHF and Block Copolymers

Sample	T_m, °C	T_g °C (PS Block)
PTHF	42.7	—
ST–1	41.9	86.1
–2	41.9	89.4
–3	41.8	92.2
–4	40.2	94.2
STS–1	41.3	86.2
–3	41.1	92.1

graphs of block copolymers ranges from 12 to 40 nm, which is almost identical to the lamellar thickness observed in the fractured surface of bulk crystallized PTHF (6).

It appears that the crystalline lamellar thickness in the block copolymers does not change as much with PS content. This agrees with the observation of the low melting point depression of block copolymers compared with that of the homo-PTHF since the melting point is determined not only by the end interfacial free energy of crystallite but also by the crystallite thickness.

The glass transition temperatures (T_g) of the PS block as determined by dilatometers are plotted against mole % PS in Figure 13. Both ST and STS copolymers show the same increase in T_g with increasing PS content. However, closer inspection of Figure 13, together with the PS molecular weights given in Table I, suggests that the T_g is governed by a molecular weight dependence rather than by block copolymer composition. Comparison of our data with the data of molecular weight dependence of T_g of PS reported by Fedors (13) supports the above conclusion. Crystal, Erhardt, and O'Malley (14, 15) also reached the same conclusion for styrene–ethylene oxide block copolymers.

The glass transition temperatures of the PTHF block were estimated from dynamic mechanical measurements. Measurements of the storage and loss component of the rigidity modulus (E', E'') at 138 Hz are depicted in Figure 14. Data for the homo-PTHF agree well with those reported by Wetten (12).

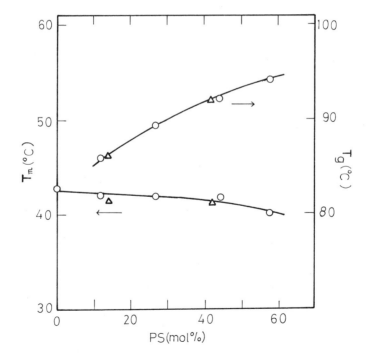

Figure 13.   Transition temperatures (T_m *and* T_g) *vs.* (mole %)
of PS content
O, *ST and* △, *STS*

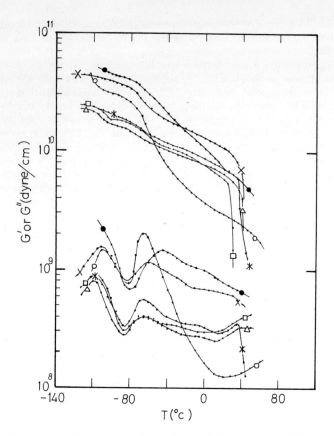

Figure 14. Temperature dependence of the storage and loss components of the rigidity modulus of PTHF, ST, and STS copolymers

O, *PTHF;* ●, *ST-1;* ×, *ST-2;* △, *ST-3;* □, *ST-4; and* ★, *STS-3*

Both the PTHF and the block copolymers exhibit a main relaxation region associated with the onset of micro-Brownian motion of the main chain which is similar to that of other polymers. The secondary mechanical γ process, which is considered characteristic of a chain containing four or more methylenes, is also observed. In the melting region, catastrophic drops in modulus appear for both PTHF and block copolymers. The main glass transition peaks in the homo-PTHF and the block copolymers are at *ca.* −64°C, which is somewhat higher than the T_g of PTHF (−86°C). However, the main glass transition peaks and thus the T_g values are essentially independent of PTHF content in the block copolymers. These results are expected since the molecular weights of the PTHF blocks are almost constant. In conclusion, the presence of two glass transition temperatures corresponding to the T_g of the homopolymers also indicates microphase separation.

Surface Activity of Block Copolymers. Modification of surface properties of a homopolymer by incorporating a small proportion of block copolymer has been studied for several block copolymer–homopolymer blends. LeGrand

and Gains (*16*) demonstrated that (AB)$_n$ type poly(bisphenol-A carbonate)–polydimethylsiloxane copolymer is concentrated at the surface of bisphenol-A carbonate homopolymer. Gains and Bender (*17*) have reported that polystyrene$_{75}$–polydimethylsiloxane$_{77}$ is surface active when incorporated in small amounts in homopolystyrene, and the time-dependent surface tension lowering of the blend is clearly indicated. In these block copolymers, however, the surface tension difference between block chains and also homopolymers is as large as 17 dynes/cm. It is interesting to examine whether a block copolymer, which consists of block segments with surface tension differences as small as 2 or 3 dynes/cm, can serve as surface active block copolymer when incorporated in a homopolymer.

Surface tension *vs.* temperature for PS and PTHF are shown in Figure 15. The data of Gains and Bender (*17*) on PS surface tensions (γ_{oPS}) agree well with ours. However, our value for the surface tension of PTHF (γ_{oPTHF}) is about 4.5 dynes/cm higher than Roe's value (*18*). The reason for this discrepancy is not clear. The γ_{oPTHF} is always smaller than the γ_{oPS} by $\Delta\gamma = 3$ dynes/cm, which is much smaller than the surface tension difference between the blocks of PS and polydimethylsiloxane. The time-dependent surface tensions of four blends (ST–PS, ST–PTHF, PS–PTHF, and PTHF–PS) were measured. To prepare the blends, the block or homopolymers were added in small amounts (0.3–1 wt %) to the homo-PS or PTHF. The mixture was completely dissolved in benzene, the solutions were quickly frozen by a dry ice-acetone mixture, and the samples were freeze dried.

The surface tension–time curves obtained are shown in Figures 16, 17, 18, 19, and 20. For the ST-1–PS and ST-3–PS blends, surface tensions first fall

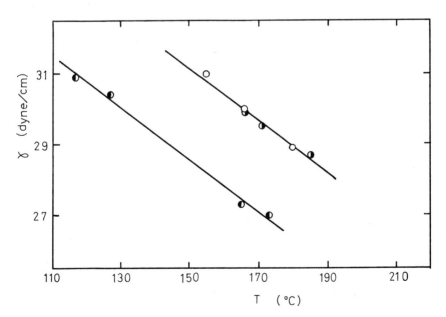

Figure 15. Surface tension vs. *temperature plots for homo-PS* (M$_n$ = *1* × *10^4*) *and homo-PTHF* (M$_n$ = *5.4* × *10^3*)
O, *PS;* ◐, *PS (from Ref. 17); and* ◑, *PTHF*

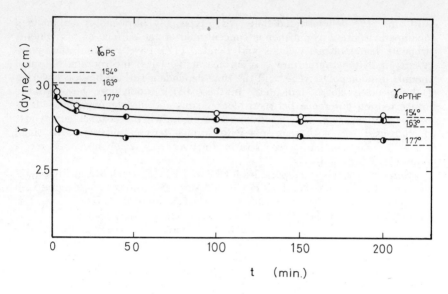

Figure 16. Surface tension vs. *time plots for ST-1 copolymer–PS blend (ST-1 conc.*
= 1 wt %)

rapidly and then gradually approach that of molten PTHF. This means that
PTHF blocks are adsorbed at the surface of the molten blends. It appears
that the rate of surface tension lowering decreases with increasing PS content—*i.e.*, this is caused by the effect of the molecular weight of the PS block

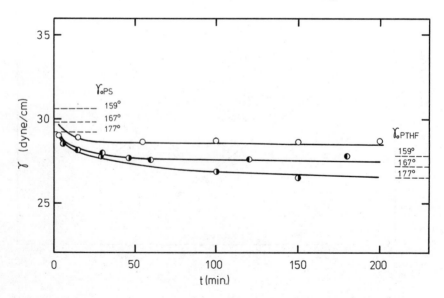

Figure 17. Surface tension vs. *time plots for ST-3 copolymer–PS blend (ST-3 conc.*
= 1 wt %)

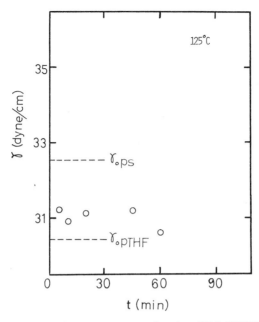

*Figure 18. γ vs. time plot for ST-3–PTHF
(ST-3 conc. = 0.3 wt %)*

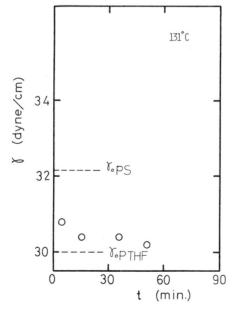

*Figure 19. γ vs. time plot for homo-PS in
homo-PTHF (PS conc. = 1.1 wt %)*

in the copolymer. On the other hand, the PTHF–PS blend showed only a small lowering of surface tension (Figure 20). Although the surface tensions of both ST–PTHF and PS–PTHF blends decrease, the initial surface tensions approach that of molten PTHF and never that of molten PS.

From these results, it is concluded that for the ST–PS blend, both the surface tension difference and the incompatibility between PTHF block and homo-PS assist in the adsorption of the PTHF block, which has lower surface tension than molten homo-PS. On the contrary, it is clear that incompatibility alone is not sufficient to force the PS block chain to the surface of the ST–PTHF blend against the increase of surface free energy. Therefore, quite an important conclusion has been reached from these studies, that is, an AB block copolymer, in which the surface tension difference between the two blocks $\Delta\gamma = \gamma_A - \gamma_B$ is as small as 3 dynes/cm, is still sufficiently surface active when it is added in low concentration to a homopolymer corresponding to block A as long as $\gamma_A > \gamma_B$.

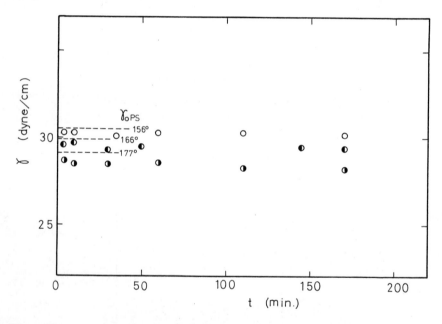

Figure 20. γ vs. time plots for the homo-PTHF–homo-PS mixture (PTHF conc. = 1.3 wt %)

Comparison of the PTHF–PS and the ST–PS blends indicates the critical importance of block structure. Thus, adsorption of the PTHF block at the homo-PS surface may represent the surface analog of domain formation in the bulk. Electron micrographs of cast films of ST-5 and STS-5, in which the concentration of PTHF block is ca. 10%, seem to support this idea. However, the data on surface tension lowering at constant block copolymer concentration are not sufficient to warrant any conclusion about the conformation of the block copolymers at the surface, and more detailed study is in progress.

Acknowledgments

We are pleased to recognize the preparative and the technical assistance of H. Wakabayashi, H. Kainuma, and K. Nagata. Special thanks are due to T. Kuroda and R. Takahashi of the Plastics Research Laboratory, Mitsubishi Petrochemical Industries, Ltd., for the electron photomicrographs and the measurements by Vibron.

Literature Cited

1. Aggarwal, S. L., Ed., "Block Copolymers," Plenum, New York, 1970.
2. Allport, D. C., Janes, W. H., "Block Copolymers," Applied Science, London, 1973.
3. Yamashita, Y., Hirota, M., Matsui, H., Hirao, A., Nobutoki, K., *Polymer J.* (1973) **2**, 43.
4. Yamashita, Y., Nobutoki, K., Nakamura, Y., Hirota, M., *Macromolecules* (1971) **4**, 548.
5. Roe, R. J., *J. Phys. Chem.* (1968) **72**, 2013.
6. Takahashi, A., *Polym. Symp. Jap. 21st*, Kyoto, Nov., 1972.
7. Inoue, T., Soen, T., Hashimoto, T., Kawai, H., *J. Polymer Sci. Part A-2* (1969) **7**, 1283.
8. Hoffmann, M., Kampf, G., Kroner, H., Pampus, G., in "Multicomponent Polymer Systems," ADVAN. CHEM. SER. (1971) **99**, 351.
9. Mandelkern, L., "Crystallization of Polymers," McGraw-Hill, New York, 1964.
10. Flory, P. J., *J. Chem. Phys.* (1949) **17**, 223.
11. Flory, P. J., *Trans. Faraday Soc.* (1955) **51**, 848.
12. Wetten, R. E., *Brit. Polym. Phys. Group Meetg.*, Shrivenham, 1965.
13. Fedors, R. F., *J. Polym. Sci. Part C* (1969) **26**, 189.
14. Crystal, R. G., Erhardt, P. F., O'Malley, J. J., in "Block Polymers," S. L. Aggarwal, Ed., p. 179, Plenum, New York, 1970.
15. O'Malley, J. J., Crystal, R. G., Erhardt, P. F., in "Block Copolymers," S. L. Aggarwal, Ed., p. 163, Plenum, New York, 1970.
16. LeGrand, D. G., Gains, Jr., G. L., *Amer. Chem. Soc., Div. Polym. Chem., Prepr.* **11** (2), 442 (Chicago, Sept. 1970).
17. Gains, Jr., G. L., Bender, G. W., *Macromolecules* (1972) **5**, 82.
18. Roe, R. J., *J. Colloid Interface Sci.* (1969) **31**, 228.

RECEIVED March 18, 1974. Work supported by grant 50189 of the Ministry of Education, Japan.

25

Domain Structure and Domain Formation Mechanism of ABA and AB Block Copolymers of Ethylene Oxide and Isoprene

E. HIRATA,[1] T. IJITSU, T. SOEN,[2] T. HASHIMOTO, and H. KAWAI

Department of Polymer Chemistry, Faculty of Engineering, Kyoto University, Kyoto 606, Japan

The domain structure and crystalline texture of AB and ABA type block copolymers of ethylene oxide (EO) and isoprene (Is) are studied, and the effects of the casting solvents and the fractional compositions of each block segment are determined. The domain structures of EO–Is copolymers are essentially identical to those of EO–Is–EO copolymers, but they strongly depend on the fractional compositions and the casting solvents. The role of casting solvent in the different domain formation mechanisms is interpreted in terms of an interrelation of two binodal surfaces that represent the critical concentration for crystallization of the EO segment and the critical concentration for micelle formation of the incompatible EO and Is segments.

Thermodynamic incompatibility of the A and B block segments of amorphous AB and ABA block copolymers involves microphase separation at a critical micelle concentration (*1, 2*). The micelles formed at this concentration are essentially maintained through the solvent evaporation process to produce the domain structures observed in solid state (*2, 3, 4*). The shape and size of the micelles, and consequently of the domains, depend on the incompatibility, molecular weight, and fractional composition of the block copolymers and on the casting solvent and the temperature (*3–10*).

When one component of an amorphous block copolymer is replaced by a crystalline polymer, the domains or crystalline texture formed by the solvent cast should depend at least on two factors: (a) crystallization of the crystalline block segment and (b) microphase separation resulting from the incompatibility of the A and B blocks. The crystalline texture observed in solid film is considered strongly dependent on the relative contributions of the two phase

---

[1] Present address: Toyama Plant, Mitsubishi Rayon Co., Ltd., Toyama, Japan.

[2] Present address: Department of Textile Engineering, Faculty of Fiber Technology, Kyoto University of Industrial Arts and Fiber Technology, Kyoto, Japan.

separations. We investigated the effects on solid texture of the casting solvent and the fractional composition of ethylene oxide (EO) and isoprene (Is) block copolymers of the EO–Is and EO–Is–EO types.

Test Specimens and Experimental Procedures

Preparation of AB and ABA Type Block Copolymers of Ethylene Oxide and Isoprene. The block copolymers were synthesized by Szwarc's anionic polymerization technique (*11*) with cumyl potassium and sodium naphthalene as the initiators for EO–Is diblock copolymers and EO–Is–EO triblock copolymers, respectively, and the ethylene oxide and isoprene contents varied from 0 to 100%. The detailed procedure is being published (*12*).

Characterization of Block Copolymers. The copolymers were characterized by velocity ultracentrifugation, osmotic pressure, and ordinary element analysis in order to determine the molecular weight monodispersity of the copolymers and the degree of contamination with the corresponding homopolymers, the number average molecular weight $(\overline{M}_n)$, and the fractional compositions of the block copolymers, respectively.

The sedimentation pattern was determined in benzene solution (0.3–0.5 g/l) at 20°C in a Spinco model E ultracentrifuge at 56,100 rpm. The Schlieren patterns for most of the copolymers had a relatively sharp, single peak; thus the synthesized copolymers were genuine AB or ABA type block copolymers with a sharp distribution of molecular weights. However, some copolymers had twin peaks which indicated possible contamination with homopolyisoprene; fractionation to remove the homopolymer was then repeated with relatively dilute (<5%) benzene solutions and *n*-hexane or methanol as precipitant until the sedimentation patterns had a single peak (*12*). The number average molecular weights were determined by high speed membrane osmometer in a toluene solution at 37.0°C.

The weight fraction of each block sequence was determined by ordinary element analysis and was verified by the infrared absorption band at 1683 cm^{-1} characteristic of polyisoprene (*13*) in CCl$_4$ solution (0.5 g/l). The findings agreed within experimental error, and the arithmetic average was used as the weight fraction for the EO and Is block segments.

The degree of crystallinity and the melting point of the EO crystals were measured on film specimens cast from benzene solutions by differential scanning calorimetry (DSC) with the Perkin-Elmer model DSC-1B operated at an increasing temperature rate 10°C/min (*12*). The degree of crystallinity was determined by measuring the area under the endothermic peak (ranging from *ca.* 40° to 70°C) and using the reported heat of fusion of the homopoly(ethylene oxide) (PEO) crystal, 45 cal/g (*14*); the determined degree of crystallinity is not the overall crystallinity of the copolymers but rather the value for copolymers with an EO fraction of 100%. The melting point was determined by the position of the endothermic peak. The findings are presented in Table I.

Procedures for Microscopic Observations. Ultrathin film specimens (*ca.* 700–1000 A thick were prepared for electron microscopy by placing a drop of 0.1% benzene or ethylbenzene solution on a microscope mesh coated with collodion and carbon and then evaporating the solvent as gradually as possible at 30°C. Ethylbenzene as a solvent was selectively good for Is but poor for EO segments whereas benzene was a good solvent for both segments. Similar studies were made recently by Kovacs *et al.* (*15, 16*), Kawai *et al.* (*17*), Skoulious *et al.* (*18, 19, 20, 21*), and Crystal *et al.* (*22, 23*).

For the electron microscopy, the specimens were stained by osmium tetroxide vapor, which selectively stains the Is phase, in order to reveal with deep contrast the detailed submicroscopic domain structure that results from the microphase separation of incompatible EO and Is segments. Alternatively, the

Table I. Characterization of AB and ABA Type Block Copolymers of Ethylene Oxide and Synthesized Isoprene

| Polymer | Ethylene Oxide, wt % | M_n, × 10^{-4} | | | Degree of Crystallinity, % | Melting Point, °C |
		Total	Is Block	EO Block		
EO–Is 1	87.0	135.8	17.7	118.1	79.5	62.2
EO–Is 2	76.0	98.8	23.7	75.1	81.0	64.0
EO–Is 3	72.0	65.2	18.3	46.9	72.7	63.3
EO–Is 4	60.0	53.0	21.2	31.8	61.5	60.0
EO–Is 5	23.0	24.9	19.2	5.7	10.1	50.8
EO–Is–EO 1	78.4	31.5	7.1	12.2	88.0	62.5
EO–Is–EO 2	74.6	21.9	5.6	8.2	86.4	61.5
EO–Is–EO 3	67.4	29.9	9.7	10.1	83.6	61.0
EO–Is–EO 4	52.4	31.5	14.9	8.3	60.0	60.5
EO–Is–EO 5	14.0	34.8	30.0	2.4	—	—
EO–Is–EO 6	11.8	68.9	61.1	3.9	35.0	55.5
EO–Is–EO 7	8.8	48.5	44.2	2.1	—	—
PEO 1	100	21.0	0.0	21.0		
PEO 2	100	3.8	0.0	3.8		
PIs 1	0.0	7.8	7.8	0.0		

specimens were shadowed by Pt–Pd in order to reveal the morphology of the crystalline PEO texture.

It was not possible to prepare ultrathin sections with an ultramicrotome, even after osmium tetroxide fixation (24). The EO segment is hydrophilic, and therefore the specimens were too ductile for ultrathin sectioning. Furthermore, there was no suitable liquid which is a common poor solvent for both segments on which to float the ultrathin sections in order to mount them on the microscope mesh.

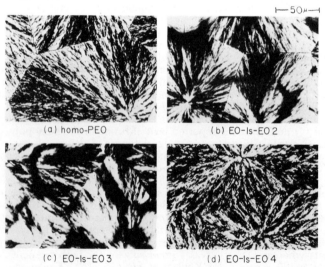

├─50µ─┤

(a) homo-PEO   (b) EO–Is–EO 2

(c) EO–Is–EO 3   (d) EO–Is–EO 4

Figure 1. Cross polarized photomicrographs of films of EO–Is–EO block copolymers with relatively large fractions of EO segments revealing less perfect development of spherulitic crystalline texture as the Is content increases

Films cast from 1% benzene solutions at 30°C

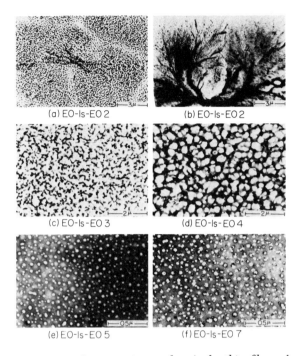

Figure 2. Electron micrographs of ultrathin films of EO–Is–EO block copolymers

Films cast from 0.1% benzene solutions at 30°C; a, EO–Is–EO 2 stained by OsO₄; b, EO–Is–EO 2 shadowed by Pt–Pd; c, EO–Is–EO 3 stained by OsO₄; d, EO–Is–EO 4 stained by OsO₄; e, EO–Is–EO 5 stained by OsO₄; and f, EO–Is–EO 7 stained by OsO₄

Domain Structure of EO–Is–EO Triblock Copolymers Cast from Benzene Solutions

Crosspolarized photomicrographs of PEO 1, EO–Is–EO 2, EO–Is–EO 3, and EO–Is–EO 4 films cast from 1% benzene solutions at 30°C are presented in Figure 1. The spherulitic texture with negative birefringence became less perfect and led to a less clear Maltese cross as the fraction of amorphous Is segment increased. When the EO fraction constituted less than 50%, the texture was not clearly resolved by light microscopy.

The submicroscopic structure of spherulites, especially the location of the Is phase within the spherulites as a function of EO content, was examined by transmission electron microscopy. The fine structure of the EO–Is–EO triblock copolymers cast from 0.1% benzene solutions at 30°C is revealed in Figure 2. These figures represent typical textures of copolymers with EO segments of more than 75 wt % (Figures 2a and 2b), of 40–75 wt % (Figures 2c and 2d), and of less than *ca.* 40 wt % (Figures 2e and 2f).

The stained specimen of EO–Is–EO 2 (Figure 2a) suggests that the dark spherical domains of *ca.* 1000 A are dispersed in the bright matrix. The dark and bright phases correspond to the Is and EO phases which are selectively stained and unstained by osmium tetroxide. The EO matrix is clearly spherulitic (Figure 2b). The spherical Is domains are neither uniform in size nor

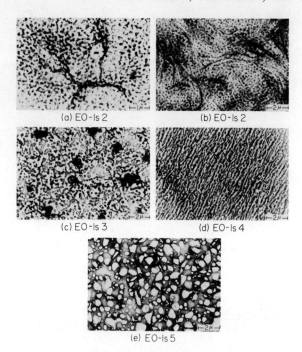

(a) EO-Is 2   (b) EO-Is 2

(c) EO-Is 3   (d) EO-Is 4

(e) EO-Is 5

Figure 3.   Electron micrographs of ultrathin films of
EO–Is block copolymers

Films cast from 0.1% benzene solutions at 30°C; a, EO– Is 2
stained by OsO₄; b, EO–Is 2 shadowed by Pt–Pd; c, EO–Is 3
stained by OsO₄; d, EO–Is 4 stained by OsO₄; and e, EO–Is 5
stained by OsO₄

regularly arranged in the spherulitic EO matrix as in amorphous block copolymers (*9, 10*) because the EO segment is crystallizable.

When the Is content is increased the spherical Is domains become interconnected and form rod-like Is domains dispersed in the spherulitic EO matrix
(Figure 2c). When the Is content is further increased, the rod-like Is domains
become interconnected and both the Is and EO phases are then continuous
(Figure 2d). The EO phase, however, is still spherulitic although disordered
(*see* Figure 1d).

Finally, a phase inversion occurs and the spherical EO domains of *ca.*
400 A are dispersed in the amorphous Is matrix (Figure 2e). The spherical
EO domains are crystallized by wide angle x-ray diffraction (*cf.* diblock copolymers in Figure 4) and DSC thermograms of thin film specimens. The size
of the dispersed EO domains was *ca.* 400 A in diameter for EO–Is–EO 5, *ca.*
750 A for EO–Is–EO 6, and *ca.* 400 A for EO–Is–EO 7. These values did not
necessarily agree with the theoretical values for amorphous AB and ABA type
block copolymers (*3, 4, 5, 6*), probably also because the EO segment is
crystallizable.

Domain Structure of EO–Is Diblock Copolymers Cast from Benzene Solutions

Crosspolarized photographs of EO–Is 1–4 (all cast from 1% benzene
solutions at 30°C) revealed a similar trend (*cf.* Figure 1), *i.e.*, the spherulitic

texture, which becomes less perfect as the content of amorphous Is increases, was volume-filling when the fraction of EO segments was greater than *ca.* 50%.

The fine structure of diblock copolymers cast from 0.1% benzene solutions at 30°C is revealed in electron micrographs (Figure 3) which are quite similar to those for the triblock copolymer series (*cf.* Figure 2). The electron micrograph of stained EO–Is 2 (Figure 3a) indicates that the Is spherical domains of *ca.* 1300 A are dispersed in the EO matrix which is clearly spherulitic (Figure 3b). When the Is content is increased, the spherical Is domains become interconnected and form irregular, rod-like domains dispersed in the spherulitic EO matrix (Figure 3c). When the Is and EO contents are similar, the EO and Is segments both form continuous phases (Figure 3d) with the EO phase spherulitic. When the Is fraction is much larger, phase inversion occurs (Figure 3e) with the spherical EO domains dispersed in the amorphous Is matrix. Thus, the texture of the diblock copolymer series was quite similar to that of the triblock copolymer series. Texture depends primarily on the fractional composition of the block copolymers and hardly at all on the sequence arrangements; as with amorphous AB and ABA type block copolymers also (25).

X-ray diffraction patterns for a series of EO–Is diblock copolymers including homopolyisoprene (homoPIs) and homopoly(ethylene oxide) (homoPEO) are presented in Figure 4. The crystal structure of the EO segments of the block copolymers is identical to that of the homoPEO, even for a block copolymer with little EO crystallinity and low melting point (*e.g.*, EO–Is 5).

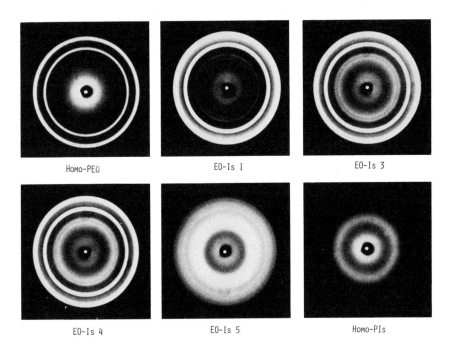

Homo-PEO   EO-Is 1   EO-Is 3

EO-Is 4   EO-Is 5   Homo-PIs

Figure 4.   X-ray diffraction diagrams of EO–Is block copolymers, homopoly(ethylene oxide), and homopolyisoprene
Films cast from benzene solutions

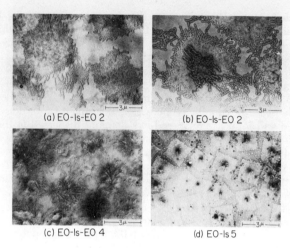

(a) EO-Is-EO 2   (b) EO-Is-EO 2

(c) EO-Is-EO 4   (d) EO-Is 5

Figure 5. Electron micrographs of ultrathin films

*Films cast from 0.1% ethylbenzene solutions at 30°C; a, EO–
Is–EO 2 shadowed by Pt–Pd; b, EO–Is–EO 2 stained by OsO₄;
c, EO–Is–EO 4 shadowed by Pt–Pd; and d, EO–Is 5 shadowed
by Pt–Pd*

Domain Structure of EO–Is–EO and EO–Is Type Block Copolymers Cast from Ethylbenzene Solutions

For films cast from 0.1% ethylbenzene solutions at 30°C, electron microscopy of both stained and shadowed specimens revealed a structure characteristic of a single, crystal-like lamella (hedrite) having a square shape a few microns in size, regardless of the EO content if it was greater than about 20 wt % (Figure 5). Disc-like domains of Is segments appear on the free surface of the EO single crystals. This Is phase is considered to be segregated into the interlamellar region, as proposed by Kovacs *et al.* (*15, 16*) and Kawai *et al.* (*17*). However, when the Is and EO contents are nearly equal, the development of crystalline structure is unstable and either a sheaf-like structure (Figure 5c) or a square hedrite of a few microns is formed.

Wide angle x-ray diffraction and small angle x-ray scattering patterns were recorded with the x-ray beam normal and parallel to the film surfaces of EO–Is–EO 2 with the film surfaces oriented horizontally (Figure 6). The specimens were cast from 3% ethylbenzene solution at ~60°C, and then the temperature was gradually lowered to 30°C. The patterns reveal preferential orientation of paratropic crystal planes such as (120) plane perpendicular to the film surfaces and, in addition, planar orientation of crystal lamellae parallel to the film surfaces. The first and second order meridional scattering peaks at 33.6 and 67.2 min (the latter peak appearing as a shoulder) arise from a single interlamellar spacing of 157 A (from Bragg's law). When the densities of the crystalline and amorphous phases of the EO segment and of the amorphous phase of the Is segment are taken as 1.234, 1.130, and 0.913 g/cm³, respectively, and when the composition and the crystallinity of the EO–Is–EO 2 copolymer are used, the layer thicknesses of the crystalline and amorphous EO phases and of the Is phase are estimated at 96, 10, and 51 A, respectively. This thickness of the crystalline EO phase agrees fairly well with that reported for homopoly(ethylene oxide) (*14*).

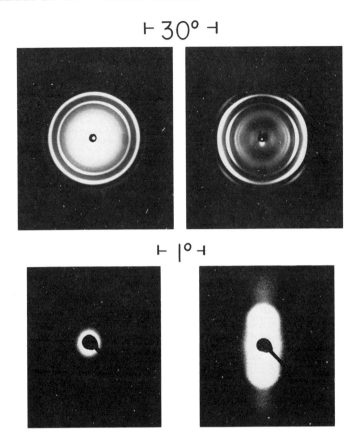

*Figure 6.   Wide angle x-ray diffraction patterns (top) and small
angle x-ray scattering patterns (bottom) of EO–Is–EO 2 film*

Films cast from 3% ethylbenzene solution at 60°C with the temperature
then gradually lowered to 30°C; left, through radiation; and right, edge
radiation (film normal is in vertical direction)

Discussion

The domain formation mechanism and structure of these copolymers were
similar to those of amorphous AB and ABA type block copolymers of styrene
and isoprene when benzene, a nonselective solvent, was used as the casting
solvent except that one phase was crystallizable and the domain structures
were distorted with resultant irregularities in size, shape, and mutual arrangement of domains. The irregularities are attributable to the development of
crystalline texture. This is probably valid also for a selective solvent that is
good for the crystallizable EO segment but poor for the noncrystallizable Is
segment (*see* below).

On the other hand, the crystalline texture of the EO component and the
submicroscopic domain structure of the Is component obtained by casting with
ethylbenzene, a selective solvent good for Is but poor for EO segments, were
very different from those obtained by casting with benzene. When ethylbenzene is used, crystallization of the EO segments must occur prior to micelle

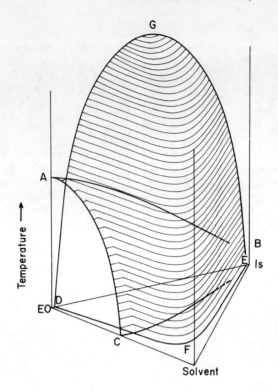

*Figure 7.   Schematic of phase diagram for a given
solvent with two kinds of binodal surfaces*

ACB, a curved surface for the critical concentration for
crystallization of EO segments; and DFEG, a curved surface for the critical concentration for micelle formation

formation and in association with the relatively less entangled segments of the
EO component. Consequently, the EO segments crystallize as square-shaped
hedrites as large as a few microns. The Is segments may be segregated to form
Is layers in between the EO hedrites. With benzene, on the other hand, the
situation may be just the opposite, with crystallization resulting in a dendritic
or spherulitic texture for the EO-rich copolymers and a spherical EO domain
dispersed in a matix of Is for the Is-rich copolymers.

The origin of this phenomenon may be qualitatively interpreted in terms
of a relation between two kinds of binodal surfaces in a phase diagram (Figure
7). In the simplified diagram, the curved surface ACB represents a sort of
binodal surface for crystallization of EO segments and the curved surface
DFEG represents a sort of binodal surface for micelle formation (27) resulting
from the microphase separation of the EO and Is block segments. In the
phase diagram at a given temperature, as illustrated by a bottom section of
the trigonal prism, the curve DFE represents the critical micelle concentration
(C_m) while CB represents the critical concentration (C_c) at which crystallization
of the EO segment starts.

C_m depends on the fractional composition, incompatibility, and molecular
weight of the copolymer as well as on the casting solvent and temperature;

empirically it is known to be about 5–10 wt % at room temperature (*1, 2, 26, 27*). C_c depends on the crystallization temperature, solvent, and fractional composition of the copolymer. At 30°C, the experimental temperature, the value of C_c should be considerably higher than that of C_m if the solvent is good for the EO segment, but considerably lower than that of C_m if the solvent is poor for the EO segment.

When benzene is the solvent, C_c is therefore expected to be higher than C_m (*see* Figure 8a) whereas when ethylbenzene is the solvent the situation should be modified as depicted in Figure 8b. Hence when a copolymer having the composition represented by Point d is cast from benzene, microphase separation occurs at first at the concentration corresponding to Point c prior to crystallization of the EO segment; this is followed by crystallization at the higher concentration b in the crystallizable EO phase. The Is domains or matrix formed at the critical micelle concentration may interrupt crystal growth and cause disorders in the EO crystal (*see* Figure 9b). The domain structure is less regular in terms of size, shape, and mutual arrangement than that of the amorphous AB and ABA block copolymers. Furthermore, the size of the domains, especially that of the EO domains, does not necessarily agree with that theoretically expected for amorphous block copolymers. These effects are considered to be a result of crystallization of the EO component.

On the other hand, when ethylbenzene is used as the casting solvent, for EO–Is–EO 2 block copolymer for example, the EO segment crystallizes into single crystal-like texture at Point b in Figure 8b prior to the microphase separation of the incompatible EO and Is segments. The Is component is segregated into the interlamellar region (Figure 9a). Therefore, crystallization is the predominant factor which controls the solid texture, and phase separation has little effect. This occurs with a copolymer which has an EO fraction as small as 20 wt % (EO–Is 5 in Figure 5). It should be noted, however, that the texture is exceptionally ill-defined, being either single crystal-like or sheaf-like in copolymers with nearly equal fractions of EO and Is segments (*e.g.*, EO–Is–EO 4). This may be attributed to the closeness of C_c and C_m as depicted in Figure 8b (*see* Line ab′c′d′).

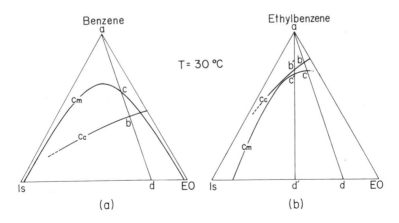

Figure 8.   Schematic of the triangular phase diagram at 30°C
Cast solvent: a, benzene, and b, ethylbenzene

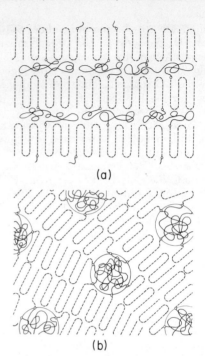

(a)

(b)

Figure 9.   Schematic of the textures of EO–Is–EO block copolymers
Cast solvent: a, ethylbenzene, and b, benzene

The behavior of similar block copolymers (styrene–ethylene oxide and styrene–propylene oxide) in solution in selective solvents for both block chains was studied intensively by Skoulios and co-workers (*18, 19, 20, 21*). At high concentrations, they found highly organized, mesomorphic structures which could be detected by analyzing small angle x-ray scattering patterns. These structures, 100–500 A in size, were lamellar, rod-like, or spherical.

The organization found by Skoulios *et al.* may correspond to the spherical, rod-like, or lamellar type domain structure formed by microphase separation when benzene is used as the solvent. On the other hand, when the selective solvent good for the noncrystallizable block segment is used, crystallization of the EO segment occurs at very low concentrations with a resultant single crystal-like texture. In this case, the mesomorphic organization found by Skoulios *et al.* may correspond to the dispersed single crystal-like texture in solutions.

Our observations on the domain structure and domain formation mechanism are also in rough accord with those of Crystal *et al.* (*22, 23*). However, they did not necessarily fully explore the detailed, submicroscopic domain structures, and application of the osmium tetroxide staining method to their styrene–ethylene oxide copolymers still leaves some doubts.

Literature Cited

1. Sadron, C., *Angew. Chem.* (1963) **75**, 472.
2. Vanzo, E., *J. Polym. Sci. Part A-1* (1965) **4**, 1727.
3. Inoue, T., Soen, T., Hashimoto, T., Kawai, H., *J. Polym. Sci. Part A-2* (1969) **7**, 1283.
4. Soen, T., Inoue, T., Miyoshi, K., Kawai, H., *J. Polym. Sci. Part A-2* (1972) **10**, 757.

5. Meier, D. J., *J. Polym. Sci. Part C* (1969) **26**, 81.
6. Meier, D. J., *Amer. Chem. Soc., Div. Polym. Chem., Prepr.* (Chicago, September, 1970).
7. Matsuo, M., Sagae, S., Asahi, H., *Polymer* (1969) **10**, 79.
8. Leary, D. F., Williams, M. C., *J. Polym. Sci. Part A-2* (1973) **11**, 345.
9. Campos-Lopez, E., McIntyre, D., Fetters, L. J., *Macromolecules* (1973) **6**, 415.
10. Hashimoto, T., Nagatoshi, K., Todo, A., Hasegawa, H., Kawai, H., *Macromolecules* (1974) **7**, 364.
11. Richards, D. H., Szwarc, M., *Trans. Faraday Soc.* (1959) **56**, 1944.
12. Hirata, E., Ijitsu, T., Soen, T., Hashimoto, T., Kawai, H., *Polymer (London),* in press.
13. Zbinden, R., "Infrared Spectroscopy of High Polymers," p. 240, Academic, New York, 1964.
14. Mandelkern, L., "Crystallization of Polymers," p. 78, McGraw-Hill, New York, 1964.
15. Kovacs, A. J., Lotz, B., *Kolloid Z. Z. Polym.* (1966) **209**, 97.
16. Lotz, B., Kovacs, A. J., Bassett, G. A., Keller, A., *Kolloid Z. Z. Polym.* (1966) **209**, 115.
17. Kawai, T., Shiozuka, S., Sonoda, S., Nakagawa, H., Matsumoto, T., Maeda, H., *Macromol. Chem.* (1969) **128**, 252.
18. Skoulious, A., Fianz, G., Parrod, J., *C.R. Acad. Sci.* (1960) **251**, 739.
19. Skoulious, A., Fianz, G., *C.R. Acad. Sci.* (1961) **253**, 265.
20. Skoulious, A., Fianz, G., *J. Chim. Phys.* (1962) **59**, 473.
21. Skoulious, A., Tsoulande, G., Franta, E., *J. Polym. Sci. Part C* (1963) **4**, 507.
22. Crystal, R. G., Erhardt, P. F., O'Malley, J. J., "Block Copolymers," S. L. Aggarwal, Ed., p. 179, Plenum, New York, 1970.
23. Crystal, R. G., "Colloidal and Morphological Behavior of Block and Graft Copolymers," G. E. Molau, Ed., p. 279, Plenum, New York, 1971.
24. Kato, K., *Polym. Eng. Sci.* (1967) **7**, 38.
25. Uchida, T., Soen, T., Inoue, T., Kawai, H., *J. Polym. Sci. Part A-2* (1972) **10**, 101.
26. Sadron, C., Gallot, B., *Macromol. Chem.* (1973) **164**, 301.
27. Inoue, T., Soen, T., Hashimoto, T., Kawai, H., *Macromolecules* (1970) **3**, 87.

RECEIVED September 23, 1974. Work was supported in part by a grant from the Scientific Research Funds (Kagaku Kenkyu-hi, 743020-1973) of the Ministry of Education, Japan, and in part by the Bridgestone Tire Co., Ltd. of Tokyo, Japan and the Japan Synthetic Rubber Co., Ltd. of Tokyo, Japan.

26

Poly(α-methylstyrene–dimethylsiloxane) Block Copolymers. The Effects of Microstructure on Properties

ANDREW H. WARD, THOMAS C. KENDRICK,[1] and JOHN C. SAAM

Dow Corning Corp., Midland, Mich. 48640

A highly regular microstructure of rod-like microdomains of poly(α-methylstyrene) in a polydimethylsiloxane matrix forms in extruded plugs or films. When compared with solution cast films, the extruded films have anisotropic mechanical properties and increased oxygen permeability. Extruded and blown films show no definable structure, whereas solution-cast films have a morphology of islands of poly(α-methylstyrene) in the siloxane matrix in some regions and the inverse structure in others.

Considerable attention has been devoted to the effects of molecular structure on properties and morphology of block copolymers where hard glassy blocks bound soft rubbery blocks. Studies of the effects of sample history and fabrication, however, have so far been limited to cases where polystyrene is the hard block, and polybutadiene or polyisoprene are the rubber components (*1*, *2*, *3*). These studies have shown that the level of shear applied during melt fabrication has profound effects on both morphology and mechanical properties in the fabricated sample.

The present study seeks to correlate sample history and fabrication with morphology and subsequent effects on modulus and permeability to oxygen in an analogous system where the rubber block is polydimethylsiloxane (*4*). This is of interest not only because of the significantly larger disparity in solubility, but also because of the uniquely high gas permeability of polydimethylsiloxane. Permeability as well as mechanical properties might be expected to be sensitive to changes in microstructure brought about during fabrication. Permeability in related poly(sulfone–siloxane) block copolymers was unaffected by solvent history in solution-cast films (*5*), but the effects of mechanical shear during fabrication have not been examined in block copolymers based on siloxanes.

The system chosen for study has poly(α-methylstyrene) as its glassy component; it differs from typical block copolymers in that the hard and soft

[1] Present address: Dow Corning Ltd., Barry, Wales CF6 7YL, U.K.

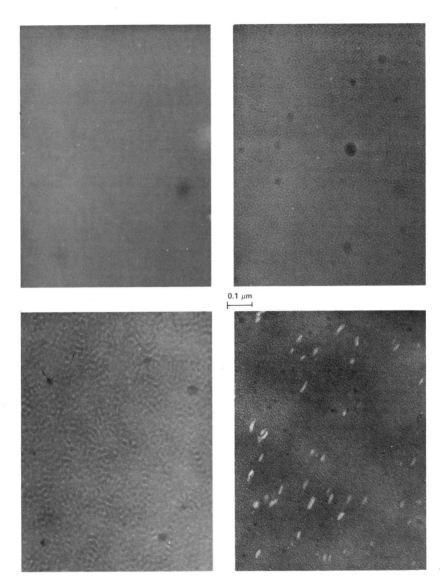

0.1 μm

Figure 1. Electron micrographs showing effect of block size on morphology
A, B, *and* C: *surface of thin films cast from toluene of block copolymers containing 40% poly(α-methylstyrene)* [A, *poly(α-methylstyrene) block—4000 g/mole;* B, *8000 g/mole; and* C, *18,000 g/mole*]; D: *thin section perpendicular to the surface, composition as in* B *but thicker film*

blocks alternate six to eight times and are terminated with the siloxane blocks. The hard blocks in this system, as in others that were based on polystyrene (6), are necessarily small (molecular weight 4000–15,000) compared with typical triblock copolymers. Larger block sizes in this system result in materials that have poor strength and are difficult to fabricate from the melt. An incidental advantage is that staining of the rubbery phase is not necessary in the electron microscopy (7).

Experimental

With the exception of the sequence shown in Figure 1, all data were obtained on a single block copolymer of the formula $[BAB]_{7.3}$ where A represents poly(α-methylstyrene) blocks of molecular weight 6000, and B polydimethylsiloxane blocks of molecular weight 4500. Synthesis and methods of characterization have already been reported (4, 6).

Tensile modulus was measured on an Instron under the conditions given in Table I. Electron micrographs were obtained with a Hitachi HS-75 electron microscope. Solution-cast films were prepared by removal of solvent at room temperature from 10% toluene solutions followed by final solvent removal at 70°C in vacuum. Samples were prepared for the micrographs as indicated in Figures 1, 2, and 3. No special staining techniques were required. Permeabilities of 0.5-mm solution-cast films and 0.25-mm extruded films were measured by a technique based on the mass spectrometer (8).

Table I.   Effect of Sample History on Mechanical Properties[a]

Method of Fabrication	Initial Modulus,[b] psi	Tensile Properties at Break[b]	
		Stress, psi	Strain, %
Solution-cast from toluene	4350	1150	930
Sample annealed at 150°C	1300	1850	825
Extruded, measured parallel to flow[c,d]	4500	1600	660
Sample annealed at 135°C	3800	1500	370
Extruded, measured perpendicular to flow	490	990	260
Sample annealed at 135°C	480	1700	420
Extruded and blown[e]	16500	—	—

[a] Data obtained on polymer described in Figure 2.
[b] Run at 10 in./min.
[c] Data subject to error due to high degree of dependence on the angle of applied stress (Equation 1); maximum values reported.
[d] Brabender extruder, head temperature 230–250°C; draw ratio 4:1, extruded as tubing, 0.010 in.-thick wall.
[e] Draw and blow ratios 4:1; 0.002-in. film.

Morphology

Figures 1A, 1B, and 1C represent a sequence of electron photomicrographs of films cast from toluene. The copolymers were of the same composition but increasing block size. The dark domains correspond to polydimethylsiloxane, the lighter domains to poly(α-methylstyrene). The copolymer with smallest block size had what can best be described as a texture observable at the limits of resolution. Copolymers with larger block size, however, had a discrete morphology reminiscent of poly(styrene–butadiene) triblock copolymers (2) despite the differences in composition and architecture. The poly(α-methylstyrene) tended to form irregular islands in some areas while in other areas it apparently composed the continuous phase. Often the islands appeared to merge at their borders. Domain size tended to increase with molecular weight of the blocks, indicating a higher degree of aggregation in the domains as molecular weight increased.

Figure 1D is a thin section taken perpendicular to the surface of a toluene-cast film of a polymer similar to that in Figure 1B. The morphology was similar which confirms the island or sphere type morphology in thicker, solution-cast samples.

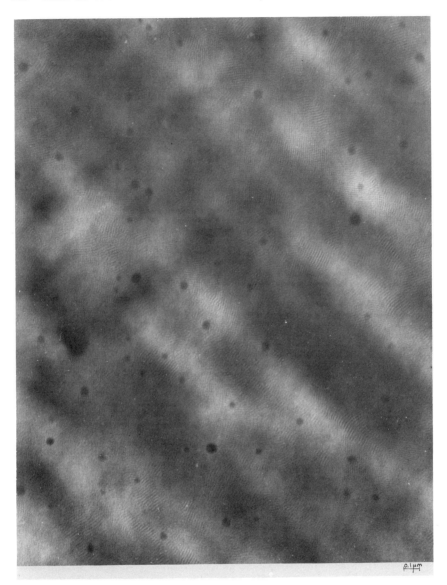

*Figure 2.   Electron micrograph of extruded copolymer freeze-sectioned perpendicular
to direction of extrusion [copolymer, 40 wt % poly(α-methylstyrene)]*

Melt extrusion and melt extrusion followed by drawing produced a new,
highly organized microstructure. When viewed perpendicular to the direction
of extrusion (Figure 2), locally regular hexagonal arrays of circular domains
composed of poly(α-methylstyrene) surrounded by a background matrix of
polydimethylsiloxane was visible in some regions of the field. Observation
parallel to the direction of extrusion revealed regular bands alternately com-

Figure 3. Electron micrograph of same copolymer as in Figure 2, but sectioned parallel to direction of extrusion

posed of polydimethylsiloxane and poly(α-methylstyrene) (Figure 3). The distance between bands corresponds to the distance between circles in Figure 2.

The light bands were approximately 60 A wide, and they were separated by ~65 A of intervening siloxane. These dimensions correspond roughly to the unperturbed chain end-to-end distances calculated for homopolymers of each block.

The observations suggest an ordered microstructure consisting of arrays of poly(α-methylstyrene) rods in a polydimethylsiloxane matrix. The rods were oriented in the direction of extrusion. Many of the rods in Figure 3 appear to terminate and be displaced by their neighbors. This and similar irregularities were probably caused by the freeze-sectioning.

Little, if any, distinctive morphology was observed when samples of block copolymer were biaxially stressed while at an elevated temperature. This was done by the bubble process of film blowing in which an extruded tube is simultaneously drawn and expanded radially. Both draw and expansion ratios were 4:1.

Mechanical Properties

The effects of sample history on mechanical properties are summarized in Table I. In all but the extruded sample that was measured perpendicular

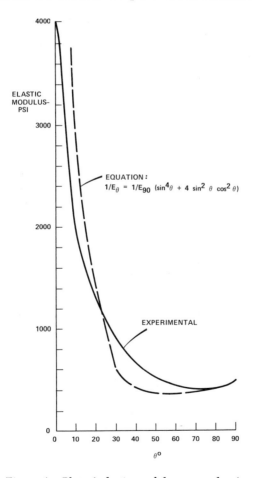

EQUATION:
$$1/E_\theta = 1/E_{90} (\sin^4\theta + 4 \sin^2 \theta \cos^2 \theta)$$

Figure 4. Plot of elastic modulus vs. angle of applied stress

Table II. Effect of Draw Ratio on Extruded Rods[a]

Draw Ratio	Initial Modulus, psi
2.0	2040
1.8	1750
1.4	690
1.2	480

[a] Same polymer as in Figure 2 extruded through a 1/4 in.-diameter circular die at 230–250°C.

to the direction of flow, a decrease in initial modulus was observed when the samples were annealed.

This decrease was especially large in the solution-cast films. These films can be dissolved after annealing and re-cast to give the original properties; this procedure was repeated up to three cycles. The drop in modulus was accompanied by increased tensile properties.

The initial modulus of extruded and drawn film had a high angular dependence. These films tended to be stiff when stressed in the direction of extrusion and very rubbery perpendicular to the extrusion direction (see Table I).

Figure 4 shows the same effect in more detail. Annealed samples fell on the same curve. Similar effects were noted by Folkes and Keller (1) who were able to fit their data to an equation derived by considering the deformation of a rubber matrix filled with glassy rods.

$$1/E_\theta \; = \; 1/E_{90°} \, (\sin^4 \theta \; + \; 4 \sin^2 \theta \cos^2 \theta) \tag{1}$$

where θ is the angle of applied stress relative to the direction of extrusion, $E_{90°}$ is the modulus perpendicular to the direction of flow, and E_θ is the modulus at angle θ. The broken line in Figure 4 is the plot of Equation 1 using the value for $E_{90°}$ that we found with our system. The samples used for Figure 4 were obtained at a draw ratio of 3:1. Other samples drawn at 1.5:1 fell on a similar curve, but the modulus was significantly lower (at $\theta = 0$, $E_\theta = 1490$ psi). The effect of draw ratio on modulus was similar when samples were extruded as rods (see Table II).

The initial modulus of films drawn biaxially, i.e. blown films, increased markedly (see Table I). Proper balance of draw ratio and bubble diameter gives isotropic materials.

Table III. The Effect of Sample History on Permeability to Oxygen in a Poly(α-methylstyrene–dimethylsiloxane) Block Copolymer[a]

Method of Fabrication	Permeability at 25°C, ml in./100 in.[2] 24 hr atm
Polydimethylsiloxane,[b] compression-molded and lightly crosslinked	1.2×10^5
Poly(α-methylstyrene)[c]	1.4×10^2
Block copolymer	
solution-cast	1.7×10^4
extruded film	—
extruded and blown film	3.3×10^4

[a] Same copolymer as in Figure 2.
[b] See Ref. 9.
[c] An estimate based on polystyrene (10).

Permeability

Blown film had a permeability to oxygen roughly twice that of the same film cast from toluene. In these studies, permeability was measured only perpendicular to the extrusion direction. The data are summarized in Table III.

Discussion of Results

The semi-empirical model discussed by Takayanagi *et al.* (*11*) is useful in collating observations of modulus and permeability in our system. When considering modulus in a composite, the model is such that stresses are distributed so that one field of force passes through only the rubber phase while a second passes through both phases. Equation 2 is appropriate for such a model when the rubber phase is continuous.

$$1/E = \frac{\phi}{(1 - \lambda)\,E_r + \lambda\,E_g} + \frac{1 - \phi}{E_r} \tag{2}$$

The modulus of the glassy phase, E_g, exceeds that of the rubbery phase, E_r, by three decades in our copolymer. Therefore the observed modulus of the copolymer might be expected to be more responsive to changes in the glassy phase. The mixing parameters, ϕ and λ, approach equality in finely dispersed systems. The numerical value of ϕ in this discussion indicates the degree to which a single field of force passes through both phases—the degree of series mixing. The value of λ indicates the degree of parallel mixing.

Mixing parameters can also be determined from observations of oxygen permeability by the application of an analogous model where pressure gradient and flow replace a force field and strain. In this model, flow in the rubber phase is influenced by both phases while flow in the glassy phase is unique. Permeability in the rubbery phase, P_r, exceeds flow in the glassy phase, P_g, by three decades, and observations of permeability, P, are very highly responsive to changes in the rubbery phase. Therefore, to obtain meaningful values for the mixing parameters, ϕ' and λ', it is necessary to assign the role of discrete phase to the rubber arbitrarily.

$$1/P = \frac{\phi'}{(1 - \lambda')\,P_g + \lambda'\,P_r} + \frac{1 - \phi'}{P_g} \tag{3}$$

Table IV.   Effect of Sample History on Mixing Parameters[a]

Method of Sample Fabrication	Modulus Measurements[b]		Permeability Measurements[c]	
	ϕ	λ	ϕ'	λ'
Solution cast	4.46	.090	1.63	.37
Above sample annealed	8.05	.050	—	—
Extruded, measured parallel to flow	4.40	.091	—	—
Above sample annealed	4.77	.084	—	—
Extruded, measured perpendicular to flow	13.21	.030	—	—
Extruded and blown	2.32	.17	.992	.605

[a] By the approach of Takayanagi *et al.* (*11*).
[b] Calculated presuming the rubbery phase continuous, $E_r = 200$ psi, and $E_g = 200,000$ psi.
[c] Calculated presuming the glassy phase continuous, and using data in Table III.

The significantly higher values of ϕ and ϕ' over λ and λ' in Table IV infer a higher degree of series mixing. This is interpreted as aggregation of the microdomains as opposed to a uniform distribution. Essentially this is also what is observed in electron photomicrographs of solution-cast as well as extruded films viewed perpendicular to the direction of extrusion. Extruded film stressed parallel to the extrusion direction approaches solution-cast film in mechanical equivalence. This is apparent from Table IV where ϕ and λ values for extruded and solution-cast films are similar.

The extremely high values of ϕ noted for the extruded sample tested perpendicular to the direction of extrusion is ascribed to a nearly ideal state of series mixing. This is substantiated by electron micrographs (Figures 2 and 3) that show the arrays of long rods which in this case run perpendicular to the field of applied stress. In such an arrangement, the applied field of force must run through both domains—the glassy rods and the rubbery matrix. Blown films, on the other hand, have little detectable microstructure. They would be expected to have values of ϕ and λ approaching one another more closely than do those for solution-cast films because of better dispersion.

Although we do not attempt to explain the driving force leading to the formation of the various observed microstructures, it seems clear that, in addition to block dimensions and insolubility of the phases, shearing forces play a major role. The magnitude of fabrication-induced changes in basic properties reported here is certainly remarkable. Similar changes and anisotropic effects probably effect other properties as well.

Literature Cited

1. Folkes, J. M., Keller, A., *Polymer* (1971) **12**, 222.
2. Aggarwal, S. L., Lavigni, R. A., Marker, L. F., Dudek, T. J., "Block and Graft Copolymers," J. J. Burke and V. Weiss, Eds., p. 157, Syracuse University, Syracuse, 1973.
3. Beecher, J. F., Marker, L., Bradford, R. N., Aggarwal, S. L., *J. Polym. Sci. Part C* (1969) **26**, 117.
4. Saam, J. C., Ward, A. H., Fearon, F. W. G., ADVAN. CHEM. SER. (1973) **129**, 239.
5. Roberson, L. M., Nosha, A., Matzner, M., Merriam, C. N., *Angew. Makromol. Chem.* (1973) **29/30**, 47.
6. Saam, J. C., Gordon, D. J., Lindsey, S. L., *Macromolecules* (1970) **3**, 1.
7. Saam, J. C., Fearon, F. W. G., *Ind. Eng. Chem. Prod. Res. Develop.* (1971) **10**, 10.
8. Dow Chemical Co., "Modern Methods of Research and Analysis," Interpretive Analytical Services (1973) p. 66.
9. Flaningham, O., private communication.
10. Brandup, J., Immergut, E. H., "Polymer Handbook," Interscience, New York, 1966.
11. Takayanagi, M., Harima, H., Iwata, Y., *Mem. Fac. Eng. Kyushu Univ.* (1963) **23C1**, 1.

RECEIVED June 11, 1974.

Block Copolymers of Methyl Methacrylate

RAYMOND B. SEYMOUR, GLENN A. STAHL, DON R. OWEN,[1] and
HUBERT WOOD

University of Houston, Houston, Texas 77004

Polymerization in poor solvents is a two-step process in which the length of the precipitating macroradical coil is controlled by its solubility in the solvent. The length of the growing chain in the coils is diffusion-controlled. Hence, the molecular weight of poly(methyl methacrylate) prepared in hexane is essentially independent of the concentration of initiator. The following block copolymers were prepared by the addition of selected monomers to macroradicals formed by polymerizing methyl methacrylate (MMA) in poor solvents: poly(MMA-b-acrylic acid), poly(MMA-b-acrylonitrile), poly(MMA-b-ethyl methacrylate), poly(MMA-b-styrene), poly(MMA-b-vinyl acetate), poly(MMA-b-vinylpyrrolidone), poly(MMA-co-styrene-b-acrylonitrile), and poly(MMA-b-styrene). These block copolymers were characterized by yield data, solubility, pyrolysis gas chromatography, and turbidimetric titration.

It is generally accepted that the most versatile procedure for the preparation of block copolymers is the addition of monomers to living ionic polymers (*1*). Accordingly, poly(methyl methacrylate-*b*-alkyl acrylates) and the corresponding block alkyl methacrylates have been prepared by anionic techniques (*2*). However, attempts to prepare poly(methyl methacrylate-*b*-styrene) by anionic techniques were not successful (*3*).

Poly(methyl methacrylate-*b*-acrylonitrile) has also been readily prepared by adding acrylonitrile to methyl methacrylate (MMA) living macroradicals (*4, 5*), and this technique is also a versatile procedure for the preparation of many other block copolymers (*6*).

Investigations of MMA macroradicals in solvents (*7, 8*) and in the vapor phase (*9*) have been reported. It was shown that these MMA macroradicals are stable in solvents which have solubility parameter values (δ) below 7.4 or above 11.0 hildebrand units (H) (*10, 11*).

In this investigation, block copolymers were prepared by adding vinyl monomers to macroradicals formed by free radical initiation of MMA at temperatures below 50°C in hexane ($\delta = 7.4$ H) and in 1-propanol ($\delta = 11.9$ H). Attempts to prepare blocks from this macroradical by adding isobutyl

---

[1] Present address: University of Southern Mississippi, Hattiesburg, Miss. 39401.

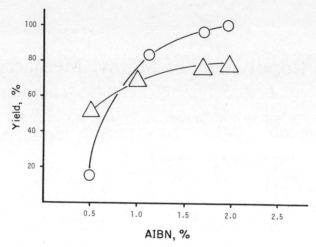

*Figure 1. Yield of PMMA after 24 hrs at 50°C in hexane
(○) and in benzene (△)*

methacrylate, isoprene, or vinyl bromide at 50°C were not successful. There
was some evidence of short blocks when butyl acrylate, hydroxyethyl methacrylate, and α-methylstyrene were added to MMA macroradicals in 1-propanol
at 50°C.

Homoblocks of MMA were readily obtained when MMA monomer was
added to MMA macroradicals in hexane or 1-propanol at 50°C. Blocks of significant size were formed when selected vinyl monomers were added to MMA
macroradicals in hexane or 1-propanol at temperatures below 50°C.

Experimental

Freshly distilled MMA was polymerized at 25°C in the absence of oxygen
by the ultraviolet light (UV) irradiation of a 10% solution of MMA monomer
in hexane which also contained 1% di-*tert*-butyl peroxide (based on MMA).
MMA macroradicals were also prepared by heating monomer solutions in
hexane or 1-propanol in the presence of 2.5% azobis(isobutyronitrile) (AIBN)
for 48 hrs at 50°C.

The rate of polymerization was monitored by gas chromatography (GC)
of residual monomer and initiator, by precision dilatometry, and by the yield of
macroradicals in aliquot samples. Block copolymers were prepared by transferring the slurry of MMA macroradicals in an inert atmosphere to small bottles
containing additional vinyl monomer. These mixtures were heated in sealed
bottles at 50°C for 72 hrs. The end products were characterized by GC
analysis of residual monomer, yield of solvent-washed product, pyrolysis GC
(PGC), and turbidimetric titration (TT).

Viscometric data were obtained by averaging five effluent times measured
to ±0.1 sec at 25.00°C in a no. 1 Ubbelohde viscometer. GC retention times
were obtained on a model A100C aerograph (Wilkens) equipped with a
Servo-Ritter II recorder (Texas Instrument) and packed with acid-washed
Chromosorb W (Johns Manville Corp.) and 20% SE-20 (General Electric Co.).

This equipment was also used to measure the retention times of the off-
gases of the pyrolyzate. The latter was obtained by placing a solution of the
sample on a rhenium tungsten code 13-002 coil (Gow-Mac Instrument Co.),

evaporating the solvent, and pyrolyzing the residual film for 5 sec. The relative amounts of components were estimated by comparison with the areas of PGC peaks of samples of known composition.

Turbidity data were obtained by titrating a selected nonsolvent continuously using a microsyringe. The nonsolvent was injected into a magnetically stirred, extremely dilute solution of the polymer in a square glass cell. This solution was illuminated by a parallel beam of light, and the intensity of the scattered light was measured by a photoelectric cell and recorded continuously.

Results and Discussion

Essentially quantitative yields of polymer (PMMA) were obtained when MMA was polymerized for 24 hrs at 50°C in hexane in the presence of 2.0% AIBN (Figure 1). Yields in this heterogeneous polymerization system were at least 80% when AIBN concentration exceeded 1%. In a homogeneous polymerization system, however, yields were less than 80% with 2% AIBN when benzene was the solvent.

Since the macroradicals precipitated when their molecular weight exceeded the solubility limit in hexane, initiator concentration affected the rate of formation but had little effect on the molecular weight of these macroradicals (Figure 2). In benzene, molecular weight decreased as AIBN concentration increased.

Although UV initiation of MMA polymerization was discontinued after eight hours, polymerization continued (Figure 3). The increase in molecular weight of the product with time after removal of the UV source demonstrated that the increased yield resulted from addition of MMA monomer to the previously produced MMA macroradicals.

Homoblocks of MMA were also produced from AIBN-initiated MMA macroradicals at 50°C in hexane. Both yield and reduced viscosity increased

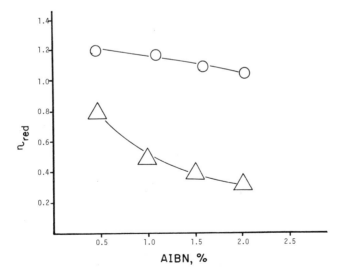

Figure 2. Reduced viscosity of PMMA solutions after 24-hr polymerization of MMA at 50°C in hexane (○) and in benzene (△)

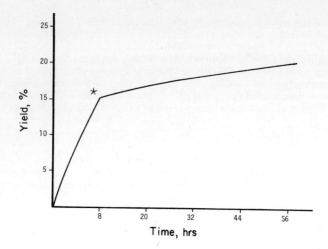

Figure 3.   Rate of UV-initiated polymerization of MMA in
1-propanol at 35°C
UV radiation was discontinued after 8 hrs (*)

with time (Figures 4 and 5). Since the hexane slurry contained no residual
AIBN, the increase in weight resulted from the addition of monomer to the
MMA macroradicals. This conclusion was supported by the findings from
viscometry and turbidimetry.

The polymerization of MMA in a poor solvent occurred in two steps. The
critical chain length of the precipitated MMA macroradical was governed by
its solubility in the poor solvent, and the number of precipitated macroradical
coils was a function of initiator concentration. The rate of diffusion of monomer
into the precipitated coils in the second step was governed by the ratio of

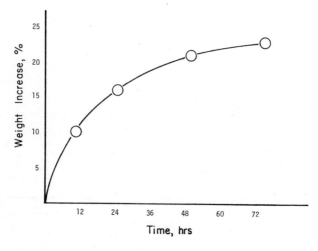

Figure 4.   Rate of formation of homoblocks of MMA
macroradicals at 50°C

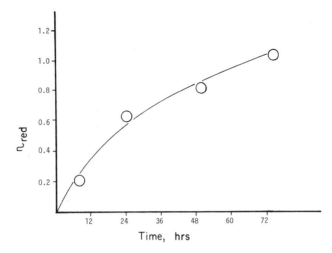

Figure 5.   Effect of increasing size of MMA homoblock on reduced viscosity of MMA macroradicals

$\overline{MMA}$ monomer to the number of coils and by the diffusibility of the monomer into the coils (*11*).

PGC data (Figure 6) revealed that a block copolymer in which the block constituted 24% of the macromolecule was obtained when acrylic acid (AA) was added to a MMA macroradical. The MMA macroradical was prepared by heating MMA monomer with 2.5% AIBN in hexane 48 hrs at 50°C. The block copolymer was prepared by heating the second monomer and the MMA macroradical in hexane 96 hrs.

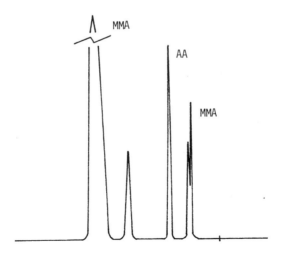

Figure 6.   Gas chromatographic pyrogram of poly(MMA-b-AA) (76:24)

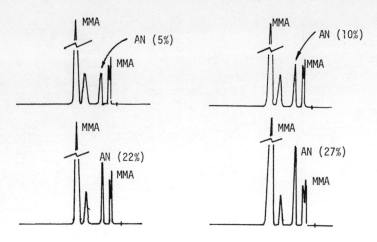

Figure 7.   Gas chromatographic pyrograms of poly(MMA-b-AN)

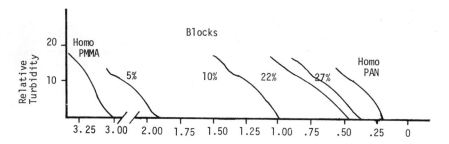

Figure 8.   Turbidimetric titration of poly(MMA-b-AN) in DMF

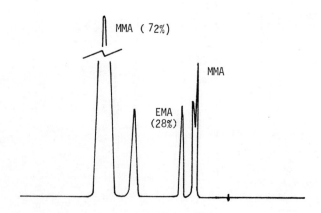

Figure 9.   Gas chromatographic pyrogram of poly(MMA-b-EMA)

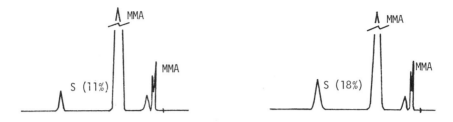

Figure 10. Gas chromatographic pyrograms of poly(MMA-b-S)

The preparation of block copolymers of MMA and acrylonitrile (AN) was reported (*12*). Findings from PGC (Figure 7) and TT (Figure 8) demonstrated that poly(MMA-*b*-AN) was prepared by heating AN and MMA macroradicals in hexane 96 hrs at 50°C; the blocks consisted of 5, 10, 22, and 27% copolymer. These block copolymers were soluble in both acetone and dimethylformamide (DMF).

Because of the low glass transition temperature (T_g) of poly(ethyl methacrylate) (PEMA), only a short block (10%) was obtained in hexane at 50°C; termination resulted from increased segmental motion of the EMA block. However, a block with a weight corresponding to 28% copolymer was obtained at 40°C. PGC data are presented in Figure 9.

Poly(MMA-*b*-styrene) formed in both hexane and 1-propanol. The styrene block (S) constituted 11 and 18% of the copolymer prepared in hexane and in 1-propanol, respectively (Figures 10 and 11).

The T_g for poly(vinyl acetate) (PVAC) is only 28°C. However, as shown by the PGC data (Figure 12), a copolymer with a VAC block equal to 13% of the macromolecule was obtained in 1-propanol.

PGC (Figure 13) and TT (Figure 14) revealed that a block consisting of 14 wt % vinylpyrrolidone (VP) was obtained when VP was heated with MMA macroradical in 1-propanol 96 hrs at 50°C. An ABC type block copolymer was obtained by adding AN to MMA-VP block copolymer (Figure 15).

It was shown that blocks of styrene did not add to AN macroradicals at moderate temperatures unless small amounts of a solvent for polyacrylonitrile were present (*12, 13*). However, AN does form a block with styrene macroradicals, and, as shown by PGC (Figure 16), poly(MMA-*co*-S-*b*-AN)

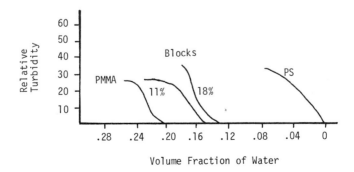

Figure 11. Turbidimetric titration of poly(MMA-b-S) in acetone

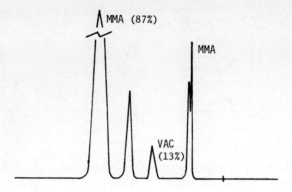

Figure 12. Gas chromatographic pyrogram of poly-
(MMA-b-VAC)

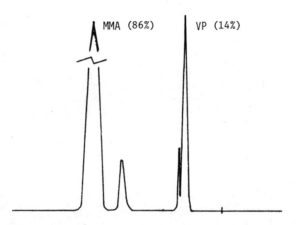

Figure 13. Gas chromatographic pyrogram of poly-
(MMA-b-VP)

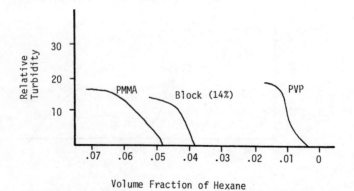

Figure 14. Turbidimetric titration of poly(MMA-b-VP)

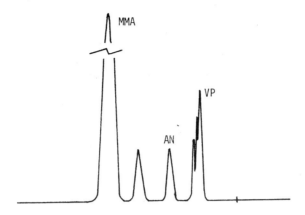

*Figure 15. Gas chromatographic pyrogram of poly-
(MMA-b-VP-b-AN)*

(43:37:20) was obtained when AN was heated at 50°C with poly(MAA-co-S) (54:46) macroradicals in hexane.

PGC (Figure 17) also demonstrated that S formed blocks with copolymers of MMA and AN. The size of the S block increased as the proportion of MMA in the copolymer macroradical increased; thus the S block constituted 12, 22, and 35% of the macromolecule when S was heated at 50°C in hexane with AN copolymers consisting of 50, 66, and 80% MMA respectively.

Likewise, PGC (Figure 18) revealed that an S block equal to 12% of the macromolecule was produced when S was heated at 50°C with poly(MMA-b-AN) (72:28) in 1-propanol in the presence of small amounts of DMF (which was miscible with 1-propanol) as well as in its absence. However, the S block obtained with this macroradical in hexane equaled only 6% of the macromolecule; nevertheless, a larger block of 15% S was obtained in the presence of relatively small amounts of DMF.

DMF was immiscible in hexane, and hence it could preferentially swell the macroradical and thereby increase the rate of diffusion of the S monomer.

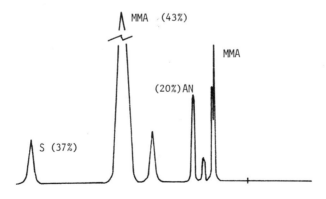

*Figure 16. Gas chromatographic pyrogram of poly(MMA-
co-S-b-AN)*

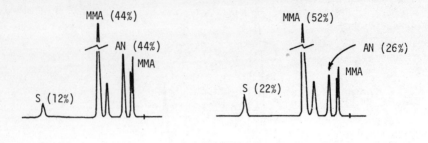

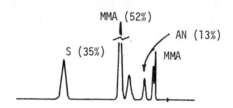

Figure 17.   Gas chromatographic pyrograms of poly(MMA-co-AN-b-S)

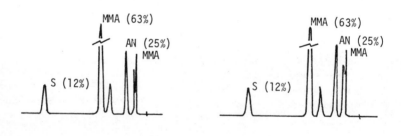

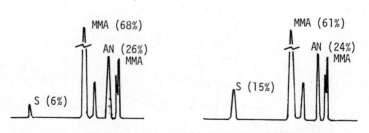

Figure 18.   Gas chromatographic pyrograms of poly(MMA-b-AN-b-S)

It was anticipated that S (δ = 9.3 H) would diffuse very slowly into poly-acrylonitrile macroradicals (δ = 12.5 H) and into macroradicals consisting of AN blocks.

Conclusion

Whereas the molecular weight of PMMA prepared in homogeneous solution was inversely proportional to the concentration of initiator, the molecular weight of PMA prepared in poor solvents was essentially independent of initiator concentration. MMA homopolymers with increasing molecular weights were prepared by adding MMA monomer to MMA macroradicals.

The following block copolymers were prepared and characterized by yield data, solubility, PGC, and TT: poly(MMA-b-AA), poly(MMA-b-AN), poly-(MMA-b-EMA), poly(MMA-b-S), poly(MMA-b-VAC), poly(MMA-b-VP), poly(MMA-b-VP-b-AN), poly(MMA-co-S-b-AN), and poly(MMA-b-AN-b-S).

Literature Cited

1. Lenz, R. W., "Organic Chemistry of Synthetic High Polymers," p. 722, Interscience, New York, 1967.
2. Graham, R. K., Panchak, J. R., Kampf, M. J., *J. Polym. Sci.* (1960) **44**, 411.
3. Graham, R. K., Dunkelberger, D. L., Goode, W. E., *J. Amer. Chem. Soc.* (1960) **82**, 400.
4. Slavnitskaya, N. N., Semchikov, Y. D., Ryabov, A., Bort, D. N., *Vysomol. Soedin, Ser. A* (1970) **12**(18), 1956; *Chem. Abstr.* (1970) **73**, 99240.
5. Seymour, R. B., Owen, D. R., Stahl, G. A., *Polymer* (1973) **14**, 324.
6. Seymour, R. B., Owen, D. R., Kincaid, P. D., *Chem. Technol.* (1973) **3**(9), 549.
7. Norrish, R. G. W., Smith, R. R., *Nature* (1942) **150**, 336.
8. Atkinson, B., Cotten, G. R., *Trans. Faraday Soc.* (1958) **54**, 877.
9. Melville, H. W., *J. Chem. Soc.* (1941) 414.
10. Burrell, H., "Polymer Handbook," J. Brandrup, E. H. Immergut, Eds., chap. 4, Interscience, New York, 1965.
11. Seymour, R. B., Kincaid, P. D., Owen, D. R., *J. Paint Technol.* (1973) **45**(580), 33.
12. Seymour, R. B., Owen, D. R., Stahl, G. A., Wood, H., Tinnerman, W. N., *Amer. Chem. Soc., Div. Polym. Chem., Prepr.* **14**(2), 658 (Chicago, August, 1973).
13. Minoura, Y., Ogata, Y., *J. Polym. Sci. Part A-1* (1969) **7**, 2547.

RECEIVED April 1, 1974. This investigation was supported in part by a grant from the Robert A. Welch Foundation.

28

Polymerization of Cyclic Imino Ethers IX: Surface Properties of Block Copolymers

MORTON H. LITT and TAKEHISA MATSUDA

Department of Macromolecular Science, Case Western Reserve University, Cleveland, Ohio 44106

A series of AB block coplymers of 2-lauroyl oxazoline ($X_n = 10$) with 2-ethyloxazoline ($X_n = 8, 23, and 68$) were synthesized and then characterized by surface tension and surface orientation. In glycol, the CMC was about 1.0 gm/l with $\gamma_{CMC} = 27.5–25$ dynes/cm. In methanol/water (80/20), the CMCs were 0.1–0.06 gm/l for the series with $\gamma_{CMC} = 24–23$ dynes/cm. Using the Gibbs adsorption equation, the surface area per non-polar residue was calculated at 14–16 Å^2. The low γ at the CMC and the low area per residue imply that the non-polar portion of the polymer is oriented as in the crystal phase with a surface of close-packed methyl groups. Moreover, after several days the polymers in emulsions crystallize at the oil–water interface.

B lock copolymers having different segments within the same molecule show composite behavior (*e.g.*, surface activity) when the physical properties of two segments contrast violently, in addition to the inherent properties characteristic of each homopolymer. Block copolymers of poly(N-lauroylethylenimine) and poly(N-propionylethylenimine) have the same polymer backbone but side chains with long and short hydrocarbon tails. The hydrophobic blocks of the copolymer consist of crystalline segments [poly(*N-lauroylethylenimine*), $T_m = ca.$ 110°–111°C, water insoluble]; the hydrophilic blocks consist of amorphous segments [poly(N-propionylethylenimine), $T_g = ca.$ 70°C, water soluble (1)]. The block copolymers were investigated with reference to molecular organization of the polymer side chains in bulk and at the interface.

Experimental

Synthesis of Block Copolymers. Since the polymerization of 2-oxazolines is nonterminating, after polymerization a polymer chain has an oxazolinum cation as an active chain end which can initiate the second stage of polymerization (1). Polymerization procedures were almost the same as those described previously (2). Since the chain transfer to monomer occurs in the alkyloxazo-line system to about 1/300 (3), the degree of polymerization was kept very low. The poly(N-lauroylethylenimine)–poly(N-propionylethylenimine) AB block copolymers were made by sequential cationic ring-opening polymerization

Table I.   Conditions for Block Copolymer Synthesis

| | Component, M/I^a | | Yield, wt % | Component, wt % | | Polymerization[c] Time, min | | Melting |
Copolymer	A^b	B^b		A	B	A	B	Point, °C
B-1	—	22.5	96	—	100	—	25	—
B-3	10.0	67.8	98	25	75	35	50	109–111
B-4	10.0	22.6	95	50	50	35	25	109–112
B-5	10.1	7.8	98	75	25	35	25	110–111
B-2	10.0	—	98	100	—	35	—	110–112

[a] M/I is the ratio of moles of monomer to moles of initiator; under the polymerization conditions, it is equivalent to the degree of polymerization.
[b] Component A is 2-undecyl-2-oxazoline [-poly(N-lauroylethylenimine)]; component B is 2-ethyl-2-oxazoline [-poly(N-propionylethylenimine)].
[c] Polymerization temperature, 125°C.

of 2-undecyl-2-oxazoline and 2-ethyl-2-oxazoline using 2-phenyl-2-oxazolinium perchlorate as the initiator. Details of polymer compositions and conditions are given in Table I. All the copolymers had 10 monomer units of poly(N-lauroyl-ethylenimine) but different chain lengths of poly(N-propionylethylenimine) segments.

The melting points of these block copolymers and homopolymer were 109°–112°C. These values are almost identical to the value (111°C) for poly(N-lauroylethylenimine) with 10 monomer residues as calculated from Flory's equation (4) using the experimental data determined previously by Litt *et al.* (2):

$$\frac{1}{T_m} - \frac{1}{T_{mo}} = -\frac{R}{\Delta Hu} \ln(1 - 2/\bar{P}_n)$$

where T_{mo} is the melting point of the crystallizable homopolymer with infinite chain length (150°C), ΔHu the heat of fusion (1,800 cal/mole), R the gas constant, and $\bar{P}_n$ the average degree of polymerization of crystallizable segment ($\bar{P}_n = 10$). This finding confirms that the average degree of polymerization of the polymers produced at the first stage of polymerization is almost equivalent to the feed molar ratio of monomer to initiator.

Critical Surface Tensions of Block Copolymers. Thin films of copolymers were deposited from 0.5% chloroform solutions onto clean glass slides. Contact angles of droplets of a number of liquids with polymer were measured with an NRL C. A. goniometer model A-100 at 21°C and 50% humidity using the sessile drop technique.

Surface Tension Measurements. Surface tensions of dilute solutions of the copolymers were measured by the ring method using the Noüy tensiometer at 22°C. The solvents used were chloroform ($\gamma_{LV} = 27.1$), ethylene glycol ($\gamma_{LV} = 48.3$) and methanol/water solution (methanol 80 vol %, $\gamma_{LV} = 27.1$).

Infrared Spectra. IR spectra were obtained with a Perkin Elmer model 521 spectrometer equipped with an internal wire grid polarizer. Internal reflection spectra were taken using an ATR Wilkes model 12 double beam attachment. Transmission and internal reflection spectra were obtained from plates composed of a germanium single crystal which were dipped into chloroform solutions of samples, air dried, heated at 100°C in vacuum, and then gradually annealed to room temperature.

Results

Surface Tension of Copolymer Solutions. The copolymers had little surface activity in chloroform which is a good solvent for both segments, but surface activity in methanol/water mixture (methanol 80 vol %) and in

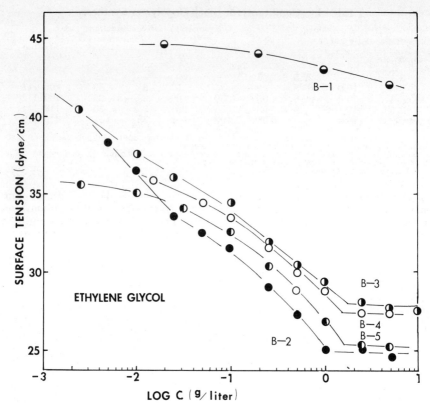

Figure 1. Surface tension vs. concentration isotherms of block copolymers in ethylene glycol

◖, *B-1, poly(N-propionylethylenimine);* ●, *B-2, poly(N-lauroylethylenimine);* ◑, *B-3;* ○, *B-4; and* ◐, *B-5*

ethylene glycol was considerable. Equilibrium surface tensions are plotted as a function of copolymer concentration in Figures 1 and 2. Surface tension decreased with increasing copolymer concentration to a limiting value, and then it remained unchanged despite further increases in copolymer concentration. Such behavior is usual for surfactants; the break in the log concentration/ surface tension curve is called the critical micelle concentration (CMC). At this concentration, the liquid surface is covered by a film of the surfactant. The surface tension at this point (low surface tension plateau) in both solvents is dependent on the copolymer composition; the higher the poly(*N*-lauroylethylenimine) content, the lower the surface tension (Figures 1 and 2 and Table II). Moreover, CMC decreases with an increase in poly(*N*-lauroylethylenimine) content in the copolymer. The low surface tension plateau for pure poly(*N*-lauroylethylenimine) occurred at 1.0 gm/l in ethylene glycol and at 0.06 g/l in methanol/water, the values being 24.9 and 23.0 dyne/cm, respectively. The latter is very close to the critical surface tension of a surface composed of close-packed methyl groups and to the surface tension of poly-(dimethylsiloxane) (~ 20 dynes/cm) (5). This may result from a favorable orientation of the lauroyl group at the interface because of its insolubility in the solvents used. For poly(N-propionylethylenimine), little surface activity was observed (*see* Figure 1).

From the slope of the surface tension *vs.* concentration curves at the CMC in Figures 1 and 2, the surface concentration of solute molecules adsorbed at the interface was calculated using the approximate form of Gibbs adsorption equation (6). Concentrations (C) were used instead of activities because the solutions were highly dilute:

$$\Gamma = \frac{1}{NA_L} = \frac{-1}{RT}\frac{d\delta}{d\ln C}$$

where Γ is the surface excess (surface concentration), A_L the limiting area (molecular area at closest packing), N Avogadro's number, δ the surface tension, R the ideal gas constant, and T the temperature (°K). The general trend is for the Gibbs limiting area to decrease as the per cent lauroyl groups increases. The calculated limiting areas in methanol/water and in ethylene glycol for poly(N-lauroylethylenimine) were 14–16 A^2/monomer unit, and the limiting areas for copolymer (B-5) containing 75% poly(N-lauroylethylenimine) in both interfaces were almost equal to those of poly(N-lauroylethylenimine). This may indicate that the monomolecular film of the macromolecules adsorbed at the interface consist mainly of poly(N-lauroylethylenimine) segments. These values are smaller than the limiting areas of monolayers of vinyl stearate or ethyl stearate (19.5–20 A^2) whose molecules stand very nearly perpendicular to the water/air interface (7).

The Surface Energy of Copolymer Films. The wettability of a surface is essentially controlled by the nature, molecular organization, and packing of

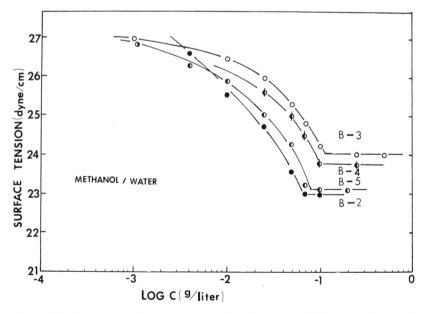

Figure 2.   Surface tension vs. concentration isotherms of block copolymers in methanol/water mixture (methanol, 80 vol %)
●, *B-2;* ○, *B-3;* ◓, *B-4; and* ◐, *B-5*

Table II. Surface Properties of Block Copolymer Solutions

Copolymer	Solvent	Concentration at Start of Plateau (CMC), g/l	Minimum Surface Tension, dyne/cm	Gibbs Limiting Area, A^2/molecule	Molecular Area Occupied per Monomer of A, A^2
B-3	methanol/	0.103 ± 0.01	24.0	240 ± 10	24 ± 1
B-4	water	0.100 ± 0.01	23.8	180 ± 10	18 ± 1
B-5	80/20	0.080 ± 0.005	23.1	162 ± 5	16 ± 0.5
B-2		0.060 ± 0.005	23.0	160 ± 10	16 ± 1.0
B-3	ethylene	1.07 ± 0.1	27.5	175 ± 5	17.5 ± 0.5
B-4	glycol	1.04 ± 0.1	27.0	160 ± 10	16 ± 1.0
B-5		1.05 ± 0.1	25.2	150 ± 10	15 ± 1.0
B-2		1.00 ± 0.1	24.9	142 ± 5	14 ± 0.5

the outermost groups on the surface. Table III lists the critical surface tension (γ_c) of the annealed block copolymers as determined from Zisman plots (Figure 3). The γ_c values of poly(N-lauroylethylenimine) given in Table III for methanol/water solutions and a series of hydrocarbons are much closer to the γ_c value, 22 dyne/cm, of a close-packed methyl group surface than to the values of organized methylene groups such as paraffin or polyethylene single crystal (8, 9). This indicates that the methylene and tertiary amide groups are not exposed at the interface, and that close-packed methyl groups form the surface (10).

The critical surface tensions of block copolymers with 25–75% poly(N-lauroylethylenimine) was always 22 dyne/cm, the value of the more surface-active segment, poly(N-lauroylethylenimine). This agrees with previous findings for block copolymers of poly(N-lauroylethylenimine) and poly(N-methyl-trimethylenimine) (2). Therefore, for block copolymers whose different sub-chains have different cohesive energies, there is a concentration of the lower cohesive energy segments at the interface, as also occurs with block copolymers of poly(dimethylsiloxane) and poly(bis-phenol-A carbonate) (11).

Molecular Orientation of Copolymer Film Surface. Attenuated total reflection (ATR) IR spectroscopy is a useful tool for determining molecular structure and orientation of surfaces, especially when polarized monochromatic light is used (12). The IR transmission and ATR spectra of annealed films of block copolymer B-4 and poly(N-lauroylethylenimine) deposited on the germanium plate are presented in Figures 4 and 5. Both ATR spectra closely resemble the corresponding transmission spectra except that the peak intensities are more distinct and sharper in the ATR spectra. This effect is generally attributable to a high degree of order of orientation. The carbonyl stretching peak at 1630 cm⁻¹ shows little difference in absorbance between both polarized spectra, that is, the absorbance of the parallel and perpendicular components

Table III. Critical Surface Tensions (γ_c), in dynes/cm, of Block Copolymers[a]

Copolymer	Methanol/Water[b]	Hydrocarbons
B-2	21	22
B-3	—	22
B-4	22	—
B-5	22	22

[a] Determinations were valid to about ± 0.5 dyne/cm.
[b] Methanol/water solutions with high methanol content injure the surfaces of block copolymer with high poly(N-propionylethylenimine) content.

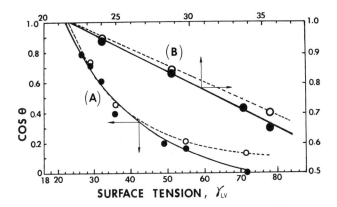

Figure 3. Zisman type plots by methanol/water solutions (bottom and left axes, A), and by hydrocarbon liquids (top and right axes, B)

Solid line, poly(N-lauroylethylenimine); and dashed line, block copolymer B-4

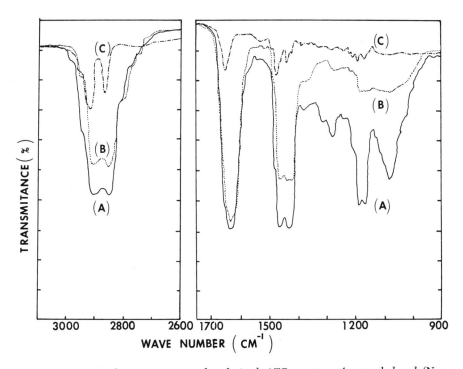

Figure 4.   Infrared transmission and polarized ATR spectra of annealed poly(N-lauroylethylenimine) film

A, ATR, parallel; B, ATR, perpendicular; and C, transmission

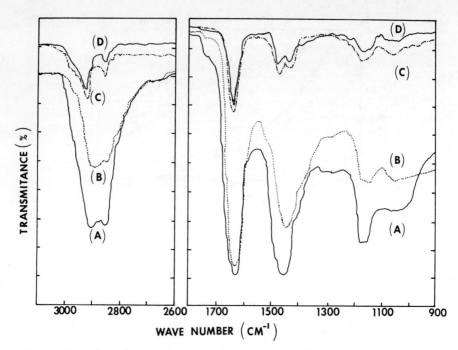

Figure 5. Polarized infrared ATR and transmission spectra of annealed block co-polymer (B-4) film

A, *ATR, parallel*; B, *ATR, perpendicular*; C, *transmission, parallel*; and D, *transmission, perpendicular*

of the light. This means that the carbonyl band has no special orientation in the XZ or XY planes. The characteristic peaks of methylene groups at 2920 (CH_2 assym), 2850 (CH_2 symm), and about 1480 cm^{-1} indicate higher absorbance of the parallel component of the light than of the perpendicular component. Moreover, the 1000–1200 cm^{-1} region is highly polarized in both spectra. The absorption of C—N stretching vibration appears in this region where, however, other bands also usually appear (13). These results suggest that the hydrocarbon tails and especially the C—N bond are oriented perpendicular to the XY plane.

Discussion

From the spectroscopic and surface wettability data, we may conclude that the outermost layer of the annealed copolymer surface is covered by the long hydrocarbon side chains of polymer with methyl groups forming a close-packed surface. According to the crystallographic study on a series of poly(N-acylethylenimine)s by Litt *et al.* (1), crystalline poly(N-lauroylethylenimine) has triclinic unit cell with two monomer units per unit cell [a = 4.9 A, b = 27.9 A, c (fiber axis) = 6.4 A, $\alpha = 90°$, $\beta = 62°$, and $\gamma = 98$ A]. In the proposed structure, the molecules assume a planar configuration with the main chain not fully extended; the lateral groups alternate on each side of the main chain, and they are tilted about 36° from the perpendicular to the chain axis in a parallel configuration. Therefore, if the condensed monomolecular film adsorbed at the interface is lying flat or horizontal on the interface, one monomer unit occupies an estimated area of about 89 A^2. On the other hand, when the adsorbed polymer segment is floating vertically on the interface, the cross sectional area of one monomer unit along the chain axis parallel to the interface is 14.8 A^2.

The limiting areas (the close-packed area) of the segment adsorbed at the interface calculated from the approximate Gibbs adsorption equation are 14–16 A^2/monomer unit (*see* above). These values are very close to that of the latter model. This suggests that the surface of block copolymer solutions at the CMC is composed of two-dimensional crystalline films with the poly(N-lauroylethylenimine) segments lifted out of the interface (Figure 6). The polymer backbone is parallel to the interface, while the side chains are tilted at the normal 36° angle from the perpendicular as in the bulk crystalline phase. The individual side chain is approximately perpendicular to the interface, and it is organized into a close-packed two-dimensional crystal where the organizing force is crystallization of the long hydrocarbon chain. Presumably, the hydrophilic, soluble poly(N-propionylethylenimine) segments anchor in the liquid phase where the segments can be solvated. Thus, the monomolecular film adsorbed at the solution/air interface arranges itself to have the lowest possible energy content, which produces the low surface energy. The larger area per

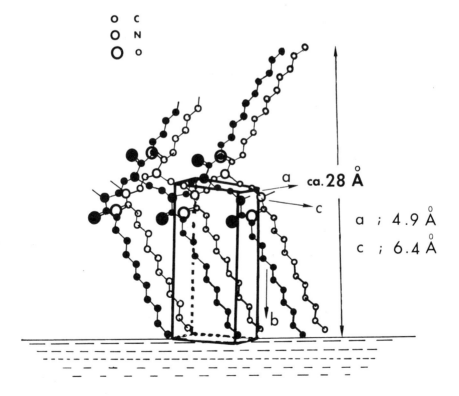

Figure 6.   Schematic representation of packing of poly(N-lauroylethylenimine) molecule at the solution/air interface

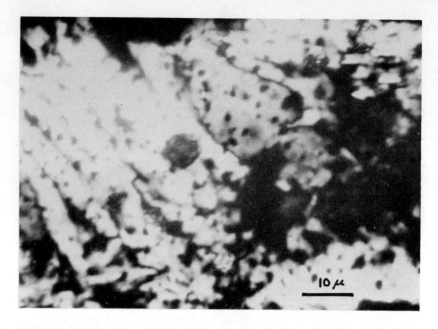

Figure 7. Most common morphology found for crystallized block copolymer (viewed between crossed polaroids)

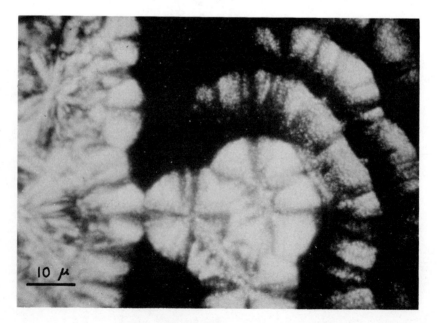

Figure 8. Spherulitic (two-dimensional) morphology of crystallized block copolymer (viewed between crossed polaroids)

molecule as the polar phase increases (Table II) is then caused by the repulsion of the water-soluble fraction of the polymer chains which prevents close packing of the crystalline phase.

Note on Interface Crystallization. Emulsions of water with a hydrocarbon can be prepared easily from these block copolymers by warming to 60°C with stirring. All form water-in-oil emulsions which are stable for several days but which eventually collapse. Optical microscopy of the fragments of interface of copolymer B-5 is very interesting; it is reported here as a preliminary note. The major reason for the collapse of the emulsion is that the polymers crystallized in three dimensions.

Figures 7 and 8 show two typical morphologies viewed under crossed polarized filters. The more common type (Figure 7) appears to be extended sheets of crystalline material with no spherulitic texture. The orientation was relatively uniform over large areas; in some cases sharp extinctions of regions were observed. The second type of morphology (Figure 8) is spherulitic (it must be mainly two-dimensional). The rings at the right seem to be a continuation of the growth of the central spherulite; the dark areas are regions where the film was too thin for birefringence. The driving force for crystallization in three dimensions is sufficient to break the emulsion.

Literature Cited

1. Litt, M. H., Rahl, F., Roldan, L. G., *J. Polym. Sci. Part A-2* (1969) **7**, 463.
2. Litt, M. H., Herz, J., Turi, E., in "Block Polymers," S. L. Aggarwal, Ed., pp. 313-319, Plenum, New York, 1970.
3. Litt, M. H., unpublished data.
4. Flory, P. J., in "Principles of Polymer Chemistry," pp. 568-576, Cornell University, New York, 1953.
5. Fox, H. W., Taylor, P. W., Zisman, W. A., *Ind. Eng. Chem.* (1947) **39**, 1401.
6. Adamson, A. W., in "Physical Chemistry of Surfaces," 2nd ed., John Wiley & Sons, New York, 1967.
7. Fort, Jr., T., Alexander, A. E., *J. Colloid Interface Sci.* (1959) **14**, 190.
8. Zisman, W. A., in "Contact Angle; Wettability and Adhesion," ADVAN. CHEM. SER. (1964) **43**, 1.
9. Shafrin, E. G., Zisman, W. A., *J. Colloid Sci.* (1952) **7**, 166.
10. Litt, M. H., Herz, J., *J. Colloid Interface Sci.* (1969) **31**, 248.
11. LeGrand, D. G., Gaines, Jr., G. L., *Amer. Chem. Soc., Div. Polym. Chem., Prepr.* **11**, 442 (Chicago, Sept., 1970).
12. Barr, J. K., Fluornoy, P. A., in "Physical Methods in Macromolecular Chemistry," B. Carroll, Ed., Vol. 1, Marcel Dekker, New York and London, 1969.
13. Bellamy, L. J., in "The Infrared Spectra of Complex Molecules," 2nd ed., Methuen, London, 1958.

RECEIVED June 7, 1974.

29

Block and Random Copolymerization of Episulfides

A. ROGGERO, A. MAZZEI, M. BRUZZONE, and E. CERNIA

Snam Progetti S.p.A., Polymer Research Laboratories, S. Donato Milanese, Milan, Italy

The polymerization and copolymerization of episulfides using anionic catalysts obtained by metalation of sulfoxides and sulfones with alkali metals, their hydrides, amides, or alkyl derivatives is reported. Depending on the nature of the sulfoxides (or sulfones) used, both mono- and dianionic catalysts were obtained; the latter gave particularly interesting results regarding catalyst activity and the molecular weight of the obtained polyepisulfides. The living nature of these polymerizations is discussed, and some technological properties of the AB, ABA, and random copolymers are reported, with particular attention to the oil resistance of these elastomers.

Although polymerization of cyclic sulfides has been studied much less than that of the corresponding oxides, much information is available about the synthesis and the properties of the polymers obtained. Among cyclic sulfides, the thiiranes have certainly been studied the most. The use of anionic type catalysts in their polymerization has been reported for the homopolymers (*1, 2, 3, 4, 5*) and for block copolymers (*2, 4, 5, 6, 7, 8*). Fewer data are available on their statistical copolymerization (*2, 9*).

The synthesis of particular anionic initiators obtained by the metalation of dimethyl sulfoxide (*10*) and their use in the polymerization of vinylic monomers (*11, 12, 13*) and epoxides (*14, 15, 16, 17*) were reported recently. Similar research on sulfone metalation (*18, 19, 20, 21*) has led to well-defined products which were rarely used (*22*) as anionic initiators. In any case, the last reference is to compounds which do not contain hydrogen which can be metalated.

In this work we continued the study of the metalation of several sulfoxides and sulfones. We used them for the first time in the polymerization (*23, 24*) of episulfides such as ethylene sulfide (ES), propylene sulfide (PS), isobutylene sulfide (IBS), and allyl thioglycidyl ether (ATE). In relation to the sulfur contained in the main chain, these polymers show an interesting combination of oil resistance and a low glass transition temperature. On the other hand, the best known polythiiranes (such as ethylene sulfide and propylene sulfide) have a limited thermal stability because of the presence of carbon–sulfur bonds as well as the nature of the groups near the sulfur atoms.

This initial study attempts to clarify these facts by taking into consideration the statistical and block copolymers of ES, PS, and IBS. Such a study also required the synthesis of polymers containing limited quantities of an unsaturated episulfide (ATE) with the aim of evaluating on vulcanized products the dynamic and mechanical properties of the different synthesized structures.

Experimental

Catalysts. Sulfoxides and sulfones were purchased from Fluka or prepared by standard oxidations of the corresponding sulfides. $CH_3—(CH_2)_{12}\overset{\overset{\displaystyle O}{\|}}{—S}—CH_3$

and $CH_3\overset{\overset{\displaystyle O}{\|}}{—S}—(CH_2)_6\overset{\overset{\displaystyle O}{\|}}{—S}—CH_3$ were obtained by reaction of methylsulfinyl carbanion with lauryl chloride and 1,4-dibromobutane, respectively.

Metalation was performed using sodium hydride, sodium amide, butyllithium (LiBu), lithium diethylamide, etc., and the resulting organometallic compounds were added to $(C_6H_5)_2CO$, C_2H_5Br, and CH_3I in order to clarify their structures

Monomers. ES and PS were Fluka products. IBS and ATE were synthesized by reaction of KSCN with the corresponding oxides. All monomers (purity >99% as determined by gas liquid chromatography) were dried over freshly pulverized CaH_2 and distilled under nitrogen prior to use.

Polymerizations. All polymerizations were carried out in glass vessels, and all transfers were made under nitrogen.

For statistical copolymerizations, solvents (tetrahydrofuran and hexamethylphosphoramide) and monomers were introduced and cooled at $-30°C$, and catalyst (0.05–0.2% with respect to the monomers) was then added. After a few minutes the temperature was raised to $25°C$, and the polymerization was allowed to proceed for 2–10 hrs. C_2H_5Br (dried over phosphorus pentoxide) was used to terminate the polymerization to avoid the unstable mercaptan chain end. The copolymers were precipitated with methanol, then filtered and dried 24 hrs *in vacuo* at $45°C$. All yields were nearly quantitative.

In the synthesis of block copolymers, the first monomer was allowed to be quantitatively polymerized, and subsequently the second monomer was introduced.

Analyses. ^1H-NMR spectra were obtained with a Varian HA-100 spectrometer. The solution contained $\sim3\%$ (w/v) copolymer in o-dichlorobenzene. Tests were run at $80°C$ with hexamethyl disiloxane as an internal standard. All ^{13}C-NMR spectra were run at $80°C$ on 12-mm samples containing $\sim15\%$ solution (w/v) of copolymer in C_6D_6-o-dichlorobenzene (50/50 v/v). The Varian XL 100 spectrometer was operated at 25.14 MHz in the Fourier Transform mode.

Glass transition temperatures (T_g) and crystalline melting points were measured by differential scanning calorimetry (DSC) on a DuPont model 900 instrument (heating rate, $10°C/min$).

Number average molecular weights were determined in toluene at $30°C$ using a Mechrolab membrane osmometer. UV number average molecular weights were determined by evaluating naphthyl end group content using a Hitachi Perkin-Elmer EPS-3T spectrophotometer.

Results and Discussion

Catalysts. The metalation of sulfoxides and sulfones by substituting the hydrogen atoms situated α to the —SO— and —SO_2— groups with alkaline

Table I.   Catalysts Obtained by Metalation of Sulfoxides and Sulfones[a]

Monoanions

1.[b] $CH_3-\overset{\overset{\displaystyle O}{\|}}{S}-CH_2^{(-)}\ M^{(+)}$

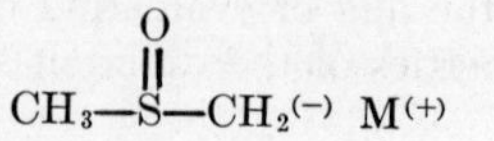

2. $CH_3-(CH_2)_{12}-\overset{\overset{\displaystyle O}{\|}}{S}-CH_2^{(-)}\ M^{(+)}$

3. $\overset{\overset{\displaystyle O}{\|}}{S}-CH_2^{(-)}\ M^{(+)}$

4.[c] $CH_3-\overset{\overset{\displaystyle O}{\|}}{\underset{\underset{\displaystyle O}{\|}}{S}}-CH_2^{(-)}\ M^{(+)}$

5.[c] $^{(-)}\ M^{(+)}$

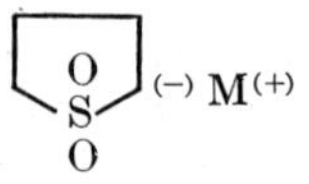

6. 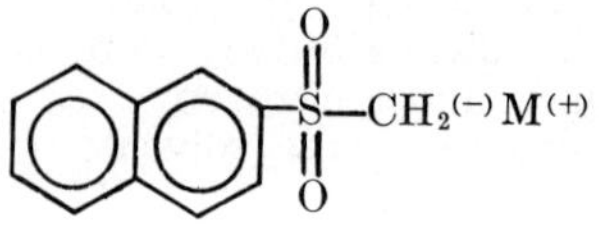 $\overset{\overset{\displaystyle O}{\|}}{\underset{\underset{\displaystyle O}{\|}}{S}}-CH_2^{(-)}\ M^{(+)}$

7. 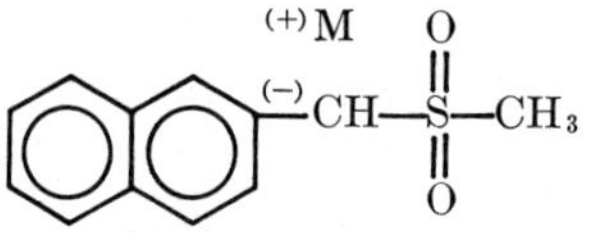 $\overset{(+)M}{\underset{(-)}{C}}H-\overset{\overset{\displaystyle O}{\|}}{\underset{\underset{\displaystyle O}{\|}}{S}}-CH_3$

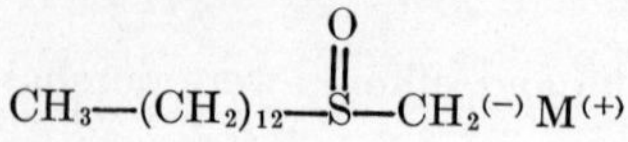

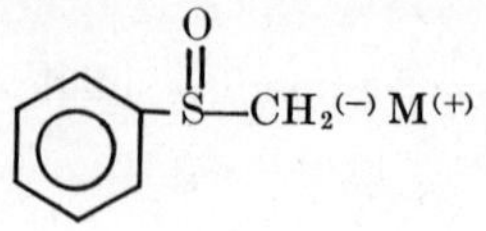

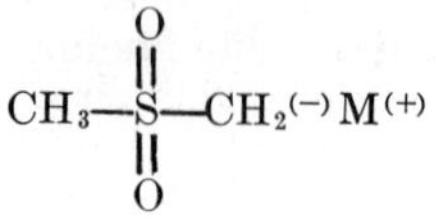

Dianions

8. 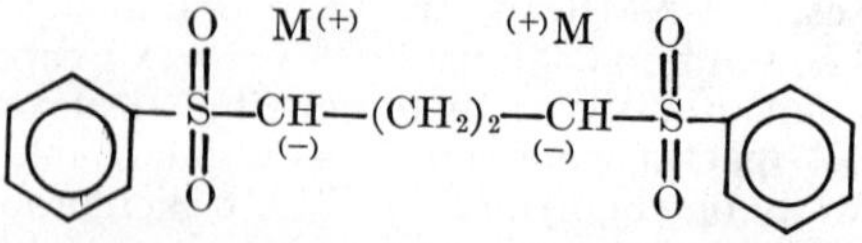 $\overset{\overset{\displaystyle O}{\|}}{\underset{\underset{\displaystyle O}{\|}}{S}}-\overset{M^{(+)}}{\underset{(-)}{C}}H-(CH_2)_2-\overset{(+)M}{\underset{(-)}{C}}H-\overset{\overset{\displaystyle O}{\|}}{\underset{\underset{\displaystyle O}{\|}}{S}}$

9. 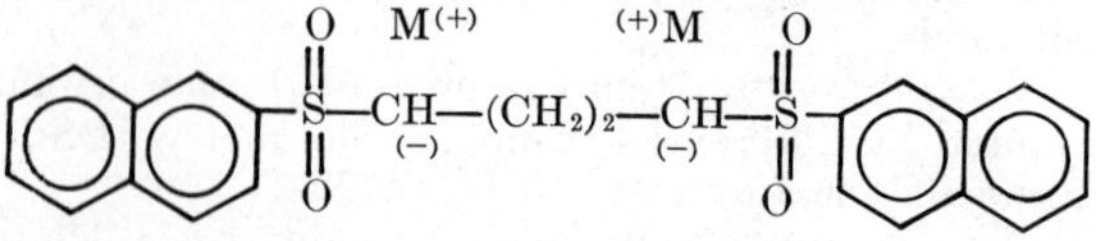 $\overset{\overset{\displaystyle O}{\|}}{\underset{\underset{\displaystyle O}{\|}}{S}}-\overset{M^{(+)}}{\underset{(-)}{C}}H-(CH_2)_2-\overset{(+)M}{\underset{(-)}{C}}H-\overset{\overset{\displaystyle O}{\|}}{\underset{\underset{\displaystyle O}{\|}}{S}}$

10. 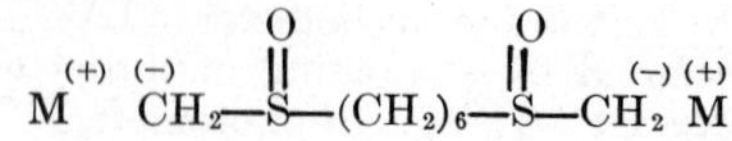 $\overset{(+)\ (-)}{M}\ CH_2-\overset{\overset{\displaystyle O}{\|}}{S}-(CH_2)_6-\overset{\overset{\displaystyle O}{\|}}{S}-CH_2\ \overset{(-)\ (+)}{M}$

[a] M = Li or Na.
[b] *See* Reference *10*.
[c] *See* References *18, 19*.

metals (*10, 18, 19, 20, 21*) or with magnesium (*21*) is known from the literature. From monosulfoxides and monosulfones it was possible to obtain mono- or
dianions, depending on the ratio of metalating agent to compound, even if the
formation of dianionic species was not particularly selective (*25, 26*). Among
the metalating species, results were best if $LiNH_2$, $NaNH_2$, and $LiNR_2$ were
used rather than LiBu.

We have extended the study of the metalation of sulfoxides and sulfones
in order to obtain carbanions with various structures. By using compounds
containing two activating groups (—SO— or —SO_2—) in the same molecule,
it was possible to obtain dianions more selectively, even if the functionality
was not rigorously univocal. Together with the dimetalated compounds, small
quantities of products having different degrees of metalation were always
present. Such selectivity is of fundamental importance for the univocal
synthesis of block polymers.

The synthesized catalysts are listed in Table I. They were condensed with
different electrophile agents; the resultant condensation products are tabulated
in Table II, together with NMR and mass spectra data.

Polymers. Preliminary studies on the homopolymerization of episulfides
with the catalysts listed in Table I demonstrated that it is possible to obtain
products that have very high crystallinity [poly(ethylene sulfide) and poly(isobutylene sulfide)] as well as essentially amorphous products [poly(propylene
sulfide)].

In the latter case, it was possible to obtain information on the configura-
tional microstructure by ^{13}C- and ^{1}H-NMR studies using poly(propylene-2-d_1
sulfide). The findings demonstrated that poly(propylene sulfide) contains the
same quantities of isotactic and syndiotactic sequences and that appreciable
quantities of irregular head-to-head, tail-to-tail enchainments were not present.

In copolymerization, use of the above-mentioned catalysts led, depending
on the procedure, to the principal formation of either statistical or block polymers. Thus, AB block polymers were formed when monoanionic catalysts were
used and polymerization was conducted in two stages; ABA blocks were formed
with dianionic catalysts.

STATISTICAL, CO-, AND TERPOLYMERS. Some data on ES–PS copolymers
obtained with the catalysts of Table I are reported in Table III.

With limited quantities of catalyst (0.05 mole % with respect to the monomers), conversion was total in relatively short times (several hours). When
the ES content of the copolymers did not exceed 45 mole %, crystallinity was
not evident on x-rays. The copolymers were soluble in $CHCl_3$ and in *o*-dichloro-
benzene, even when the ES content was around 50 mole %. Since poly(ethylene sulfide) is completely insoluble in these solvents, the copolymer nature
of the products obtained was evident. X-ray data confirmed this statement.

ES–PS copolymers were studied by ^{13}C-NMR. Complete interpretation
of the spectrum (*27*) provided information on the distribution of the monomeric sequences in terms of dyads and triads. The triad number, in per cent
monomeric unit, was calculated assuming a statistical distribution, and the
experimental value was determined by evaluating the various peak intensities
for two copolymers having different ES content (*see* Table IV). A comparison
of theoretical and experimental values demonstrates the mainly statistical
nature of the copolymers examined.

Data on ES–IBS copolymers are presented in Table V. Both homopolymers
are extremely crystalline, and therefore when the copolymer had a composition

Table II.   Condensation Products

Catalyst[a]	Electrophile	Condensation Product
1	$\phi_2\,C{=}O$	$CH_3{-}\overset{\displaystyle O}{\overset{\|}{S}}{-}CH_2{-}\underset{}{C}{\Big\langle}\overset{OH}{\underset{\phi}{\phi}}$
2	$\phi_2\,C{=}O$	$CH_3{-}(CH_2)_{12}{-}\overset{\displaystyle O}{\overset{\|}{S}}{-}\overset{\alpha'}{CH_2}{-}C{\Big\langle}\overset{OH}{\underset{\phi}{\phi}}$
2'ᵉ	CH_3I	$CH_3{-}(CH_2)_{12}{-}\overset{\displaystyle O}{\underset{\displaystyle O}{\overset{\|}{\underset{\|}{S}}}}{-}\overset{\alpha'}{CH_2}{-}\overset{\beta'}{CH_3}$
3	$\phi_2\,C{=}O$	$\phi{-}\overset{\displaystyle O}{\overset{\|}{S}}{-}CH_2{-}C{\Big\langle}\overset{OH}{\underset{\phi}{\phi}}$
4	$\phi_2\,C{=}O$	$CH_3{-}\overset{\displaystyle O}{\underset{\displaystyle O}{\overset{\|}{\underset{\|}{S}}}}{-}CH_2{-}C{\Big\langle}\overset{OH}{\underset{\phi}{\phi}}$
5	$\phi_2\,C{=}O$	ring: β and α carbons in a ring containing $\overset{O}{\underset{O}{S}}$ with side $C{\Big\langle}\overset{OH}{\underset{\phi}{\phi}}$
6	C_2H_5Br	naphthyl${-}\overset{\displaystyle O}{\underset{\displaystyle O}{\overset{\|}{\underset{\|}{S}}}}{-}CH_2{-}CH_2{-}CH_3$
7	C_2H_5Br	naphthyl${-}\overset{CH_2{-}CH_3}{CH}{-}SO_2{-}CH_3$
8	C_2H_5Br	$\phi{-}\overset{\displaystyle O}{\underset{\displaystyle O}{\overset{\|}{\underset{\|}{S}}}}{-}\overset{CH_2{-}CH_3}{CH}{-}CH_2{-}CH_2{-}\overset{CH_3{-}CH_2}{CH}{-}\overset{\displaystyle O}{\underset{\displaystyle O}{\overset{\|}{\underset{\|}{S}}}}{-}\phi$

of Catalysts with Electrophiles

	^1H-NMR Spectra	
Solvent	*Chemical Shifts,[b] ppm*	*Mass Spectra*

Solvent	Chemical Shifts	Mass Spectra (m/e)					
CDCl$_3$	7.30(ϕ,m)3.55(CH$_2$,m) 2.55(CH$_3$,s)^c	105 100^d 197 70 77 43	91 32 183 15 198 11	106 8 51 8 M$^+$·260 1			
CH$_2$Cl$_2$	7.30(ϕ,m)5.77(OH,s)3.43(α'CH$_2$m) 3.70(αCH$_2$,m)1.62(βCH$_2$,m) 1.15(other CH$_2$)0.80(CH$_3$,t)	197 100 105 58 183 20	198 18 91 17 77 16	180 16 410 13 M$^+$ 428 0			
CDCl$_3$	2.9(α,CH$_2$,α'CH$_2$,m)1.78(βCH$_2$,m) 1.31(β'CH$_3$,t)1.16(other CH$_2$,m) 0.80(CH$_3$,t)	95 100 57 44 43 40	55 30 41 39 71 28	94 23 69 20 M$^+$·276 2			
CDCl$_3$	7.30(ϕ,m)6.60(OH,s) 3.54(CH$_2$,m)	105 100 197 92 77 41	91 33 198 14 78 9	106 8 183 8 M$^+$·322 01			
—	7.35(ϕ,m)4.96(OH,s) 3.97(CH$_2$,m)2.35(CH$_3$,s)^f	183 100 105 59 77 28	184 14 121 11 199 7	91 7 51 7 M$^+$·276 1			
—	7.30(ϕ,m)4.35(OH,s) 3.20(αCH$_2$,αCH,m) 2.10(βCH$_2$,m)^f	183 100 105 44 77 21	184 19 147 6 55 6	51 5 225 4 M$^+$·302 3			
CDCl$_3$	8.5–7.5($\phi\phi$,m)2.14(αCH$_2$,t) 1.74(βCH$_2$,m)0.96(CH$_3$,t)	127 100 128 72 M$^+$·234 65	43 32 192 24 144 22	41 19 115 16 126 16			
CCl$_4$	8.3–7.3($\phi\phi$,m)4.78(CH,m) 1.7–2.6(CH$_2$,m)2.23(SO$_2$CH$_3$,s) 0.85(CH$_3$,t)	169 100 141 51 153 16	170 14 128 12 152 11	154 10 129 8 M$^+$·248 6			
CDCl$_3$	7.9–7.4(ϕ,m)2.76(CH,m) 1.95–1.35(CH$_2$,m)0.84(CH$_3$,t)	143 100 69 80 253 57	55 57 41 38 77 36	111 29 83 19 M$^+$·394 3			

$$9 \quad C_2H_5Br$$

$$\text{(naphthyl)}-\overset{O}{\underset{O}{\overset{\|}{\underset{\|}{S}}}}-\overset{CH_2-CH_3}{\overset{|}{CH}}-CH_2-CH_2-\overset{H_3C-H_2C}{\overset{|}{CH}}-\overset{O}{\underset{O}{\overset{\|}{\underset{\|}{S}}}}-\text{(naphthyl)}$$

$$10 \quad C_2H_5Br$$

$$CH_3-CH_2-CH_2-\overset{O}{\overset{\|}{S}}-(CH_2)_6-\overset{O}{\overset{\|}{S}}-\overset{\alpha'}{CH_2}-\overset{\beta'}{CH_2}-CH_3$$

a See Table I.
b m = multiplet, t = triplet, and s = singlet.
c See Reference 10.

Table III.   Ethylene Sulfide–Propylene Sulfide Copolymers

Copolymer Composition[a]					
ES, mole %	PS, mole %	$[\eta]$,[b] dl/g	T_g,[c] °C	X-ray	^{13}C-NMR
23	77	2.94	−45	amorphous	nearly random[d]
35	65	2.61	−44	amorphous	nearly random[d]
45	55	3.18	−44	nearly amorphous	nearly random
54	46	4.25	−42	low ES crystallinity	not analyzed

a ^{1}H-NMR data in o-dichlorobenzene at 80°C.
b From CHCl₃ solution at 30°C.
c From DSC analysis.
d See Table IV.

Table V.   Ethylene Sulfide–Isobutylene

Copolymer Composition[a]			
ES, mole %	IBS, mole %	X-ray	T_g,[b] °C
> 60	< 40	low ES crystallinity	−35
55	45	nearly amorphous	−35
46	54	nearly amorphous	−36
34	66	nearly amorphous	−35
21	79	low IBS crystallinity	

| CDCl₃ | 8.4–7.3(φφ,m)3.0(CH,m)
 1.8(CH₂,m)0.87(CH₃,t) | $\frac{m}{e}$: | 128
 303
 129 | 100
 28
 14 | 127
 69
 41 | 46
 26
 14 | 193
 55
 M⁺·494 | 44
 26
 4ᵍ |

| | 2.85(αCH₂,α′CH₂,m)
 1.80(βCH₂βCH₂,m)
 1.40(γCH₂,m)0.95(CH₃,t) | $\frac{m}{e}$: | 55
 83
 175 | 100
 35
 23 | 41
 27
 223 | 75
 29
 23 | 43
 63
 M⁺·266 | 42
 26
 01 |

ᵈ Relative intensity.
ᵉ CH₃—(CH₂)₁₂—SO₂—CH₂⁽⁻⁾ M⁽⁺⁾.
ᶠ *See* Reference 18.
ᵍ Probably slightly thermally decomposed.

Table IV.   Sequence Distribution of Ethylene Sulfide–Propylene Sulfide Copolymers

	Triad Number, %			
	23 mole % ES		35 mole % ES	
Triad Type	*Exptl.*	*Calcd.*	*Exptl.*	*Calcd.*
PPP + PPE	58.6	59.3	41.5	42.2
EPP + EPE	18.3	17.7	23.4	22.8
PEP	11.3	13.6	12.0	14.8
PEE	5.7	4.1	7.6	8.0
EEP	4.9	4.1	8.8	8.0
EEE	1.2	1.2	6.7	4.2

Sulfide Copolymers

THF Extract			CHCl₃ Extract			
wt %	ES,ᵃ mole %	η,ᶜ rel	wt %	ES,ᵃ mole %	η,ᶜ rel	RES, wt %
48.5		1.159	16.3			35.2
75.0	53	1.152	21.9	55		3.1
71.4	46	1.154	27.1	45	1.141	1.5
87.0	34	1.125	8.7		1.138	4.3
68.2	21	1.096	28.0	20	1.134	3.8

ᵃ ¹H-NMR data in *o*-dichlorobenzene at 80°C.
ᵇ From DSC analysis.
ᶜ From CHCl₃ solution at 30°C, (C = 0.1 g/dl).

Table VI.   Propylene Sulfide–Isobutylene Sulfide Copolymers

Copolymer Composition[a]				
PS, mole %	IBS, mole %	X-ray	$[\eta]$,[b] dl/g	T_g,[c] °C
88	12	amorphous	1.39	−41.5
84	16	amorphous	0.99	−38.0
76	24	amorphous	1.08	−37.5
61	39	amorphous	1.01	−35.0
49	51	amorphous	1.02	−29.0
36	64	nearly amorphous	0.95	−27.0
30	70	low IBS crystallinity	1.31	−20.0
20	80	medium IBS crystallinity		−16.0
13	87	high IBS crystallinity		−10.5

[a] ¹H-NMR data in o-dichlorobenzene at 80°C.
[b] From o-dichlorobenzene solution at 30°C.
[c] From DSC analysis.

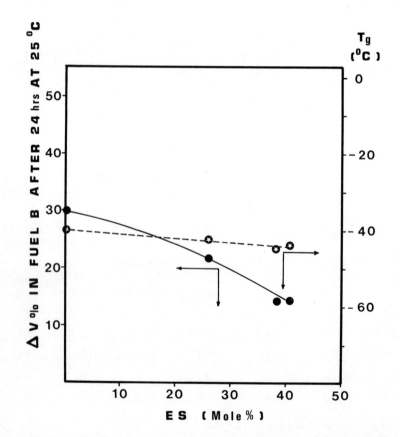

Figure 1.   Swelling and glass point of ES–PS elastomers vs. ES content

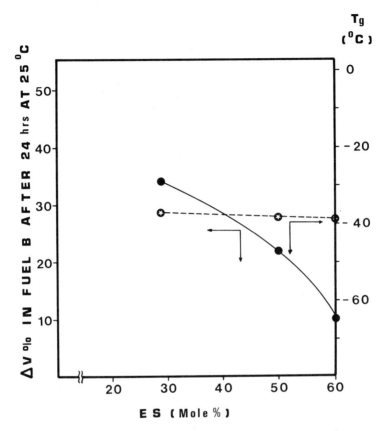

Figure 2.   Swelling and glass point of ES–IBS elastomers vs. ES content

which differed greatly from equimolarity, the crystallinity of the monomer in excess always appeared. Copolymers were therefore mostly amorphous when the molar composition of ethylene sulfide was 35–55%.

The copolymeric nature of the products was also revealed by the extraction data [remember that poly(ES) is insoluble in THF and CHCl$_3$ and poly(IBS) is insoluble in THF] and the x-ray data. Since composition and viscosity were equal, the fact that some copolymer was extracted from CHCl$_3$ and not from THF could be attributed to different microstructures.

The monomeric sequence distribution in these copolymers is still being studied by NMR. Information at the dyad level obtained so far indicates deviation from a statistical distribution and a certain blocking tendency.

Some data on the PS–IBS copolymers are presented in Table VI. These copolymers were essentially amorphous over a wide composition range (0–70 mole % IBS), and all were completely soluble in THF. The T_g of the copolymers increased regularly as IBS content increased. By extrapolation, the T_g of poly(IBS) was calculated as $\sim -10°$C. This is in good agreement with the value reported (28) for the homopolymer with a moderate crystallinity.

Although it is beyond the scope of this work to go into detail about the synthesis of unsaturated terpolymers based on episulfides, we believe that it is

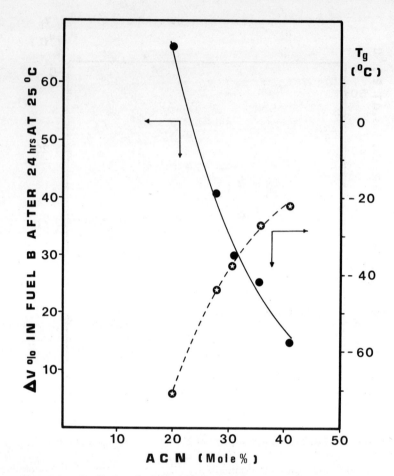

Figure 3. Swelling and glass point of a conventional NBR elastomer vs. ACN content

useful to report some relevant data on the properties of ES-, PS-, and IBS-based elastomers in which vulcanization was reached by introducing low unsaturation levels. For ATE in quantities around 5 mole %, the data on swelling of the vulcanized products in a partially aromatic solvent (ASTM

Table VII.   Molecular Weights in Propylene Sulfide Polymerization

C,[a] mmoles	PS, mmoles	M_{theor}, g/mole	$\overline{M}_{osm}$, g/mole	$\overline{M}_V$,[b] g/mole	$\overline{M}_{UV}$,[c] g/mole
0.18	50	29,000		32,000	32,500
0.04	50	101,000	109,000	110,000	103,000

[a] Seeds ($\overline{M}_{UV}$ = 8500) performed by using

$$\overset{O}{\underset{O}{\overset{\|}{\underset{\|}{S}}}}-CH_2(^-)\ M(^+).$$

[b] Calculated from $\eta = 1.8 \cdot 10^{-5} \cdot M^{0.89}$.
[c] From naphtyl group determination by UV technique.

fuel B) and on the glass point are depicted in Figures 1 and 2. Comparison with a conventional oil-resistant elastomer (Figure 3) [nitrile–butadiene rubber (NBR) with varying acrylonitrile (ACN) content] indicated the superiority of the vulcanized products in oil resistance and glass point. It is interesting that oil resistance improved as ACN content increased, despite the sharp rise in the glass point. With polyepisulfides, however, an increase in ES content improved both oil resistance and glass point (the latter was less marked).

BLOCK COPOLYMERS. Some preliminary tests on the homopolymerization of PS using the catalyst

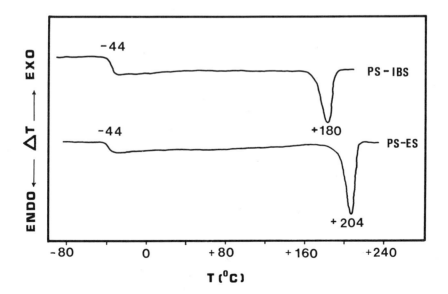

are reported in Table VII. There was good agreement between the theoretical and experimental values for $\overline{M}$ (the latter being determined by osmometry, viscometry, and UV). This agrees with the living nature of this type of polymerization (*1*). It is also evident from the presence of the naphthyl group in the polymer that, in the initiation reaction, transfer processes do not intervene to modify the nature of the active centers.

Similar results were obtained with the catalyst

Figure 4.   DSC analysis of block copolymers

Definitive quantitative data are not yet available.

The synthesis of block polymers of the AB and BAB types (dianionic initiator) (A = PS and B = ES or IBS) is further proof of the living nature of these polymers.

By using monoanionic catalysts, the following types of copolymers were obtained:

$$-(PS)_n-(ES)_m-$$

$$-(PS)_n-(IBS)_m-$$

By DSC analysis it was possible to evidence (Figure 4) both the T_g of the amorphous part (PS) and the T_f of the crystalline part (ES, IBS). The copolymers so obtained were less soluble or even insoluble in those solvents in which the corresponding statistical copolymers were soluble.

The use of dianionic catalysts did not produce such homogeneous results as those obtained with monoanionic catalysts because, as was mentioned above, it was not possible to obtain catalysts with a univocal functionality of 2.0. We recently studied new types of catalysts which enable us to resolve this problem.

Literature Cited

1. Sigwalt, P., *IUPAC Int. Symp. Macromol. Chem.*, Budapest, 1969, p. 251.
2. Sigwalt, P., "Ring Opening Polymerization," p. 191, M. Dekker, New York, 1969.
3. Lautenschlaeger, F., *J. Macromol. Sci. Chem.* (1972) **A6** (6), 1089.
4. Korotneva, L. A., Belonovskaya, G. P., *Russ. Chem. Rev.* (1972) **41** (1), 83.
5. Morton, M., Kammereck, R. F., Fetters, L. J., *Brit. Polym. J.* (1971) **3**, 120.
6. Gourdenne, A., "Block Polymers," S. Aggarwal, Ed., p. 277, Plenum, New York, 1970.
7. Nevin, R. S., Pearce, E. L., *J. Polym. Sci. Part B* (1965) **3**, 487.
8. Mackillop, D. A., *J. Polym. Sci. Part B* (1970) **8**, 199.
9. Boileau, S., Sigwalt, P., *Makromol. Chem.* (1970) **131**, 7.
10. Corey, E. J., Chaykovsky, M., *J. Amer. Chem. Soc.* (1962) **84**, 866.
11. Trossarelli, L., Priola, A., Guaita, M., Saini, G., *Int. Symp. Macromol. Chem.*, Prague, 1965, p. 578.
12. Priola, A., Trossarelli, L., *Makromol. Chem.* (1970) **139**, 281.
13. Molau, G. E., Mason, J. E., *J. Polym. Sci. Part A-1* (1966) **4**, 2336.
14. Bawn, C. E. H., Ledewith, A., McFarlane, N., *Polymer* (1969) **10**, 633.
15. Blanchard, L. P., Hornof, V., Moinard, J., Tahiani, F., *J. Polym. Sci. Part A-1* (1972) **10**, 3089.
16. Blanchard, L. P., Dinh, K. T., Moinard, J., Tahiani, F., *J. Polym. Sci. Part A-1* (1972) **10**, 1353.
17. Banks, P., Peters, R. H., *J. Polym. Sci. Part A-1* (1970) **8**, 2595.
18. Tavares, D. F., Vogt, P. F., *Can. J. Chem.* (1967) **45**, 1519.
19. Kaiser, E. M., Beard, R. D., Hauser, C. R., *J. Organometal. Chem.* (1973) **59**, 53.
20. Cuvigny, T., Normant, H., *Organometal. Chem. Syn.* (1971) **1**, 237.
21. Truce, W. E., Buser, K. R., *J. Amer. Chem. Soc.* (1954) **76**, 3577.
22. Hirahata, T., Sugimura, T., Minoura, Y., *J. Polym. Sci. Part A-1* (1970) **8**, 2827.
23. Snam Progetti, Belgian Patent **783778** (1972).
24. Snam Progetti, Belgian Patent **796431** (1973).
25. Kaiser, E. M., Hauser, C. R., *Tetrahedron Lett.* (1967) 3341.
26. Kaiser, E. M., Beard, R. D., *Tetrahedron Lett.* (1968) 2583.
27. Corno, C., Roggero, A., Salvatori, T., *Eur. Polym. J.* (1974) **10**, 525.
28. Vandemberg, E. J., *IUPAC Mtg., 23rd,* Boston, July, 1971, p. 111.

RECEIVED April 3, 1974.

Synthesis and Structure–Property Relations in High Temperature Polymer Systems Based on Aromatic Diamines

GEORGE L. BRODE, JAMES H. KAWAKAMI, GEORGE T. KWIATKOWSKI, and ALBERT W. BEDWIN

Research and Development Department, Union Carbide Corp., Chemicals and Plastics, Bound Brook, N.J. 08805

A series of novel aromatic diamines were prepared which contained an ordered sequence distribution of thermally stable, but rigid and/or flexible, units. They could be prepared through a nucleophilic displacement reaction in aprotic solvent from the alkali metal salt of p-aminophenol or the salt of a bisphenol and an activated aromatic halide. Such monomers were useful building blocks for various polymeric systems, especially those which benefit from the high glass transition temperature imparted by polar or rigid moieties and the improved impact properties conveyed by ether groups. Polyimides and polyamide–imides based on these monomers exemplify such polymers; they display high heat distortion temperatures, good solvent resistance, excellent mechanical properties, high thermal and oxidative stability, and, depending on diamine structure and molecular weight, thermoplastic characteristics.

A series of new aromatic sulfone ether diamines (I) was prepared by a nucleophilic displacement reaction in aprotic solvent from the alkali metal salt of *p*-aminophenol, or optionally from the salt of a bisphenol and an acti-

vated aromatic halide (Reaction 1) (*1, 2*). The incentives for developing this class of materials included the knowledge that polysulfones display high thermal

$$NH_2 - \langle O \rangle - O - \langle O \rangle - \overset{\overset{O}{\|}}{\underset{\underset{O}{\|}}{S}} - \langle O \rangle - O - \langle O \rangle - NH_2$$

II

and oxidative stability (3, 4, 5) and the prediction that polyimides and poly-amide–imides based on such sulfone ether diamines would be tractable. The structure of the sulfone ether diamines prepared this way included 4,4'-[sulfonyl bis(p-phenyleneoxy)] dianiline (II), the polyether diamines based on bis-phenol-A (III), and the polyether diamines based on hydroquinone (IV).

$$NH_2 - \langle O \rangle - O - \left(\langle O \rangle - \overset{\overset{O}{\|}}{\underset{\underset{O}{\|}}{S}} - \langle O \rangle - O - \langle O \rangle - \left| - \langle O \rangle - O \right| \right)_n - \langle O \rangle - \overset{\overset{O}{\|}}{\underset{\underset{O}{\|}}{S}} - \langle O \rangle \quad \langle O \rangle - NH_2$$

$$NH_2 - \langle O \rangle - O - \left(\langle O \rangle - \overset{\overset{O}{\|}}{\underset{\underset{O}{\|}}{S}} - \langle O \rangle - O - \langle O \rangle - O \right)_n - \langle O \rangle - \overset{\overset{O}{\|}}{\underset{\underset{O}{\|}}{S}} - \langle O \rangle - O - \langle O \rangle - NH_2$$

IV

Synthesis of these diamines requires control of two important side reactions: rapid oxidation of the phenate salt of p-aminophenol (6, 7) (Reaction 2), and nucleophilic displacement of halide by anilide anion (8) (Reaction 3). The oxidative side reaction was eliminated by scrupulous removal of oxygen by vacuum sparging with nitrogen. Anilide anion displacement was controlled

$$
\underset{OM}{\overset{NH_2}{\langle O \rangle}} + MOH \xrightarrow{\text{DMSO}} \underset{OM}{\overset{\overset{\ominus}{N}H \overset{\oplus}{M}}{\langle O \rangle}} + H_2O \qquad (2)
$$

$$
\overset{\oplus \ominus}{M} NH - \langle O \rangle - OM + Cl - \langle O \rangle - \overset{\overset{O}{\|}}{\underset{\underset{O}{\|}}{S}} - \langle O \rangle - Cl \longrightarrow
$$

$$ (3) $$

$$
MO - \langle O \rangle - NH - \langle O \rangle - \overset{\overset{O}{\|}}{\underset{\underset{O}{\|}}{S}} - \langle O \rangle - Cl + MCl
$$

Table I. Physical Properties of Sulfone Ether Diamines

Structure	Amine Equiv. Wt		Melting Range, °C
	Calcd.	Observed	
II	216	216	191–192
III (n = 1)	437	449	127–140
III (n = 2)	658	710	130–165
IV (n = 2)	540	insoluble	222–229

by ensuring exact stoichiometric quantities of metal alkoxide *vs.* p-amino-phenol and bisphenol; in some cases, p-aminophenol was used at 0.5–1 mole % excess relative to caustic. With these precautions, diamine synthesis which provided polymer grade monomer in high yield was reproducible. When these side reactions were avoided, the deviation of amine equivalent weight from theory in certain cases was due entirely to the loss of a small fraction of lower molecular weight species during isolation procedures. The observed molecular weights were higher than the theoretical values (*see* Table I for data on diamines based on bisphenol-A). Unlike diamines II and IV which were crystalline, diamines III were amorphous.

Table II. Product Distribution Obtained with Bisphenol-A-Based Diamines (III) at Stoichiometric Values of n = 1 and n = 2

Component		Stoichiometry, %	
		n = 1	n = 2
n = 0	(II)	26.5	10.1
n = 1	(III)	29.0	20.8
n = 2	(III)	19.5	20.1
n = 3,4 . . .	(III)	25.0	49.0

A distribution of products with various molecular weights was expected with diamines of Structure I when n = 1 since these materials were prepared by condensing several reactants (Reaction 1). It was important to ascertain the percentage of each component as the polyaryl ether mer units in the diamine would be expected to affect the properties of the resultant polymers. The components were resolved by liquid–liquid chromatography on a Sephadex column (Table II). When dichlorodiphenyl sulfone was added rapidly to a mixture of the sodium salts of bisphenol-A and p-aminophenol at elevated temperatures, the distribution was significantly different from that obtained at lower temperatures with sequential addition of sulfone monomer (Table III). This is believed to result from the solubility characteristics of the disodium

Table III. Effect of Reaction Conditions on Bisphenol-A-Based Diamine Distribution at Stoichiometric Values n = 1

Component		Sulfone Addition, %	
		Rapid[a]	Sequential[b]
n = 0	(II)	26.5	28.4
n = 1	(III)	29.0	15.4
n = 2	(III)	19.5	16.6
n = 3,4 . . .	(III)	25.0	39.3

[a] $T = 160°–175°C$.
[b] $T = 100°–160°C$.

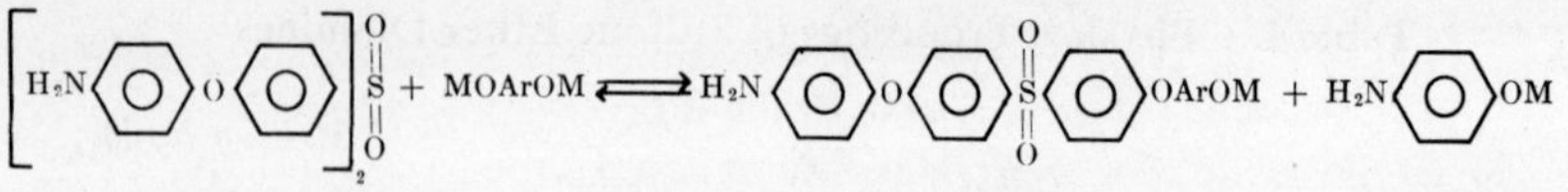

$$\left[H_2N\langle O\rangle O\langle O\rangle\right]_2 \overset{O}{\underset{O}{\overset{\|}{S}}} + MOArOM \rightleftharpoons H_2N\langle O\rangle O\langle O\rangle\overset{O}{\underset{O}{\overset{\|}{S}}}\langle O\rangle OArOM + H_2N\langle O\rangle OM \quad (4)$$

salt of bisphenol-A. In the sequential low temperature reaction, the ratio of the concentration of the *p*-aminophenol salt to that of the bisphenol-A salt was higher because of the low solubility of the latter. This results in higher yields of II, lower yields of the $n = 1$ and $n = 2$ diamines, and higher yields of the oligomer diamines $n = 3, 4. \ldots$. Since these systems probably equilibrate (Reaction 4), it is likely that the initial concentration of II was actually higher than 26.5% or 28.4%. Therefore, in order to obtain uniform quality diamines of types III and IV and their polymers, it was important to control reaction conditions.

Polymer Synthesis

The polymer synthesis program (*9, 10, 11*) initially centered on II. Reaction of II with either isophthaloyl or terephthaloyl chloride yielded, respec-

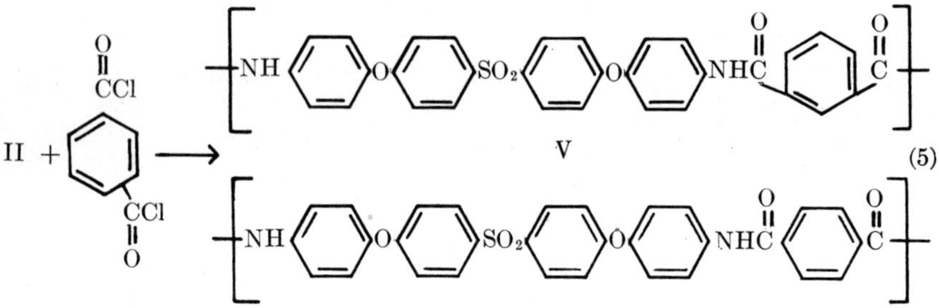

(5)

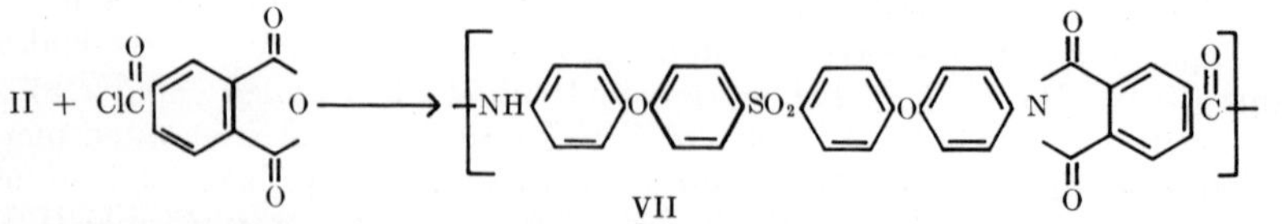

(6)

tively, polyisophthalamide (V) or terephthalamide (VI) (Reaction 5). Condensation of II with either trimellitic acid anhydride or trimellitolyl chloride yielded polyamide–imide (VII) (Reaction 6), whereas II plus benzophenone tetracarboxylic acid anhydride or pyromellitic acid anhydride yielded, respectively, polyimides VIII and IX (Reactions 7 and 8).

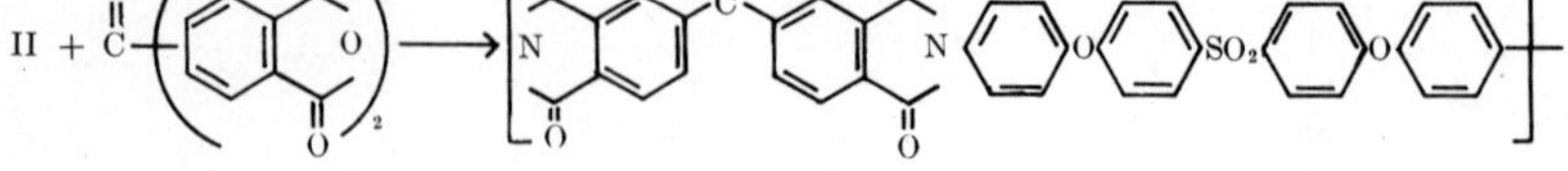

(7)

The mechanical properties and glass transition temperatures of the polymer classes prepared from II are summarized in Table IV. In general, the polymers had excellent mechanical properties. Except for terephthalamide (VI), the polymers were amorphous with glass transition temperatures ranging from a

IX

low of 230°C for isophthalamide (V) to a high of 320°C for the polyimide based on pyromellitic dianhydride (IX). Terephthalamide (VI) films cast from a 5% lithium chloride–dimethyl acetamide solution had a glass transition temperature of 260°C. There was no evidence of crystallinity in the x-ray diffraction studies; however, VI did not flow under pressure at 350°C, and its melting point (transition), as measured by differential scanning calorimetry, was 425°–430°C with a heat of fusion of 8.40 cal/g. It was interesting that the poly-amide–imide (VII) had a glass transition temperature at least equivalent to and actually somewhat higher than that of the polyimide based on benzophe-none tetracarboxylic acid dianhydride (VIII). This is pertinent, considering that the amide–imide (VII) was tractable whereas the straight imide (VIII) had a low level of crosslinking and could not be handled under thermoplastic processing conditions. This is generally true of polyimides, with intractability increasing as the aromatic diamine molecular weight decreases. Even at the relatively high molecular weight of II, this crosslinking tendency was observed for both pyromellitic dianhydride (IX) and benzophenone dianhydride (VIII) polyimides. The data also indicate that, while in every case the tensile strength, elongation, and pendulum impact properties were quite acceptable, the mechanical properties of VIII were somewhat poorer despite an extremely high molecular weight. Both VII and IX had significantly greater impact strength and elongation. The excellent mechanical properties of VII, together with the observed good solvent resistance, flame-retardant characteristics, outstanding thermo-oxidative stability and tractability (*vide infra*), led to in-depth studies of this ploymer class based on sulfone ether diamines.

Polyamide–imides were prepared from I and trimellitic acid anhydride (Reaction 9) or trimellitoyl chloride. The latter was used with either the tri-

Table IV. Mechanical Properties of 4,4′-[Sulfonyl Bis(p-Phenyleneoxy)] Dianiline-Based Polymers

Structure	T_g, °C	Reduced Viscosity[a]	Tensile Modulus, psi	Tensile Strength, psi	Elongation at Break, %	Pendulum Impact, ft lb/in^3
V	230	1.0	323,000	11,150	11	135
VI	260	—	—	—	—	—
VII	270	0.97	370,000	12,500	15	171
VIII	260	1.8	440,000	14,000	8	70
IX	320	insoluble	270,000	9,600	15	197

[a] Values were obtained using 0.2% solutions of the respective polymer in dimethylacetamide.

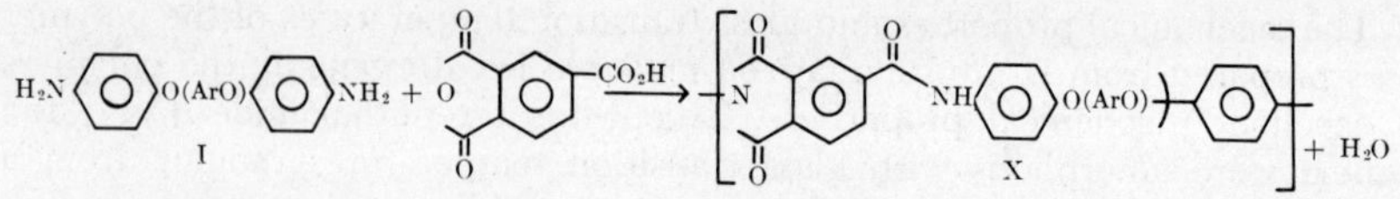

$$(9)$$

ethylamine route or the ethylene oxide route (Reaction 10). The resultant polyamide acid was then cyclized by heating or with acetic anhydride (Reaction 11). Phthalic anhydride was introduced sometimes to control molecular weight. Because of the high purity of the starting diamines and the chemical efficiency of Reaction 10, ultrahigh molecular weight polymers could be prepared. In

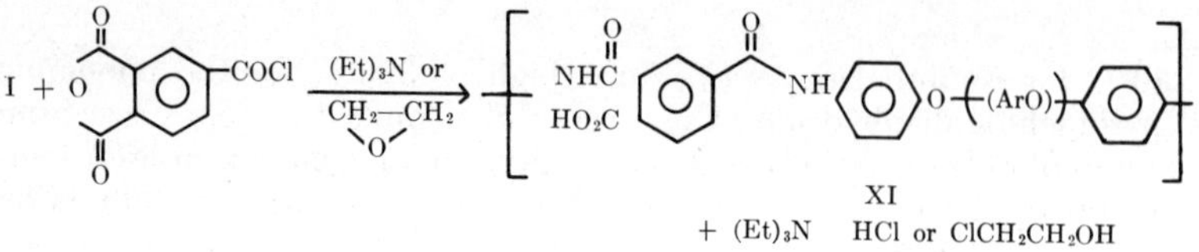

$$(10)$$

$$+ (Et)_3N \quad HCl \text{ or } ClCH_2CH_2OH$$

addition, this process could produce high molecular weight polyamide–imides from such acidic diamines as 4,4'-diaminodiphenyl sulfone (Reaction 12). Sulfone dianiline polyamide–imide (XII) was prepared at high molecular weight ($M_n \sim 30,000$) by the *in situ* chemical imidization reaction. Conventional imidization processes reportedly yield low molecular weight products with deficient mechanical properties (12).

$$XI + (CH_3CO)_2O \longrightarrow X + CH_3CO_2H \qquad (11)$$

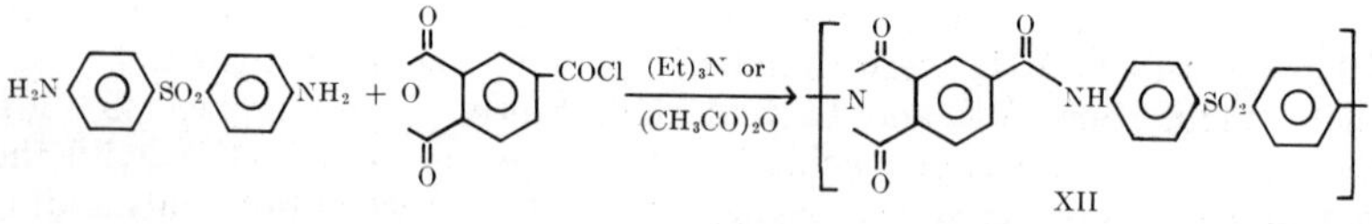

$$(12)$$

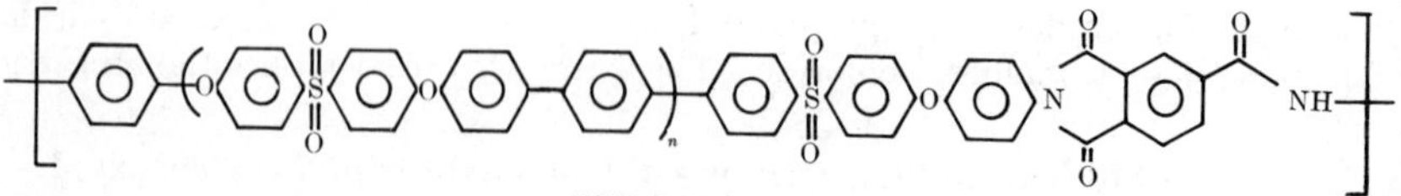

XIII, XIV

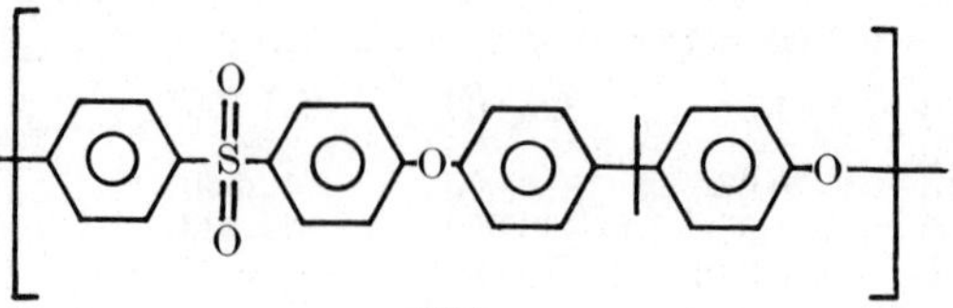

XV

Table V. Properties of Polyamide–Imides from Sulfone Ether Diamines

Structure	Reduced Viscosity,[a] dl/g	T_g, °C	Modulus, psi	Tensile Strength, psi	Elongation at Break, %
XII	0.50	350	425,000	11,300	17
VII	0.60	270	360,000	14,000	18
XIII (n=1)	0.50	225	315,000	11,650	8.5
XIV (n=2)	0.50	200	305,000	11,000	10
XV	0.50	180	270,000	10,000	50

[a] Dimethylacetamide solvent.

Properties of Sulfone Ether Diamine and Related Polyamide–Imides and Polyimides

Mechanical Properties. Properties of polyamide–imides VII, XII, XIII, and XIV from the sulfone ether diamine family were measured (Table V); for comparison, the properties of polysulfone (XV) from the polyether family are also included. Glass transition temperature was calculated by a summation of the products of the unit weight fraction and the unit contribution (13) to T_g

Table VI. Relation between Structure and T_g

Repeat Unit	Molecular Weight	Unit Contribution to T_g
	216	335
	188	360
	92	105
	442	180

(Table I) (Equation 13). The polar diphenylsulfone unit and the rigid, polar, amide–imide unit were primarily responsible for the high T_g's of these materials. The ether and isopropylidene linkages markedly reduce glass transition temperature but provide toughness and tractability. Through proper selection of the structural units, polyamide–imides can be designed with a range in T_g from about 180° to 350°C.

$$T_g = \Sigma \text{ (unit weight fraction)(unit } T_g \text{ contribution)} \tag{13}$$

Table VII. Environmental Stress Aging Characteristics of Sulfone Diamine-Based Polymers

Solvent	Stress, psi	XII	VII	XVI	XIII	XIV	VIII	IX
Trichloroethylene	4000	U	U	U	C, B	C, B	U	U
Xylene	4000	U	U	U	C, B	R (<1)	U	U
Ethanol	4000		U				U	U
Acetone	1000	C, B	C, B	C, B	R (<1)	R (i)	U	U
	2000						U	U
	4000	R (7)	R (1)	R (<1)	R (i)			

^{a} U: films unaffected after 10-min exposure at indicated stress; C, B: films crazed and brittle after 10-min exposure at indicated stress; and R: films ruptured (number in parentheses = minutes to rupture; i = instantaneous).

Environmental Stress Aging Characteristics. The sulfone-based polyamide–imides were soluble only in polar organic solvents such as amides (dimethylformamide, dimethylacetamide, and N-methyl pyrrolidone) and dimethyl sulfoxide. The bisphenol-A oligomer sulfone ether polyamide–imides were somewhat soluble in chlorinated hydrocarbons, and solubility increased with the molecular weight of the oligomer component. In general, the polyamide–imides and the polyimides had good resistance to boiling water and 5% sulfuric acid, but they were attacked after about 24 hrs of immersion in 1% sodium hydroxide.

The solvent stress crazing resistance of the sulfone diamine and sulfone ether diamine polyamide–imides or polyimides is presented in Table VII. Polymers with high amide–imide or imide weight contributions per repeat unit (VII, VIII, IX, XII, and XVI) had excellent environmental stress aging resistance to trichloroethylene, xylene, and ethanol, but fair resistance to acetone. Polyamide–imides that contain the diphenylisopropylidene structure (XIII and XIV) had only fair resistance to xylene, ethanol, and trichloroethylene, and poor resistance to acetone. The environmental stress aging characteristics of all polymer systems studied improved with increasing molecular weight.

Thermal Stability. The thermo-oxidative stability of polyamide–imides VII, XII, and XVI was assessed by thermogravimetric analysis in air and also by measuring the retention of impact strength at 220°C as a function of time.

XVI

Table VIII. Thermal Gravimetric Analysis of Polyamide–Imides in Air[a]

	Weight Loss, %		
T, °C	VII	XII	XVI
350–400	1	0	1
400–450	2	2	3.5
450–500	5	6	7
500–550	10	10	15

[a] Heating rate, 5°C/min.

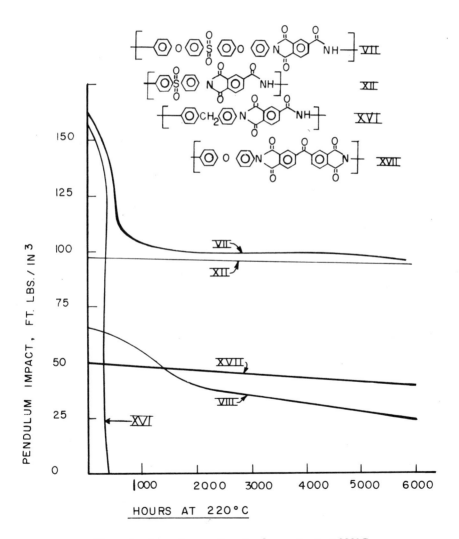

Figure 1. Impact properties of polymers in air at 220°C

The latter test demonstrated a clear relation between polymer structure and retention of performance at a given temperature; the benzophenone-based polyimide IX was also evaluated by the latter technique. The sulfone-based polyamide–imides had initial weight loss at 350°–400°C (Table VIII). The materials retained good performance properties after short exposures to 400°C, but at 500°–550°C weight loss was rapid. The polyamide–imide based on methylene dianiline (XVI) had a higher rate of weight loss which was indicative of a lower level of thermal stability.

Long-term, isothermal aging data, compiled by exposing polymer films in air at 220°C while periodically measuring impact properties, further established this trend (Figure 1). As a reference point, materials with impact strengths less than 30 ft lb/in.[3] are generally brittle. The data demonstrate the marked superiority of sulfone-based polymers over those based on the oxidatively unstable methylene linkage. The other systems were also interesting. After an initial loss of impact properties during the first 2–4 weeks of aging, VII had level performance over the ensuing 30 or more weeks of testing. XII performed consistently throughout the testing period. Finally, impact data on imides based on benzophenone tetracarboxylic anhydride and, respectively, oxydianiline (XVII) or 4,4'-[sulfonyl bis(p-phenyleneoxy)] dianiline (VIII) indicated that the former performed slightly better under these conditions.

Flow Characteristics (Tractability). Tractability was obtained with polyamide–imides based on 4,4'-[sulfonyl-bis(p-phenyleneoxy)] dianiline and the bisphenol-A-based polyether diamines (VII, XIII, and XIV) and, to a limited extent, with the polyimide based on benzophenone tetracarboxylic dianhydride (VIII). Work was most extensive with polyamide–imide VII. Polymers with a reduced viscosity of 0.43–0.50 were the most useful in molding applications. In order to obtain polymer in this molecular weight range reproducibly, phthalic anhydride was used to moderate the polymerization reaction. In addition to controlling molecular weight, this method of end-capping the polymer also gave enhanced thermal stability. As a result, recovery through a vented extruder was possible with the product being held in the molten stage at temperatures of 325°–340°C. Under such conditions, polymer of 0.49 dl/g reduced viscosity was obtained from product with an initial reduced viscosity of 0.45 dl/g. The stability of the extruded resin was also demonstrated by melt flow studies in which melt flow ratios, MF_2/MF_1, of 1.1–1.2 were obtained (MF_1 and MF_2 were measured at 350°C, the latter after heat aging at 350°C for 10 min). Compression-molded samples of product recovered from the vented extruder had properties essentially identical to those of polyamide–imides (Table IV) except that tensile strength and elongation at break were slightly greater. Thus, the desired combination of good mechanical properties, improved solvent resistance, high temperature performance, flame retardance, and tractability can be obtained with certain members of this class of materials.

Acknowledgment

The authors thank L. M. Robeson, B. L. Joesten, W. D. Niegisch, and P. Van Riper of the Union Carbide Corp. laboratories at Bound Brook, N.J., for determination of physical properties.

Literature Cited

1. Brode, G. L., Kwiatkowski, G. T., Kawakami, J. H., *Amer. Chem. Soc., Div. Polym. Chem., Prepr.* **15** (1), 764 (Los Angeles, April, 1974).

2. Kawakami, J. H., Kwiatkowski, G. T., Brode, G. L., Bedwin, A. W., *J. Polym. Sci.* (1974) **12,** 565.
3. Johnson, R. N., Farnham, A. G., Clendinning, R. A., Hale, W. F., Merriam, C. N., *J. Polym. Sci. Part A-1* (1967) **5,** 2375.
4. Hale, W. F., Farnham, A. G., Johnson, R. N., Clendinning, R. A., *J. Polym. Sci. Part A-1* (1967) **5,** 2399.
5. Brode, G. L., "The Chemistry of Sulfides," A. V. Tobolsky, Ed., pp. 133-144, Wiley and Sons, New York, 1968.
6. Konaka, R., Kurama, K., Terabe, S., *J. Amer. Chem. Soc.* (1968) **90,** 1801.
7. LuValle, J. E., Glass, D. B., Weissberger, A., *J. Amer. Chem. Soc.* (1948) **70,** 2223.
8. Dolman, O., Stewart, R., *Can. J. Chem.* (1967) **45,** 911.
9. Brode, G. L., Kawakami, J. H., U.S. Patent **3,817,921** (1974).
10. Brode, G. L., Kawakami, J. H., Kwiatkowski, G. T., Bedwin, A. W., *J. Polym. Sci.* (1974) **12,** 575.
11. Kwiatkowski, G. T., Brode, G. L., Kawakami, J. H., Bedwin, A. W., *J. Polym. Sci.* (1974) **12,** 589.
12. Androva, N. A., Bessonov, M. J., Larus, L. A., Rudakov, A. P., "Polyimides," 1969, Engl. Transl. Israel Program Sci. Transl.
13. Niegisch, W. D., Faucher, J. A., private communication.

RECEIVED March 6, 1974.

31

The Synthesis and Properties of Multicomponent Crosslinked Polymers

C. H. BAMFORD and G. C. EASTMOND

Department of Inorganic, Physical, and Industrial Chemistry,
University of Liverpool, Liverpool L69 3BX, England

This paper reviews the method used to prepare and control the structures of multicomponent polymers in which a polymer A is crosslinked by a polymer B. Recent DSC and electron microscope data on the morphologies of materials containing branches of the crosslinking polymer are correlated with previous dilatometric data, and an explanation of the influence of branches on polymer morphology is proposed. Morphologies of some polycarbonate/polychloroprene crosslinked polymers are described. The influence of one component on segmental motions of the second component of the crosslinked polymers is discussed in terms of pulsed NMR data. Some preliminary findings on the mechanical properties of multicomponent crosslinked polymers are presented.

The multicomponent crosslinked polymers with which this paper is concerned consist of chains of polymer A crosslinked by chains of a second polymer B. They contain a basic structural unit (*see* Structure I) in which both A and B

(I)

chains are long, typically $\overline{P}_n \sim 1000$. We refer to these crosslinked polymers as ABCPs. The polymeric components which can be incorporated into ABCPs include a range of chemical types.

ABCPs are prepared by generating free radicals at specific reaction sites on a preformed polymer A in the presence of a monomer. The macroradicals initiate graft polymerization of the monomer to produce growing branches of polymer B, and combination termination of the propagating radicals gives polymers with Structure I. Continued reaction involving initiation on additional A chains and at additional sites on reacted A chains produces more complex structures such as Structure II. Further reaction increases the crosslink

(II)

density, producing more complex structures and, ultimately, highly crosslinked three-dimensional infinite networks based on structural unit I.

The A and B components are normally incompatible, and, in common with other multicomponent polymers, there is an inherent tendency for microphase separation. We demonstrated previously that microphase separation is a general feature of ABCPs (*1, 2*), even in highly crosslinked networks, despite the limitations imposed on molecular reorganization by geometrical constraints in such materials. Electron microscope studies have enabled us to identify some general correlations between the morphologies and the chemical structures of ABCPs.

In ABCPs that contain branches of polymer B, in addition to the crosslinks, microphase separation is incomplete (*1, 2*), and mixed phases containing both components are present in the polymers; the formation of mixed phases can be correlated with the structures of the polymers. Broad-line NMR studies of such materials have revealed modifications of molecular motions attributable to effects of unusual environments associated with the mixed phases and in the vicinity of polymer–polymer interfaces (*2, 3*).

This paper presents, in outline, the basic techniques for preparing ABCPs of known structure and for controlling their structural parameters; a more detailed account has been presented elsewhere (*2*). Some of our earlier conclusions are summarized, and more recent data which substantiate our previous views and add to our understanding of ABCPs are presented.

Synthesis of ABCPs

In common with many other derivatives of transition metals in low oxidation states, molybdenum and manganese carbonyls in association with certain organic halides are effective sources of free radicals that are capable of initiating free-radical polymerizations. It has been established that in each case the rate-determining process for radical formation and hence for initiation is the formation of reactive complexes from the parent derivatives (*4*). For $Mo(CO)_6$ (thermal initiation at 80°C) the reactive complex is formed by ligand exchange with a donor solvent (which may be monomer), and for $Mn_2(CO)_{10}$ (photo-initiation at 25°C, $\lambda = 435$ nm) the corresponding process is photolysis of the carbonyl. The radical-forming process is then a redox reaction between the reactive species and the halide R—X which may be represented by Reaction 1

$$M° + \text{R-X} \longrightarrow M^{I}X + \text{R}\cdot \tag{1}$$

where M represents the transition metal, and X is a labile chlorine or bromine atom. Radical R· initiates polymerization. With both $Mo(CO)_6$ and $Mn_2(CO)_{10}$, Reaction 1 is completely specific (5, 6) in generating radicals R· from R—X.

Advantage is taken of the specificity of the initiating reaction to produce ABCPs of known structure by replacing the simple halide with a polymer containing reactive halogen in the side chains; this polymer becomes the A component of the ABCP. As with simple halides, the metal complex abstracts a halogen atom from the side group of the polymer to generate a macroradical which can then initiate the polymerization of a monomer present in the reaction mixture, thereby producing a propagating graft of polymer B. Combination termination of the propagating radicals yields crosslinks of polymer B and species with Structure I; disproportionation termination of the propagating radicals generates branches of polymer B. A major consequence of the specificity of the initiation reaction is that no free homopolymer B is produced. Thus, A may be any polymer having reactive chlorine or bromine in side groups, and B may be any polymer which can be produced by free radical polymerization in solution. A components which have been used include poly(vinyl trichloroacetate) (PVTCA), copolymers of styrene and p-vinyl benzyltrichloroacetate and the polycarbonate based on 1,1,1-trichloro-bis-2-(p-hydroxyphenyl)-ethane. B components of ABCPs whose properties have been examined include polystyrene (PSt), poly(methyl methacrylate) (PMMA), poly(methyl acrylate) (PMA), and polychloroprene; other B components have been used in kinetic studies (7).

The major structural parameters of the ABCPs are the degrees of polymerization of A and B chains, the crosslink density, and the crosslink/branch ratio. The degree of polymerization of the A chains is determined prior to the crosslinking reaction. The degree of polymerization of the B chains is determined by the polymerization kinetics during the crosslinking process; it can be calculated from the rate of initiation (determined from the kinetics of initiation) and the rate of polymerization or polymer yield. Crosslink/branch ratios are controlled in simple cases by the relative rates of combination and disproportionation termination (7). Additional branches of polymer B may be incorporated in a controlled manner by adding simple halides to the crosslinking reaction mixture (2, 8). The structural parameters may be adjusted to allow for chain transfer to monomer, solvent, or added transfer agent (2).

Studies of gelation kinetics have established conditions under which random crosslinking occurs (7). Under such conditions, the kinetics of the crosslinking process conform with predictions from Flory's gelation theory (9), and the gel time, t_g, the time taken to produce one crosslinked unit per weight-average A chain, is given by

$$t_g = \frac{c}{\overline{P}_w I} \left(\frac{k_{tc} + k_{td}}{k_{tc}} \right)$$

where c is the concentration of halogen-containing groups (in base mole/l), $\overline{P}_w$ is the weight-average degree of polymerization of the A chains (referred to halogen-containing groups), I is the rate of initiation, and k_{tc} and k_{td} are, respectively, the rate coefficients for combination and disproportionation termination of the propagating B chains. Deviation from ideal kinetic conditions results in nonrandom crosslinking and excessive intramolecular crosslink formation which delays gelation. The expression for t_g may be modified for the addition of a simple halide or for transfer to monomer, solvent, or other additive.

Under conditions of random crosslinking, crosslink density in an ABCP is determined by the initiation rate and the reaction time. For convenience we define the relative crosslinking index γ_r as the ratio of the actual crosslinking index in the sample to the critical crosslinking index at the gel point. Thus, $\gamma_r = 1$ at the gel point, and, under other conditions, γ_r is equal to the reaction time expressed as multiples or fractions of the gel time assuming a constant rate of initiation (corrections can be made for initiator consumption or effects of chain transfer reactions).

Under ideal conditions, the populations of various crosslinked species are determined by the statistics of random crosslinking. Quantitative determination of these populations in ABCPs is complex since both A and B components have distributions of chain lengths. Qualitatively, an ABCP with $\gamma_r < 1$ consists of an assembly of crosslinked structures, such as Structures I and II as well as more complex species, blended with unreacted A chains. For $\gamma_r > 1$, the ABCP contains infinite network structures in addition to the simpler structures. As γ_r increases, the proportion of material existing as unreacted A chains decreases. Since the A component has a distribution of chain lengths and, therefore, of functionality, the long A chains are incorporated preferentially into crosslinked structures, and the molecular weight of unreacted A polymer decreases as γ_r increases.

Microphase Separation and Polymer Structure

We previously presented a detailed analysis of dilatometric data obtained from ABCPs with $\gamma_r > 1$ consisting of PVTCA crosslinked with PSt and PMMA; these are denoted PVTCA/PSt and PVTCA/PMMA, respectively (1, 2, 8). The ABCPs, dried from gels swollen with the monomer of the crosslinking chains, have multiple glass transitions indicative of microphase separation, even at high crosslink densities (up to $\gamma_r = 10$). Representative data are presented in Table I. The major conclusions from the dilatometric studies were that, for ABCPs containing no branches of the crosslinking polymer, microphase separation was essentially complete; however, for ABCPs containing branches of the B component, only partial microphase separation occurred and a proportion of the B chains became incorporated with the A component in a mixed phase (2).

Table I. Structures and Properties of PVTCA/PSt and PVTCA/PMMA ABCPs

Crosslinking Polymer	Casting Solvent[a]	PVTCA, wt %	γ_r	$\overline{P}_n$, Cross-links	Branches: Cross-links	T_{g1}, °C	T_{g2}, °C	w
PSt	S	40.5	3	4750	0	60	97	0.12
PSt	S	55	3.5	500	0	65	90	0.05
PSt	S	30.3	1.5	1044[b]	2	72	92	0.90
PSt	S	38.3	1.3	1150[b]	1	74	89	0.45
PSt	E	56.5	2	1700[b]	2	87.5	—	—
PMMA	M	8	9	5200[c]	4	71	102	1.7
PMMA	M	21	1	11820[c]	8	74	99	2.8
PVTCA						59		
PMMA							~100	
PSt							~100	

[a] S, styrene; E, styrene + ethyl acetate (9:1 v:v); and M, methyl methacrylate.
[b] The $\overline{P}_n$'s for branches and crosslinks are equal.
[c] The $\overline{P}_n$ for branches is one-half the $\overline{P}_n$ for crosslinks.

The evidence for these conclusions was that PVTCA/PSt ABCPs containing no PSt branches had two glass transitions at temperatures T_{g1} (lower) and T_{g2} (upper) which correspond to the T_g's of the component polymers (1), whereas both PVTCA/PSt and PVTCA/PMMA ABCPs containing PSt and PMMA branches, respectively, had a modified T_{g1} (arising from the mixed phase) whose absolute value was raised by 10°–15°C (1, 2, 8). Furthermore, for ABCPs containing branches, coefficients of expansion between T_{g1} and T_{g2} were higher than expected for heterogeneous mixtures of the component polymers of the same overall composition, whereas ABCPs without branches had coefficients of expansion consistent with A and B components in separate phases. Compositions of the mixed phases were estimated from the coefficients of expansion and expressed in terms of the parameter w; w is defined as the weight of B component mixed with unit weight of the A component.

It was also reported (2, 8) that PVTCA/PSt ABCPs containing PSt branches dried from gels swollen with styrene + ethyl acetate (9:1 v:v) had only a single glass transition which corresponded to that calculated from the Kelley–Bueche equation (10) for mixtures of the component polymers with the same overall compositions, assuming additivity of volumes and intimate mixing of the polymers. It was concluded that microphase separation did not occur under these conditions. Solvent effects of this type were not observed for PVTCA/PSt ABCPs containing no PSt branches.

Table II. Structure and Properties of Some PVTCA/PSt ABCPs

Sample[a]	PVTCA, wt %	$\bar{P}_n$ (PSt)	Branches: Crosslinks	γ_r	T_{g1}, °C	T_{g2}, °C
MI (S)	45.5	1200	0	3	58	91
MII (S)	24.5	673	2	3	76	100
MIII (E)	46.0	300	2	2	68	
PVTCA	100				67	
Blend (S)	30				67	100

[a] Casting media: S, styrene; and E, styrene + ethyl acetate (9:1 v:v).

Subsequently, another series of PVTCA/PSt ABCPs (Table II) was prepared to obtain additional data on these polymers by differential scanning calorimetry (DSC) and electron microscopy. The samples used in these supplementary studies were much thicker (1 mm) than the thin films (< 0.1 mm) used in the dilatometric investigations, and difficulties in solvent removal were encountered. Consequently, quantitative data derived from these samples are less definitive than the dilatometric data.

Differential Scanning Calorimetry (DSC). DSC traces for the PVTCA/PSt ABCPs described in Table II are presented in Figure 1; they were obtained from samples previously heated at 100°C under high vacuum to constant weight. The compression-molded sample of PVTCA (Curve 1) had a single T_g at 67°C. The PVTCA–PSt blend (Curve 2) had two T_g's at temperatures corresponding to those of the homopolymers which confirms the incompatibility of the component polymers. PVTCA/PSt ABCP MI, with no PSt branches (Curve 3) had two T_g's. The absolute values of the T_g's were lower than expected from previous dilatometric data, probably because of solvent retention; however, the difference between the T_g's was approximately equal to that in the blend which probably indicates essentially complete microphase separation in the ABCP in the absence of PSt branches. PVTCA/PSt ABCPs

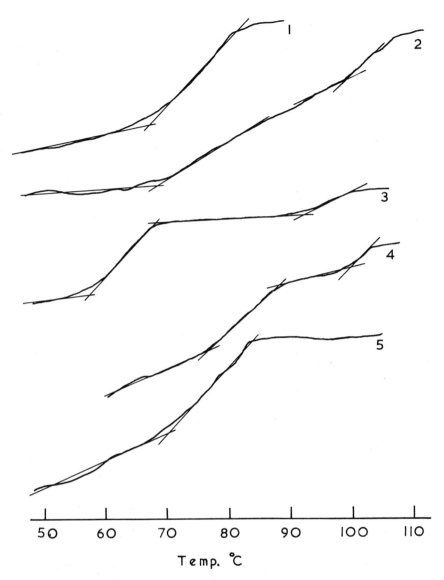

Figure 1.   DSC traces for PVTCA/PSt polymers, heating rate 16°C/min
1, *compression-molded PVTCA homopolymer; 2, PVTCA–PSt blend; 3, PVTCA/PSt ABCP MI; 4, PVTCA/PSt ABCP MII; 5, PVTCA/PSt ABCP MIII (see Table II for sample description)*

MII and MIII contain PSt branches with the same degree of polymerization as the PSt crosslinks. Polymer MII, cast from styrene, had two T_g's (Curve 4); T_{g2} was identical to that of PSt homopolymer and PSt in the blend whereas T_{g1} was higher than the T_g of PVTCA. This finding corresponds to data from dilatometric studies and supports the concept of incomplete phase separation and the formation of a mixed PVTCA–PSt phase in ABCPs containing PSt branches. Polymer MIII, which also contains PSt branches, was cast from styrene + ethyl acetate. It had only one T_g (Curve 5) with an absolute value less than expected (also probably because of solvent retention); nevertheless, this appears to support the concept of intimate mixing of the components when PVTCA/PSt ABCPs containing PSt branches are cast from a mixed solvent medium.

Electron Microscopy. Figure 2a is an electron micrograph of a PVTCA/PSt ABCP with 46 wt % PSt ($\overline{P}_n$ polystyrene = 1275, γ_r = 3.1) and no PSt branches. The micrograph shows the presence of well-defined dark domains of PVTCA with diameters of 8–16 nm which confirms microphase separation in the ABCP. It is not expected that domain diameters bear any definite relation to the size of the PVTCA chains. Previous studies on related systems (11) have shown that at high crosslink densities ($\gamma_r > 2$) domain sizes and separations are much smaller than for simpler materials with the same chain lengths but lower crosslink density ($\gamma_r < 1$) as a result of restrictions on aggregation of chains in highly crosslinked materials. In addition, if the A component forms domains, domain sizes will generally be influenced by unreacted A chains.

Figure 2b is an electron micrograph of the PVTCA/PSt ABCP MII which contains PSt branches and was cast from styrene monomer. Domain sizes are considerably larger ($\sim$ 30 nm) than in Figure 2a even though the PVTCA chain lengths are very similar. The domains are more diffuse, and sections of this sample always had poorer contrast than sections of ABCPs that contained no PSt branches. These observations are consistent with the idea of incomplete phase separation in which some PSt swells the PVTCA domains and thus reduces the inherent contrast by reducing the difference in composition between the phases. No evidence of very small scale phase separation of the components within the mixed domains was observed.

The PVTCA/PSt ABCP MIII which had only one T_g showed evidence of very small scale microphase separation (Figure 2c) with dark domains only 5–8 nm in diameter and very small domain separations. Although such materials appear homogeneous by dilatometry and DSC, electron microscopy pro-

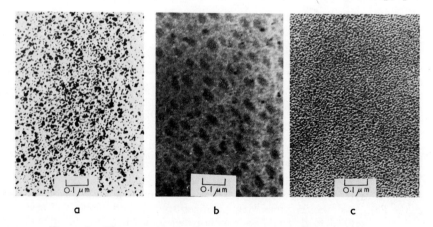

a   b   c

Figure 2.   Electron micrographs of PVTCA/PSt ABCPs (unstained)

a: *46 wt % polystyrene as crosslinks (no PSt branches), $\overline{P}_n$ polystyrene = 1275, γ_r = 3.1, cast from styrene;* b *and* c: *PVTCA/PSt ABCP MII and MIII, respectively (see Table II for sample description)*

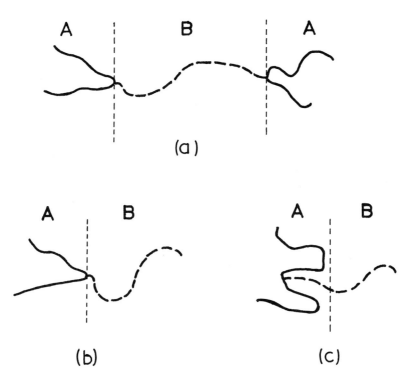

Figure 3. Schematic representation of microphase separation in ABCPs
Complete microphase separation in a, a crosslinked species of Structure I, and in b, a structure having B branches attached to A chains; c, representation of the incorporation of B branches into the A phase

vides definite evidence of microphase separation. Presumably there is too little bulk associated with a single domain or section of matrix for the phases to exhibit their own thermal properties independently. Inevitably there must be some mixing of the component polymers, if only in the interfacial regions. In addition, the presence of PSt branches may be responsible for incorporation of some PSt into the domains, thereby reducing the difference in properties of the two phases. Although the domains in Figure 2a are not much larger than those in Figure 2c, the existence of two transitions in the former sample may reflect both larger domains and a greater difference between the properties of the phases which contain virtually pure components.

Previous dilatometric data and present DSC data support the conclusion that there is a general tendency for complete microphase separation in ABCPs having no branches of the B component, but there is a tendency for incomplete microphase separation in those materials containing branches of the B component. To understand how this difference in morphology originates, consider two primary A chains attached to a B crosslink with the junction points located at the domain–matrix interfaces as depicted in Figure 3a. There are in effect two A chains attached to each end of the B chain. The A chains will be unable to assume their normal conformations especially in the vicinity of the interface, and the entropies of these chains will be less than in a corresponding free A chain. The tendency for each A chain to increase its entropy will produce a force that tends to draw the junction point into the A phase. By similar reasoning, the tendency of the B chain to increase its entropy, together with a retractive force in the B chain that results because the chain ends are permanently held apart, will tend to pull the junction in to the B phase. The forces in the A and B chains counteract each other and stabilize the junction at the interface.

When a B branch is attached to a primary A chain (Figure 3b), the same forces tend to draw the junction into the A phase. In this case, these forces are opposed only by the tendency of the B chain to increase its entropy and not by any retractive force in the B chain since one end is free. Consequently, the tendency for B branches to be incorporated into the A phase, as in Figure 4c, will be greater than for B crosslinks.

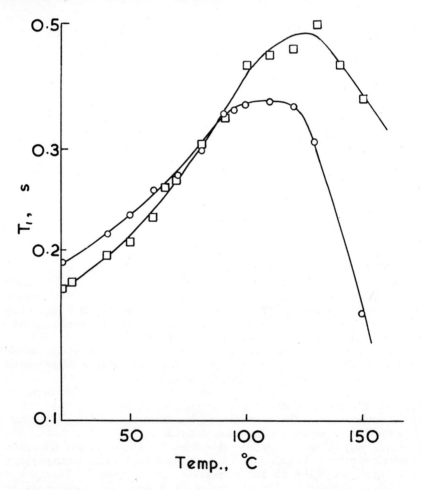

Figure 4.   Variation of spin-lattice relaxation time (T_1) *with temperature* □, *PMMA homopolymer;* O, *a PVTCA/PMMA ABCP with 50 wt % PVTCA,* $\gamma_r = 2$, $\overline{P}_n$ *PMMA crosslinks = 1300,* $\overline{P}_n$ *PMMA branches = 650, crosslink/branch ratio = 4*

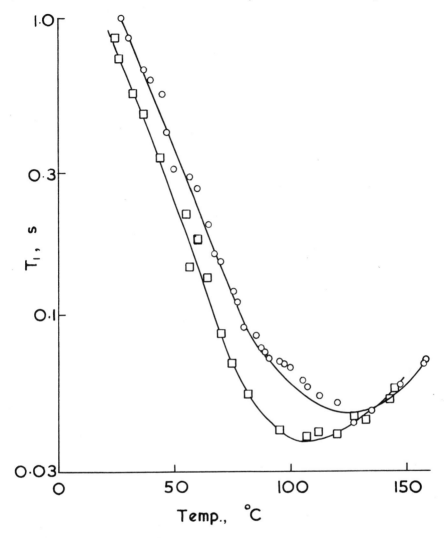

Figure 5. Variation of spin-lattice time (T₁) with temperature

□, *PMA homopolymer;* O, *a PVTCA/PMA ABCP with 20 wt % PVTCA,* γ_r = 2, $\bar{P}_n$ *PMA crosslinks = 24,000, no branches*

Molecular Motions in ABCPs

It is interesting to determine if molecular motions in polymer chains are influenced by their environment when these chains are incorporated into multicomponent polymers. We reported elsewhere (*2, 3*) that in PVTCA/PMMA ABCPs, rotations of the α-methyl groups on the PMMA chains may be enhanced or retarded with respect to rotations in PMMA homopolymer. These effects have been interpreted in terms of the morphology of the polymers, as determined by the dilatometric investigations, and they depend on the location of the PMMA chains to which the α-methyl groups are attached—whether they are in the mixed PVTCA + PMMA phase or in the PMMA phase near the domain–matrix interface.

Regarding some pulsed NMR data (12) on segmental motions in PMMA chains in PVTCA/PMMA ABCPs and in PMA chains in PVTCA/PMA ABCPs, Figure 4 presents the variation in spin-lattice relaxation time T_1 with temperature for PMMA and a PVTCA/PMMA ABCP in the vicinity of the PMMA T_g. The temperature at which the slope of the T_1 temperature curve starts to decrease with increasing temperature corresponds to the onset of segmental motion in the PMMA chains, and the rapid decrease in T_1 at slightly higher temperatures reflects the development in segmental motions in the PMMA chains. Segmental motions in the PMMA chains start at a slightly lower temperature in the ABCP than in the PMMA homopolymer, and the segmental motions in the crosslinked material develop more rapidly with increasing temperature than in the homopolymer. In the ABCP, the PMMA chains are attached to PVTCA chains which are above their T_g when segmental motions start in the PMMA. In other words, comparatively rigid PMMA chains are attached to relatively mobile PVTCA chains.

In contrast, Figure 5 illustrates T_1–temperature data for a PVTCA/PMA ABCP and for PMA homopolymer in the vicinity of the T_1 minimum associated with segmental motion in the PMA. The fact that the T_1 minimum for the ABCP occurs at a higher temperature than that for the PMA homopolymer implies that the segmental motions of the PMA chains in the ABCP are retarded with respect to the segmental motions in the homopolymer. In the PVTCA/PMA ABCP, the PMA chains (T_g for PMA = 5°C) are attached to relatively rigid PVTCA chains ($T_g \sim 60°C$). Thus, comparison of the data for the PVTCA/PMMA and PVTCA/PMA ABCPs indicates that, if chains of two chemically different polymers are attached in a multicomponent polymer, segmental motions in the relatively mobile material enhance segmental motions in the more rigid material and *vice versa*.

Morphologies of Some Polychloroprene-Containing ABCPs

We reported elsewhere (2) the results of morphological studies of ABCPs in which the A component was a copolymer (PS′) of styrene and p-vinyl benzyltrichloroacetate (containing about 10 mole % of the latter), and the B component was polychloroprene (PCp). At low crosslink densities ($\gamma_r < 1$) these materials consisted of assemblies of structures such as Structures I and II, together with unreacted PS′ chains. They contained small spherical PCp domains in a PS′ matrix. Domain diameters were approximately proportional to the square root of the average degree of polymerization of the PCp chains, as observed experimentally and predicted theoretically for linear AB block copolymers (12). The absolute domain diameters were approximately one-quarter the size predicted by Meier's treatment of AB block copolymers (12). Domain sizes in the two systems differ because in the block copolymers only one end of the B chain is necessarily located at the domain–matrix interface, whereas in the ABCP both ends of the PCp chains are located at the interface.

To extend these studies a series of ABCPs was recently prepared in which the polycarbonate derived from 1,1,1-trichloro-bis-2-(p-hydroxyphenyl)ethane (PCarb) was crosslinked with PCp (13). In these materials the number average molecular weight of the PCp chains was 97,350, and that of the PCarb was ~ 12,000. Polymer films were prepared by casting from solutions in chlorobenzene; electron micrographs of thin sections of these materials (stained with OsO_4) are presented in Figure 6.

The sequence of Figures 6a, 6b, 6c, and 6d illustrates the morphological changes which occur when crosslink density (γ_r 0.28–0.52) and PCp content

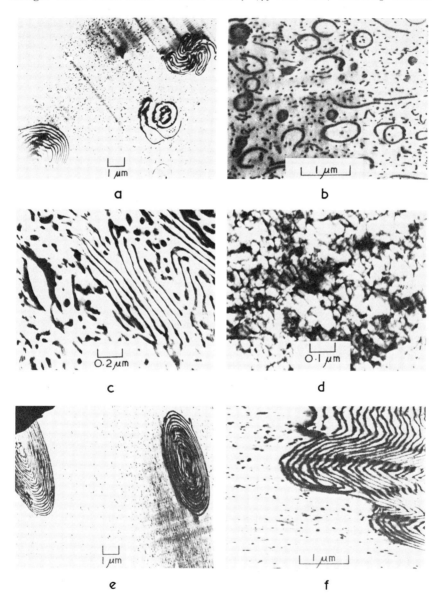

Figure 6.   Electron micrographs of PCarb/PCp ABCPs

PCarb $\overline{M}_n$ = 12,000, PCp $\overline{M}_n$ = 97, 350; a, 5 wt % PCp, γ_r = 0.028; b, 10 wt % PCp, γ_r = 0.058; c, 33 wt % PCp, γ_r = 0.25; d, 49 wt % PCp, γ_r = 0.52; e and f, ABCP with 49 wt % PCp, γ_r = 0.52 blended with PCarb to give materials containing 5 and 10 wt % PCp, respectively

(5-49 wt %) are simultaneously increased. In Figure 6a, both ribbons and small domains of PCp are visible. Domain diameters averaged about 40 nm. The ribbons (approximately 10 nm thick, they appear broadened because they are generally not perpendicular to the plane of the section) often formed loops which distorted to produce structures reminiscent of lamellar morphology. With increasing crosslink density and PCp content (Figures 6a, 6b, and 6c), the proportion of PCp in ribbon-like structures increased, and considerable order can be observed in Figure 6c.

The polymer films in Figures 6a, 6b, and 6c were optically clear. At higher crosslink densities ($\gamma_r = 0.5$, Figure 6d), the morphology became more irregular with large domains of PCarb enclosed in a fine network of PCp; the polymer film was cloudy. For materials with $\gamma_r > 1$, the morphology was similar to that shown in Figure 6d, but the PCarb domains were smaller, the network structure was finer, and the polymer films were clear.

The only real distinction between the materials with the morphologies depicted in Figures 6a, 6b, 6c, and 6d is that, as the conversion and PCp content increased, the materials contained a larger proportion of crosslinked species such as Structures I and II, and the proportion of complex crosslinked species to simpler structures increased. The average length of A chains incorporated into crosslinked structures decreased to some extent. PCarb/PCp ABCP containing 49 wt % PCp was blended with PCarb homopolymer to produce films with overall PCp contents of 5 and 10 wt % which correspond in composition to the original samples of low crosslink density. These blended materials (*see* Figures 6e and 6f for micrographs) correspond in overall composition to the materials depicted in Figures 6a and 6b respectively. While the morphologies of the blended materials did not correspond exactly to those of the original materials of similar composition, they are not unlike Figure 6a with a mixture of ordered lamellar-type structures and domains. Again the lamellae were about 10 nm thick, and they were in the form of concentric ellipsoids in the polymer that contained 5 wt % PCp (Figure 6e).

Strictly speaking, all the materials shown in Figure 6 were blends of cross-linked species and unreacted polycarbonate. Ordered structures similar to those appearing in Figures 6a, 6e, and 6f were reported by other workers (*14*) for blends of butadiene–styrene block copolymers with styrene homopolymer when the pure block copolymers have lamellar morphology and the blends contain 5–15 wt % block copolymer. Considering the composition of the simple crosslinked structures (Structure I) which were the most predominant of the multicomponent species present in the ABCPs, they were approximately 25:75 polycarbonate:polychloroprene which is approximately within the composition range for which lamellar morphologies have been predicted for linear block copolymers (*15*). It is not known how relative block lengths determine the morphology of assemblies of structures such as Structure I or II. However, it seems reasonable to suppose that the ordered lamellar-type structures in the ABCPs (containing unreacted PCarb) are predictive of lamellar morphologies in assemblies of structures such as Structures I and II of similar composition but in the absence of homopolymer.

Mechanical Properties of ABCPs

We report our findings on two materials (*16*) in which the A component is the polystyrene copolymer PS' ($\overline{M}_n = 130,000$) and the B component cross-links are PCp. The structural parameters and tensile properties of polymers

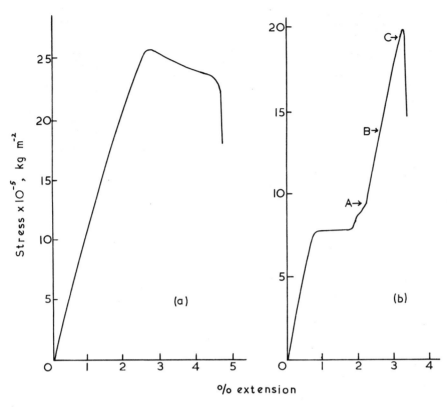

Figure 7.   Stress–strain curves for PS'/PCp ABCPs
Strain rate, 5 cm/min; temperature, 15°C; a, ABCP H1 and b, ABCP H2

are summarized in Table III. At 15°C both ABCPs underwent brittle fracture at small extensions (extensions at break were greater than for PS'), and stress–strain curves are presented in Figure 7. For the ABCPs, the 1% moduli were less than and the ultimate tensile strengths were comparable with the values for PS' (Table III). The intermediate yield, seen in the stress–strain curve for polymer H2 (Figure 7b), occurred reproducibly at ~1% extension and a stress of 7.8×10^5 kg/m². At 40°C the polymers yielded at relatively low stress (yield stresses for H1 and H2 are 2.5×10^5 and 10×10^5 kg/m²,

Table III.   Structures and Properties of PS'/PCp ABCPs

Sample	PCp, wt %	γ_r	$\overline{P}_n$, Cross-links	T, °C	1% Modulus[a] $\times 10^{-7}$kg/m²	UTS[b] $\times 10^{-5}$kg/m²	Extension at Break, %
H1	32	10	630	15	6.5	26	4.9
				40	1.5	2.7	365
H2	7.3	0.5	3300	15	7.0	20	3.5
				40	5.7	10	56
PS'				15	13	22	1.6

[a] Strain rate, 5 cm/min.
[b] UTS: ultimate tensile strength.

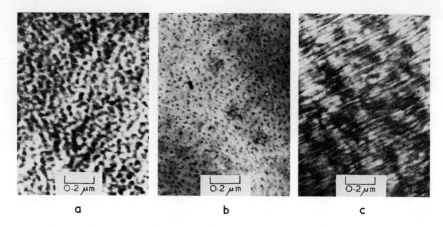

Figure 8.   Electron micrographs of PS'/PCp ABCP H1
a, unstressed; after tensile stress to fracture at 40°C, sections taken perpendicular (b) and parallel
(c) to the stress direction

respectively), and they developed considerable extensions prior to fracture, particularly polymer H1 which contained 32 wt % PCp.

The original samples contained an isotropic array of spherical PCp domains (Figures 8a and 9a); these are seen most clearly in the micrograph of the lightly crosslinked sample H2 that contained only 7.3% rubber (Figure 9a). Under stress at 15°C, sample H1 deformed, producing a short neck in the vicinity of the fracture. Electron micrographs of sections cut from the fracture zone established that deformation was associated with elongation of the domains. This effect is seen clearly in Figures 8b and 8c which are, respectively, electron micrographs of sections cut perpendicular and parallel to the stress direction from the neck produced in a sample after fracture at 40°C. Comparison of Figures 8a, 8b, and 8c indicates that the original domains under stress were pulled into thin fibrils parallel to the stress direction, and that the

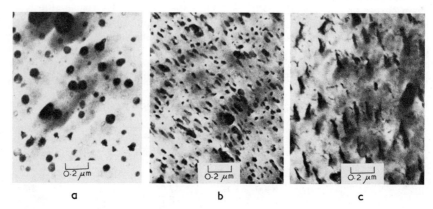

Figure 9.   Electron micrographs of PS'/PCp ABCP H2
(a) unstressed; after tensile stress to fracture at 40°C, sections taken perpendicular (b) and parallel
(c) to the stress direction

original domains preserved their original and separate identities. Similar effects were observed when sample H2 was stressed at 40°C (Figures 9a, 9b, and 9c); the domains were extended in directions approximately parallel to the stress direction.

The unusual yield observed at 1% extension when sample H2 was stressed at 15°C was investigated by examining electron micrographs of sections cut from samples stressed to points A, B, and C on the stress–strain curve (Figure 7b). After stress to point A followed by relaxation of the sample, no permanent distortion of the sample was apparent (*cf.* Figures 9a and 10a). Some permanent distortion of the PCp domains occurred after stressing to point B (Figure 10b). Distortion of the domains was not unidirectional (*see* Figure 10c); the deformation did not occur uniformly throughout the sample or even throughout

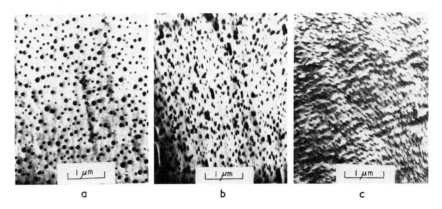

a   b   c

Figure 10.   Electron micrographs of PS'/PCp ABCP H2 after stress at 15°C to points A (a), B (b), and C (c) on the stress–strain curve in Figure 7b

a single section. At low crosslink densities as in sample H2 ($\gamma_r = 0.5$), the material must consist mainly of simple structures such as Structure I and unreacted PS' chains, so that very few PS' chains will provide interconnections between the PCp domains. Thus, the rubbery domains in H2 were effectively separate entities held together only by physical interactions between the PS' chains attached to the domains and the unreacted PS' chains in the glassy matrix. We suggest that the yield which occurred in sample H2 derived from some form of slippage of the domains within the matrix and rearrangement of the PS' chains attached to the domains and in the remainder of the matrix.

Literature Cited

1. Bamford, C. H., Eastmond, G. C., Whittle, D., *Polymer* (1971) **12,** 247.
2. Bamford, C. H., Eastmond, G. C., in "Recent Advances in Polymer Blends, Grafts and Blocks," L. H. Sperling, Ed., Plenum, New York, 1974.
3. Bamford, C. H., Eastmond, G. C., Whittle, D., *Polymer,* in press.
4. For a summary see C. H. Bamford in "Reactivity, Mechanism and Structure in Polymer Chemistry," A. D. Jenkins and A. Ledwith, Eds., Chap. 3, John Wiley and Sons, 1974.
5. Bamford, C. H., Eastmond, G. C., Robinson, V. J., *Trans. Faraday Soc.* (1964) **60,** 751.
6. Bamford, C. H., Eastmond, G. C., Whittle, D., *Polymer* (1969) **10,** 771.
7. Bamford, C. H., Dyson, R. W., Eastmond, G. C., *Polymer* (1969) **10,** 885.

370   COPOLYMERS, POLYBLENDS, AND COMPOSITES

8. Maguire, D., unpublished data.
9. Flory, P. J., "Principles of Polymer Chemistry," p. 358, Cornell University, New York, 1953.
10. Kelley, F. N., Bueche, F., *J. Polym. Sci.* (1961) **50,** 548.
11. Smith, E. G., Ph.D. Thesis, University of Liverpool, 1973.
12. Meier, D. J., *J. Polym. Sci. Part C* (1969) **26,** 81.
13. Phillips, D., unpublished data.
14. Bradford, E. B., in "Colloidal and Morphological Properties of Block and Graft Copolymers," G. E. Molau, Ed., Plenum, New York, 1971.
15. Meier, D. J., *Amer. Chem. Soc., Div. Polym. Chem., Prepr.* **11,** 400 (Chicago, September, 1970).
16. Hughes, E. O., unpublished data.

RECEIVED April 3, 1974.

32

Properties of Compatible Blends of Poly(vinylidene fluoride) and Poly(methyl methacrylate)

D. R. PAUL and J. O. ALTAMIRANO

Department of Chemical Engineering, University of Texas, Austin, Texas 78712

Evidence of the miscibility of poly(vinylidene fluoride) and poly-(methyl methacryate) is confirmed by the extensive studies of the dynamic behavior of their blends that are reported here. The blends, as well as the pure polymers, have multiple transitions. There is a single transition which shifts with blend composition as would be expected in compatible systems; however, for PVF$_2$ this is not the transition generally identified as T$_g$. The T$_g$ of the blend, the T$_m$ of the PVF$_2$, and the relative amount of crystallinity as a function of blend composition were examined by DTA. Density data indicate a lack of volume additivity caused primarily by the varying crystallinity of PVF$_2$-rich blends.

$\mathbf{B}$ecause of thermodynamics, truly miscible polymer pairs are rare (*1*). Evidence strongly suggests that poly(vinylidene fluoride), PVF$_2$, and poly(methyl methacrylate), PMMA, are miscible when melt blended (*2, 3, 4, 5*). Blends containing certain proportions of these two polymers are available commercially (*6*). Consequently, more extensive studies of blends made from this polymer pair were interesting to us. Previous efforts to establish the compatibility of this pair were based on observing the glass transition behavior of blends, mainly by thermal analysis and dilatometry (*2*). This report deals with additional thermal analysis including examination of PVF$_2$ crystallization from certain blends. Emphasis is on the transitional behavior indicated by dynamic mechanical properties which has not been reported previously. Specific volume measurements are discussed briefly.

The PVF$_2$, Kynar 301, was obtained from Pennwalt, and the PMMA, Plexiglas V(811)-100, was from Rohm and Haas. Blends were made by melt mixing in a Brabender Plasticorder 10 min at 200°C. Sheets were compression molded, and samples were annealed at 115°C for 20 min to develop maximum crystallinity. Otherwise, crystallization of some blends would occur later during testing.

Thermal Analysis of Blends

Differential thermal analysis (DTA) of blends and pure components was done cyclically by successive heating and cooling at 10°C/min between −100°

371

and 200°C. First heat analyses often differed somewhat from subsequent heats which gave identical data for as many as six repetitions. Only the later heat analyses are discussed because they are reproducible. As Noland *et al.* (2) observed, certain blends rich in PVF$_2$ have a melting endotherm characteristic of PVF$_2$ whereas blends rich in PMMA are totally amorphous.

Figure 1 shows the melting endotherm area with arbitrary units normalized for sample mass. Blends containing 50% or less PVF$_2$ have no melting endotherm; beyond this point, however, crystallinity increases rapidly with PVF$_2$ content. The dashed line connects the zero value for amorphous pure PMMA with the observed area for pure PVF$_2$. This is the peak area expected if PMMA did not interfere with PVF$_2$ crystallization but merely diluted the sample mass. Obviously, blending does interfere with crystallization. First heat areas for annealed samples were larger than these obtained with cyclical heating. Blends containing 50% PVF$_2$ showed slight crystallinity after the first heat but none after subsequent heats whereas blends containing 40% or less PVF$_2$ never showed evidence of crystallinity.

The temperature location of the melting peak, T_m, varied slightly with blend composition (*see* upper portion of Figure 2). The maximum depression in T_m was about 10°C. Thermodynamic calculations based on conventional mixture theories which account for the reduction in chemical potential that occurs on mixing (7) indicate that a miscible diluent with the molecular weight of PMMA cannot produce a depression this large. It is likely that the major effect is morphological, *e.g.* smaller or less perfect crystalline regions develop

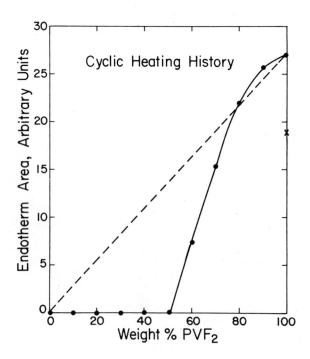

Figure 1.   Relative crystallinity of PVF$_2$–PMMA blends as determined by DTA

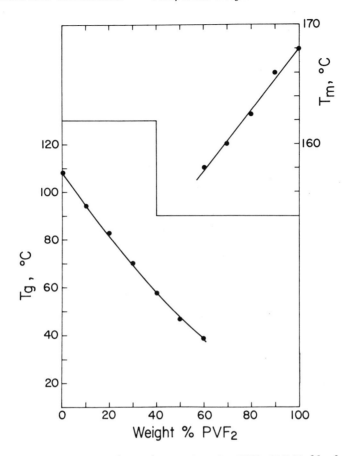

Figure 2.   DTA observed transitions for PVF₂–PMMA blends

from blends. Similar reductions in T_m have been observed in incompatible blends (8).

All the blends showed a single T_g by DTA, within the limits of detection which was quite strong for the wholly amorphous samples. The lower portion of Figure 2 shows the location of this transition as a function of blend composition. As crystallinity increases with increasing PVF_2 content, the intensity of this transition diminishes rapidly. Beyond 60% PVF_2, all indications by DTA of a T_g for blends were comparable in magnitude to instrument noise, and consequently no values are recorded in this range in Figure 2. DTA also failed to reveal an unambiguous glass transition for pure PVF_2 at cooling rates of $10°C/min$. The dependence of T_g on blend composition shown in Figure 2 is consistent with the experimental data of Noland *et al.* (2), and this confirms their conclusion of miscibility for this polymer pair, at least to PVF_2 contents where crystallinity develops. For blends with crystallinity, one might picture a two-phase structure where the crystalline regions are pure PVF_2 and the amorphous regions are a compatible mixture of PVF_2 and PMMA. This mixture composition differs from that of the total blend and is richer in PMMA because of loss of PVF_2 by crystallization. Partitioning of PVF_2 between these phases may occur on a kinetic rather than a thermodynamic basis. Some evidence for compatibility in the amorphous regions is presented below (*see* section on dynamic mechanical properties).

Noland *et al.* (*2*) encountered similar difficulty in measuring by thermal analysis or dilatometry the T_g of blends very rich in PVF_2. However, they state that their T_g data, which cover a composition range similar to that in Figure 2, are consistent with extrapolation to $-40°$ to $-46°C$ for pure PVF_2. They were able to measure T_g directly ($-46°C$) using thermal analysis for rapidly quenched PVF_2 samples. In subsequent discussions, it is important to remember that for this polymer considerable uncertainty may be associated with this observation or its interpretation because of the difficulties posed by high crystallinity (*2*). There is considerable evidence (*9–16*) to support a T_g value for PVF_2 in this range; however, as Noland *et al.* (*2*) noted, there is some controversy about this since higher values, $13°$ and $27°C$, have been suggested (*17, 18*). We mention this here because the subsequent dynamic mechanical property data are not directly consistent with this simplistic extrapolation or acceptance of the $-40°C$ region as the T_g for PVF_2. More recently Nakagawa and Ishida (*19*) reported a small specific heat jump by DSC for PVF_2 that occurs between $-50°$ and $-30°C$. More thorough examination of our PVF_2 polymer by DSC (*20*) revealed a slight base-line drift over the $-40°$ to $0°C$ range. This change was very small and indistinct in some thermal traces. Our T_g data presented in Figure 2 are not at all conclusive regarding extrapolations to pure PVF_2. If the slight curvature is real, then an intercept of $+20°C$ or higher would be possible. However, if

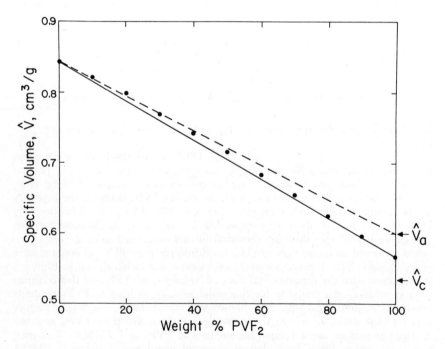

Figure 3. Specific volume of annealed PVF_2–PMMA blends

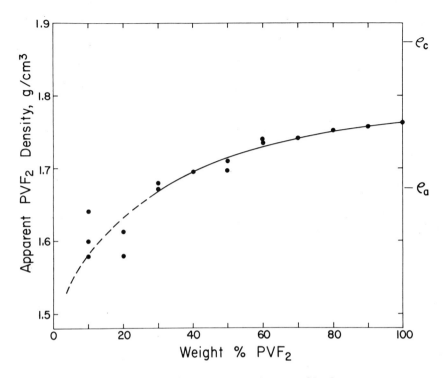

Figure 4.   Apparent PVF₂ density in blends
Markers to the right indicate density of 100% crystalline and amorphous PVF₂ in the pure state

this curvature is ignored and the extrapolation is based on the first few PMMA rich points, it will meet the PVF$_2$ axis at $-20°$ to $-25°$C.

Specific Volume of Blends

The specific volume of annealed blends was measured by a pycnometer technique. The data are plotted *vs.* blend composition in Figure 3. Estimates of the specific volume for totally crystalline, $\hat{V}_c$, and totally amorphous, $\hat{V}_a$, PVF$_2$ reported by Nakagawa and Ishida (*21*) are indicated by the arrows on the right. The $\hat{V}_a$ value differs considerably from the estimate (0.676 cm^3/g) of Dole and Lando (*22*), but it is considered more accurate. The $\hat{V}_c$ value agrees well with that reported by Dole and Lando. The experimentally determined value of $\hat{V}$ for pure PVF$_2$ lies approximately halfway between the amorphous and crystalline values which indicates 50% crystallinity for our sample. The solid line connects this point with that for amorphous PMMA. The broken line connects the latter with the $\hat{V}_a$ for pure PVF$_2$. The data for the amorphous, PMMA-rich blends lie closer to the broken line whereas those for the crystalline, PVF$_2$-rich blends lie closer to the solid line. Strict volume additivity as defined by either line should not be expected for all blends because of variation in crystallinity with blend composition. A more detailed analysis of these data can be effected by defining an apparent density for

PVF$_2$ in the blend, ρ_{PVF_2}, as:

$$\frac{1}{\rho} = \frac{w}{\rho_{PVF_2}} + \frac{(1-w)}{\rho_{PMMA}}$$

where ρ is the observed density of a blend containing a weight fraction, w, of PVF$_2$, and ρ_{PMMA} is the observed density of pure PMMA. The calculated ρ_{PVF_2} value is the density that is needed to assure volume additivity, and it is related to the partial specific volume of this component, but not simply. Figure 4 shows apparent PVF$_2$ density as a function of blend composition. At $w = 1$, the density is that observed for pure PVF$_2$, but it decreases as PMMA is added, primarily because of the reduction in PVF$_2$ crystallinity that occurs on blending (*see* Figure 1). Annealed blends showed no endotherm at all when PVF$_2$ content was 40% or less; thus one would expect that the apparent density would approach the amorphous density of PVF$_2$ (1.667 g/cm^3 according to Ref. *20*) at 40% and then would remain constant at this level for lower PVF$_2$ contents if crystallinity were the only factor. This value is not reached experimentally until PVF$_2$ content has been reduced to 30%, and it appears

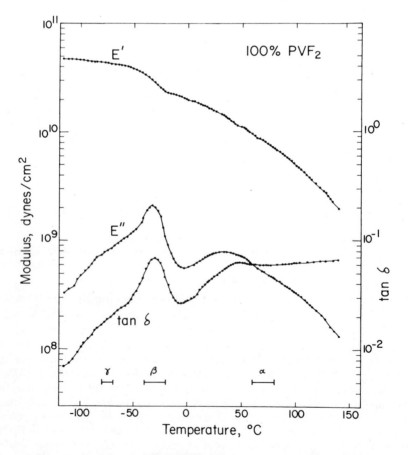

Figure 5.   Dynamic mechanical properties of annealed PVF$_2$ at 110 Hz

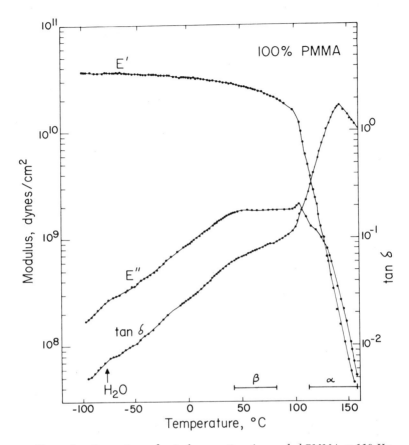

Figure 6.   Dynamic mechanical properties of annealed PMMA at 110 Hz

to continue decreasing thereafter although experimental uncertainty in the apparent density increases rapidly as w approaches zero. Lack of volume additivity in the amorphous phase (arising from molecular interactions) can also contribute to the observed apparent density. The data suggest that this effect may be operative here, with the volume changes on mixing being negative for PVF_2-rich blends and positive for PMMA-rich blends. However, the concomitant and more massive effect of crystallinity makes this conclusion uncertain.

Dynamic Mechanical Properties of Blends

The dynamic mechanical properties, E', E'', and tan δ, of annealed blends and pure components were measured at 110 Hz by a Rheovibron viscoelastometer. Data for pure PVF_2 are presented in Figure 5. Three principal relaxation regions, labeled α, β, and γ, were observed previously for this polymer type by various techniques. Markers indicate the range of peak temperatures for each transition reported in the literature for a frequency of 110 Hz. For unknown reasons, our α peak occurs at a somewhat lower temperature; however, we believe this is the same transition. Locations of the β and γ peaks

378 COPOLYMERS, POLYBLENDS, AND COMPOSITES

agree well with the literature. Most reports would be compatible with the mechanisms for these dispersion regions given by Yano (10): α, molecular motions associated with crystalline regions and their defects; β, motion of the main chain in the amorphous region which may thus be regarded as the main T_g; γ, local molecular motion in the amorphous regions. However, Peterlin and Holbrook (17, 18) assigned different meanings to these peaks (which they labeled differently). They suggest that the α peak discussed here is caused by rotation of dipoles of chains in crystal defects. They mentioned a premelting peak at higher temperatures, but they did not extend their measurements to low enough temperatures to see the γ region. They referred to calorimetric and dilatometric measurements (17) that locate T_g at 13°C and reference other dilatometric data of theirs (18) that fix this value at 27°C.

Data for pure PMMA (Figure 6) indicate two major relaxation regions labeled α and β. Markers indicate the ranges reported in the literature for these transitions at 110 Hz. The α peak is the main T_g (23), and the data locate it at 105°C by E'' and at 142°C by tan δ. This wide separation of the

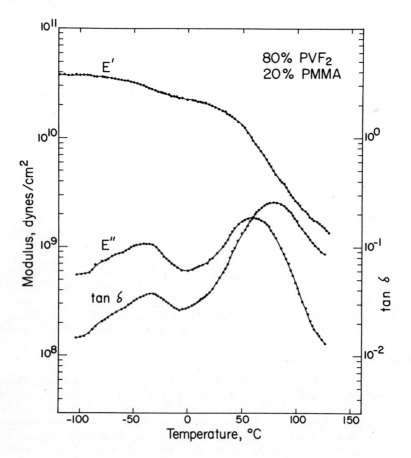

Figure 7. Dynamic mechanical properties at 110 Hz for an annealed 80 wt % PVF₂ blend

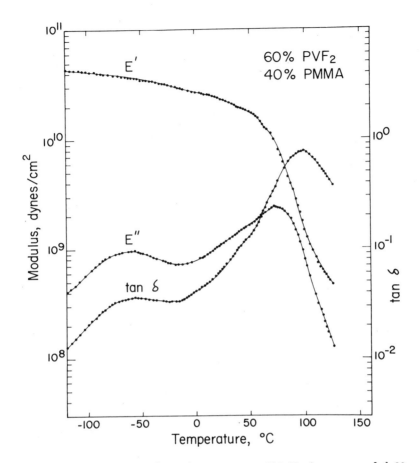

Figure 8. Dynamic mechanical properties at 110 Hz for an annealed 60 wt % PVF₂ blend

peaks is common for wholly amorphous polymers (*24*). The β region is believed to arise from rotations of the ester side group (*23*), and it appears here as a shoulder in E'' and tan δ in the vicinity of 50°C. The presence of water is known to produce a low temperature relaxation (*23*) which is weakly evident here.

The dynamic mechanical properties of various PVF₂–PMMA blends are shown in Figures 7–11. For the 80% and 60% PVF₂ blends (Figures 7 and 8) which are partially crystalline, there are two major peaks in either E'' or tan δ. The higher temperature peak is more dominant and would appear to be the T_g of the blend. As PMMA content increases, this peak becomes larger and shifts toward higher temperatures. It appears to degenerate into the α peak of pure PVF₂ shown in Figure 5 when PMMA content decreases to zero. The lower temperature peak decreases in magnitude and shifts toward lower temperatures as PMMA content increases; when all PMMA is eliminated, it appears to reduce to the PVF₂ β peak.

For the noncrystalline 40%, 20%, and 10% PVF_2 blends (Figures 9, 10, and 11), there is one major peak which occurs very near the end of the temperature range. This peak is very large, and the magnitude of tan δ is almost the same for all three materials. Its location shifts to higher temperatures as PMMA content increases. It is clearly the main T_g for the blends, and it merges directly into the pure PMMA T_g peak when all PVF_2 is eliminated. The shoulder at about +50°C is especially apparent for the 10% PVF_2 blend. This is evidently the PMMA β dispersion, and it does not shift appreciably on blending, although its magnitude diminishes as the PMMA is diluted with PVF_2. Another significant shoulder appears at or below −50°C for all three blends. This is discussed below.

Figure 12 shows peak locations as a function of blend composition for the well-defined peaks in Figures 7–11 (shoulders are not included). Since the storage modulus drops precipitously at the highest temperature transition, the tan δ and E″ peaks occur at quite different temperatures (24); consequently, both positions are indicated. Smooth curves, although sigmoidal in

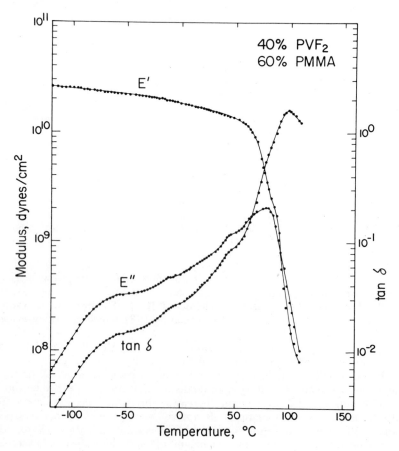

Figure 9. Dynamic mechanical properties at 110 Hz for an annealed 40 wt % PVF₂ blend

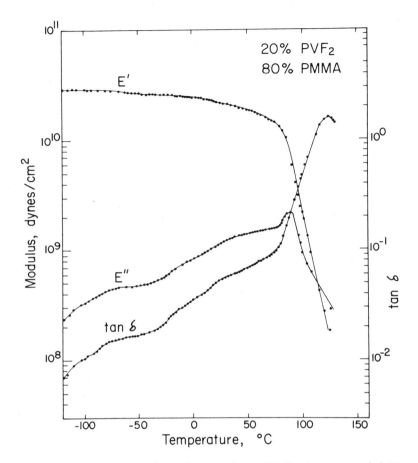

Figure 10.   Dynamic mechanical properties at 110 Hz for an annealed 20 wt % PVF₂ blend

shape, can be drawn through the upper transitions for the blends which connect with the major T_g peak of pure PMMA and the α peak of pure PVF$_2$. Only the E'' peak is shown for the lower temperature transition since tan δ peaks at substantially the same location. Only two points are shown for blends since these are the only ones that show a distinct peak. They can be connected to the pure PVF$_2$ β peak by a simple curve.

For miscible polymer blends, one expects a single T_g which depends on composition in such a way as to connect the T_g's of the two pure components (*25, 26*). To date, however, there is very little to suggest what should happen to secondary amorphous peaks or those associated with motions in crystalline regions which would include PMMA β and PVF$_2$ α and γ according to the interpretations noted above. All evidence strongly suggests that PVF$_2$ and PMMA are miscible. Thus one might expect the PVF$_2$ β peak to shift to higher temperatures as PMMA is added, and to connect with the T_g of PMMA, given that the β peak is T_g. Further, one might expect the α peak to diminish in magnitude as PMMA is added and subsequently to disappear when all crystallinity is lost, given that the α peak is associated with PVF_2 crystalline regions. However, the data for the blends, shown in Figure 13 as tan δ curves, do not conform to these expectations. Instead, when PMMA is added, the PVF_2 tan δ and E'' peaks shift as follows. The β peak diminishes in magnitude and eventually disappears (approximately when all crystallinity is lost). This peak seems to shift to lower temperatures as PMMA is added; however, this may not be real, as it could reflect the increase in relative importance of the lower temperature γ region which only appears as a shoulder in pure PVF_2. What is evidently the small α peak increases in magnitude and shifts toward higher temperatures as PMMA is added (*see* the high temperature portion of the 80% and 60% tan δ curves in Figure 13). After all crystallinity has been lost (between 60 and 40% PVF_2), this peak no longer increases in magnitude with added PMMA, but it does continue to shift toward higher temperatures.

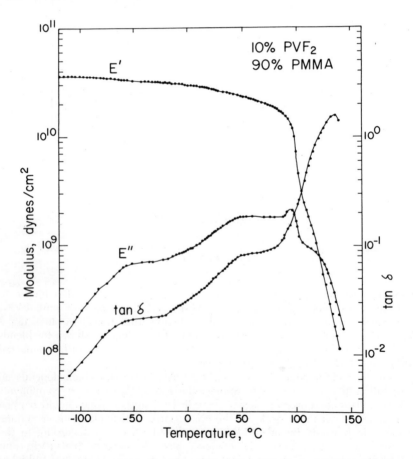

Figure 11. Dynamic mechanical properties at 110 Hz for an annealed 10 wt % PVF₂ blend

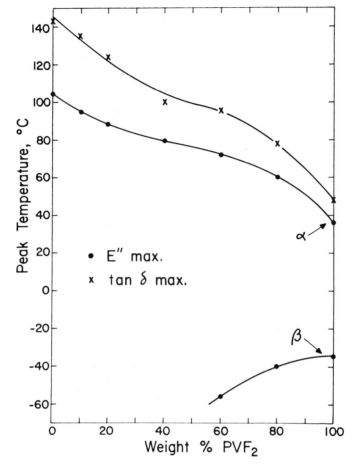

Figure 12. Effect of blend composition on temperature location of major peaks (locations of shoulders are not indicated)

These results are easily explained by the interpretation of the PVF$_2$ α and β peaks given by Peterlin and Holbrook (*17, 18*). Figure 12 plus the results discussed in earlier sections strongly support this view, and it is very tempting to draw this conclusion. However, this approach deals too lightly with the considerable evidence for the interpretations summarized by Yano (*10*). If the latter view is correct then explanation of our data is more complex, and they pose some intriguing questions about polymer blends; this explanation requires additional investigative techniques so we choose not to speculate in this direction at the present time. However, other comments about the transitional behavior of PVF$_2$ are appropriate. Nakagawa and Ishida (*16*) recently described very extensive and thorough investigations of relaxations and molecular motions in PVF$_2$. They conclude that the β region is the T_g for PVF$_2$; however, they did note a number of peculiarities by which the T_g behavior of PVF$_2$ differs from that of other polymers, and they suggest that

$-38°C$ is only an apparent T_g. They discuss these motions in terms of the size of the cooperatively rearranging region which is macroscopic at a very low temperature, T_2, and which decreases to a minimum at $T_3 = +30°C$. Starkweather (27) has published some very thorough and interesting studies on molecular motions in an alternating ethylene–tetrafluoroethylene copolymer which of course is isomeric with PVF_2. He also observed α, β, and γ regions, although the temperature locations are not exactly the same as those for PVF_2. He concludes from a variety of evidence that the α and γ relaxations reflect motions in amorphous or disordered regions, and that the β relaxation occurs in crystalline regions. He notes that these assignments parallel those made for polytetrafluoroethylene.

Summary

It is clear from our findings and the earlier work of Noland *et al.* (2) that PVF_2 and PMMA are compatible. However, no definitive mechanism has been identified that explains the specific interactions between these molecules that makes this so.

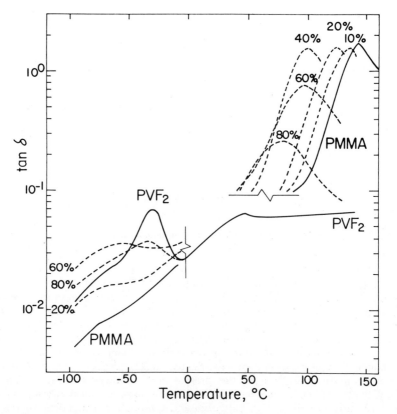

Figure 13.   Tan δ at 110 Hz for pure PVF₂ and PMMA (solid lines) and selected blends (broken lines)

For clarity, portions of curves have been omitted; no significant peaks appear in omitted regions

Several serious questions about the transitional behavior of pure PVF_2 have been raised. One might identify the α peak of PVF_2 as its amorphous phase T_g in order to explain Figure 12 in terms of previous examples of polymer compatibility that show simple connections between the T_g's of the pure components (25, 26). The sigmoidal shape of the curve in Figure 12 cannot be described by the usual type of equation (25, 26) that generally fits blend data. This, however, might easily be explained by crystallinity which changes the amorphous phase composition by removing PVF_2. The T_g from DTA (Figure 2) has a different dependence on overall blend composition, perhaps because crystallinity is lower as a result of cyclic heating than it was in the annealed blends (Figure 12). Crystallization of PVF_2 from blends rich in PVF_2 does not necessarily signal immiscibility since the evidence indicates that all the PMMA and the remaining PVF_2 form a homogeneous amorphous phase.

Acknowledgments

The authors gratefully acknowledge helpful discussions and correspondence with J. W. Barlow, C. E. Locke, Y. Ishida, and A. Peterlin.

Literature Cited

1. Paul, D. R., Vinson, C. E., Locke, C. E., *Polym. Eng. Sci.* (1972) **12**, 157.
2. Noland, J. S., Hsu, N. N. C., Saxon, R., Schmitt, J. M., "Multicomponent Polymer Systems," ADVAN. CHEM. SER. (1971) **99**, 15.
3. Koblitz, F. F., Petrella, R. G., Dukert, A. A., Christofas, A., Pennsalt Corp., U.S. Patent **3,253,060** (1966).
4. Miller, C. H., American Cyanamid Corp., U.S. Patent **3,458,391** (1969).
5. Schmitt, J. M., American Cyanamid Corp., U.S. Patent **3,459,834** (1969).
6. Dohany, J. E., private communication.
7. Flory, P. J., "Principles of Polymer Chemistry," Cornell University, Ithaca, 1953.
8. Natov, M., Peeva, L., Djagarova, E., *J. Polym. Sci. Part C* (1968) **16**, 4197.
9. Mandelkern, L., Martin, G. M., Quinn, F. A., *J. Res. Nat. Bur. Std.* (1957) **58**, 137.
10. Yano, S., *J. Polym. Sci. Part A-2* (1970) **8**, 1057.
11. Sasabe, H., Saito, S., Asahina, M., Kakutani, H., *J. Polym. Sci. Part A-2* (1969) **7**, 1405.
12. Koizumi, N., Yano, S., Tsunashima, K., *J. Polym. Sci. Part B* (1969) **7**, 59.
13. Kakutani, H., *J. Polym. Sci. Part A-2* (1970) **8**, 1177.
14. Ishida, Y., Watanbe, M., Yamafuji, K., *Kolloid Z.* (1964) **200**, 48.
15. Koo, G. P., in "High Polymer Series," Vol. **25**: "Fluoropolymers," L. A. Wall, Ed., Chap. 16, Interscience, New York, 1972.
16. Nakagawa, K., Ishida, Y., *J. Polym. Sci. Part A-2* (1973) **11**, 1503.
17. Peterlin, A., Holbrook, J. D., *Kolloid Z.* (1965) **203**, 68.
18. Peterlin, A., Elwell, J. (Holbrook), *J. Mater. Sci.* (1967) **2**, 1.
19. Nakagawa, K., Ishida, Y., *J. Polym. Sci. Part A-2* (1973) **11**, 2153.
20. Locke, C. E., private communication.
21. Nakagawa, K., Ishida, Y., *Kolloid Z.* (1973) **251**, 103.
22. Doll, W. W., Lando, J. B., *J. Macromol. Sci. Phys.* (1968) **B2**, 219.
23. McCrum, N. G., Read, B. E., Williams, G., "Anelastic and Dielectric Effects in Polymer Solids," Wiley, New York, 1967.
24. Locke, C. E., Paul, D. R., *Polym. Eng. Sci.* (1973) **13**, 308.
25. Koleske, J. V., Lundberg, R. D., *J. Polym. Sci. Part A-2* (1969) **7**, 795.
26. Krause, S., Roman, N., *J. Polym. Sci. Part A* (1965) **3**, 1631.
27. Starkweather, H. W., *J. Polym. Sci. Part A-2* (1973) **11**, 587.

RECEIVED February 20, 1974.

33

Fundamental Considerations for the Covulcanization of Elastomer Blends

MARTIN E. WOODS and THOMAS R. MASS

B. F. Goodrich Chemical Co., Avon Lake Technical Center,
Avon Lake, Ohio 44012

It has been generally reported that elastomer blends do not co-vulcanize, i.e., the blend properties are inferior to those predicted from the properties of the component elastomers. We demonstrate that the lack of covulcanization results from diffusion of the vulcanization accelerators from one phase into the other. Furthermore, this diffusion is the result of two driving forces: thermodynamic, i.e., preference of the accelerators for the more polar elastomer, and kinetic, i.e., depletion of the curatives in the phase with the greater cure rate and subsequent diffusion of more curatives into that phase. The design of cure systems to overcome the two driving forces is discussed, and their ability to covulcanize a series of elastomer blends is demonstrated.

Elastomer blend technology is widely used in the rubber industry today as it has been for many years. Typically, the science of elastomer blends has evolved much more slowly than the technology. This is evidenced by the fact that it is widely known and accepted that elastomer blends are seldom as good as one would expect empirically by interpolating between the properties of the base elastomers used in the blend (*1, 2*). This obtains even when similar elastomers are blended (*3*).

The science of elastomer blends has begun to develop more rapidly. The most notable examples are the work published by Gardiner (*4, 5*) and Zapp (*6, 7*) since 1968. Gardiner studied the problem of curative migration whereas Zapp measured the interfacial adhesion in elastomer blends. The objective of this paper is to build on the work of these authors and to demonstrate that curative diffusion occurs by two mechanisms. We will also show that true covulcanization can be obtained, *i.e.*, the blend properties will be a linear function of blend composition, when both curative diffusion mechanisms are counterbalanced by proper curative selection.

Strategy

We studied a series of blends based on the relatively nonpolar ethylene–propylene–diene monomer (EPDM) elastomer which has a solubility parameter

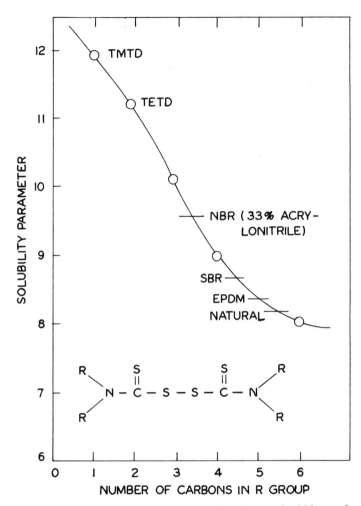

Figure 1.   Solubility parameter of tetraalkylthiuram disulfide accelerators

of 8.3 (8). Blends were made with highly polar acrylonitrile–butadiene co-polymer (NBR), with slightly polar styrene–butadiene copolymer (SBR), with nonpolar natural rubber (NR), and with nonpolar synthetic polyisoprene elastomer (SN). These elastomers have solubility parameters of 9.4, 8.5, 8.1, and 8.1 respectively (8). Since Gardiner had shown that polar accelerators such as tetramethylthiuram disulfide (TMTD) was five times more likely to diffuse into the more polar elastomer than other curatives, we studied the effects of various degrees of alkylation of similar accelerators on the properties of the elastomer blends. This is depicted in Figure 1 where the solid line represents the solubility parameter of the tetraalkylthiuram disulfide accelerators as a function of the degree of alkylation. The values were calculated using Small's method (9). Figure 1 clearly indicates that the more common methylated (TMTD) and ethylated (TETD) accelerators would have a strong tendency to diffuse into the most polar elastomer. This tendency should definitely decrease with higher degrees of alkylation. We designed the experiments to demonstrate that, when the polarity of the accelerator is reduced sufficiently, the thermodynamically induced diffusion will be overcome and covulcanization will be obtained.

We interpreted our results in terms of the vulcanization diagram shown in Figure 2; this is a three-dimensional version of the vulcanization possibilities reported by Gardiner (5). On the basis of this model, we felt that covulcanization would occur only when the following conditions were satisfied: (a) the continuous phase was vulcanized to near its optimum level, (b) the discontinuous phase was vulcanized to near its optimum level, and (c) there was excellent adhesion between the dispersed and continuous phases. The exact degree of curing depends on the main property that is desired because some properties are maximized by slightly undercuring the elastomers, others maximize at optimum cure, and still others maximize when there is some overcuring (10).

Experimental Procedure

Description of Materials. The accelerators evaluated in this study are listed in Table I. The first three accelerators are available commercially; the remainder were synthesized in our laboratory. The elastomers are described in Table II.

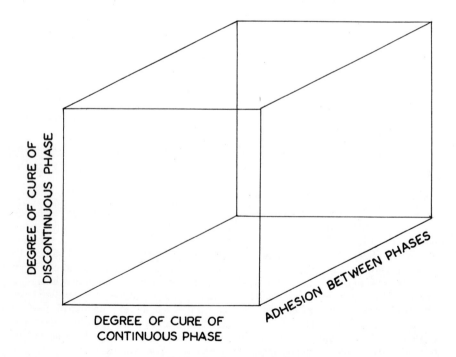

Figure 2. Vulcanization cube

Table I.   Accelerators Evaluated

Code	Chemical Description
C1	zinc dimethyldithiocarbamate
C2	zinc diethyldithiocarbamate
C4	zinc dibutyldithiocarbamate
C8	zinc diethylhexyldithiocarbamate
C10	zinc didecyldithiocarbamate
C12	zinc didodecyldithiocarbamate
C20	zinc diarachidicdithiocarbamate

Compounding. The compounds studied were all made by the base recipe in Table III. The concentration of 1.6×10^{-3} moles accelerator/100 g rubber was chosen because it represents a normal level for the low molecular weight accelerators of this type that are being used in the elastomer industry. Blend ratios were 100/0, 75/25, 50/50, 25/75, and 0/100 in order to obtain a complete picture as a function of the blend composition.

Table II.   Description of Polymeric Materials

Rubber Type	Trade Name	Mooney[a]	Supplier
EPDM	Epcar 585	50	B.F. Goodrich Chemical Co.
NBR	Hycar 1032[b]	55	B.F. Goodrich Chemical Co.
SBR	Ameripol 1502[c]	50	B.F. Goodrich Chemical Co.
NR	RSS #1	—	Rubber Trade Association of New York Dealers
SN	Ameripol SN600	80	B.F. Goodrich Chemical Co.

[a] Raw Mooney ML-4 @ 212°F.
[b] Nominal 33% acrylonitrile.
[c] 23.5% bound styrene.

The compounds were prepared using a normal Banbury milling procedure. The rubbers were charged to the Banbury and allowed to mix for 2 min; then the remainder of the ingredients were added. The batch was dumped after 5 min total mixing time. Curatives were added on a standard two-roll laboratory mill by the normal milling procedure. Monsanto rheometer curves at 320°F were obtained for the compounds. The compounds were then press cured for optimum cure times except that flexometer specimens were cured an additional 10 min because of their thickness.

Physical Testing. The physical properties were measured in accordance with the applicable ASTM procedures.

Table III.   Base Compound Recipe

Substance	Content
Rubber	100 parts
Stearic acid	2 parts
N550 (FEF)	55 parts
ZnO	5 parts
Sp sulfur	1.5 parts
Accelerator	1.6×10^{-3} moles

Results and Discussion

Blends of EPDM with NBR. Because of differences in the solubility parameters of EPDM and NBR elastomers, these blends should have the least tendency to covulcanize. This is clear in Figures 3, 4, and 5 for the ethylated

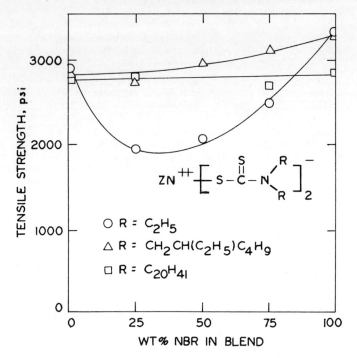

Figure 3.   Tensile strength of EPDM/NBR rubber blends

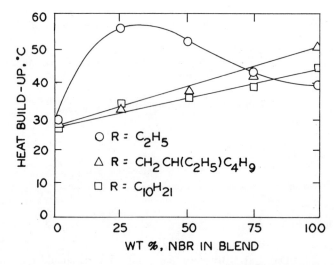

Figure 4.   Flexometer heat buildup in EPDM/NBR blends

accelerator. In each case the properties deviate greatly from the linear behavior that is indicative of covulcanization. These figures also show that when
the ethylhexyl and higher alkylated versions of the accelerators are used, tensile
stress, flexometer heat buildup, and oil swell all become linear functions of the
blend composition. This agrees directly with our hypothesis that if the accelerator is less polar, diffusion prior to vulcanization will decrease to the point
where it is insignificant and covulcanization is obtained. Similar results based
only on tensile strength measurements were reported by Mastromatteo and
co-workers (*11*).

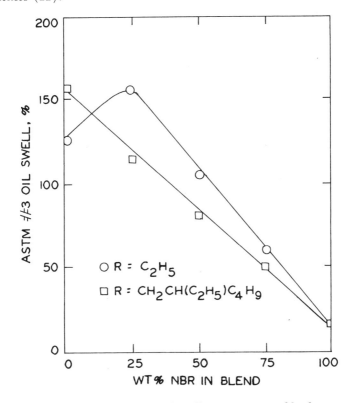

Figure 5.   ASTM #3 oil swell in EPDM/NBR blends

Linearization of the plots in Figures 3, 4, and 5 also indicates that adhesion
between the phases is good. Zapp showed that as adhesion improved, tensile
strength showed less of a dip below linearity (*7*). He did not achieve linearization, however, possibly because of carbon black distribution problems (*12*).

Furthermore, when the ethylated accelerator was used, the departure
from linearity was greatest at the 25% NBR level, *i.e.*, when the EPDM phase
was continuous. This can be explained by the rapid diffusion of the polar
accelerator from the EPDM phase into the NBR phase. Gardiner reported
that this diffusion occurs in seconds because of the fine dispersion of one
polymer in the other (*5*). Therefore, the continuous EPDM phase would
contain little accelerator and would be greatly undercured, with a resultant

substantial drop in the values of the measured properties. Consequently, as one might expect, vulcanization of the continuous phase plays a major role in determining the physical properties of elastomer blends.

The effect of the continuous phase is also demonstrated in Figure 6 where the oil swell at three blend ratios is plotted as a function of degree of accelerator alkylation. When EPDM is the continuous phase (75/25 and 50/50), the effect of alkylation on oil swell is strong. When NBR is continuous (25/75), alkylation essentially has no effect. Likewise, the improvement in tensile strength with alkylation is more pronounced when EPDM is the continuous phase (*see* Figure 7). In this case, tensile strength did not reach a plateau value because the tensile strength of the nitrile phase increased as the degree of alkylation of the accelerator increased.

Blends of EPDM with SBR. These are representative of blends of a slightly polar elastomer with a nonpolar elastomer. One would expect that these would be easier to covulcanize than the NBR/EPDM blends discussed in the previous section. Figure 8 shows that despite the small difference in polarity, the ethylated accelerator failed to covulcanize the blends. Covulcanization was achieved with the more highly alkylated accelerators (*see* Figures 8 and 9). It is apparent that the blends of EPDM with slightly polar SBR are covulcanized at lower degrees of alkylation than are the blends of EPDM with the highly polar NBR elastomers (*cf.* Figures 9 and 7). These results agree with the curative diffusion mechanism mentioned previously and demonstrate that

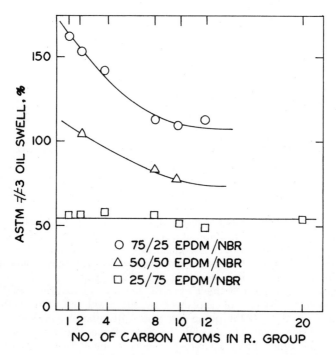

Figure 6. ASTM #3 oil swell in EPDM/NBR blends

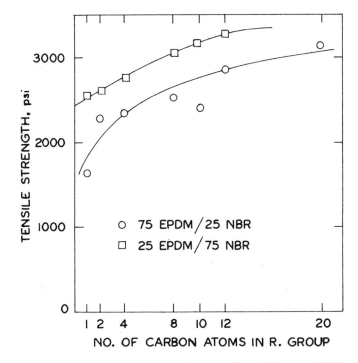

Figure 7. Tensile strength of 75/25 and 25/75 EPDM/NBR rubber blends

this diffusion is important even when there are slight differences in elastomer solubility parameters.

Blends of EPDM with Natural Rubber. These blends were included in the study to represent the case where diffusion effects should be minimal, *i.e.*, in blends of two nonpolar elastomers. The tensile strength (Figure 10) and the flexometer heat buildup data (Figure 11) were quite surprising. These blends were not covulcanized by either the ethylated or the ethylhexylated accelerator. Since the latter did covulcanize EPDM/NBR blends where the difference in solubility parameters was large, these results do not agree with the picture of curative diffusion that evolved from the studies of EPDM blends with NBR and SBR elastomers. Figure 12 shows that the effect of alkylation on the EPDM/NR blends also differs from that observed for the other two blends. After an initial increase, tensile strength improved gradually as the degree of alkylation increased. However, covulcanization was obtained only when 20-carbon atom alkyl groups were added to the accelerator. This gradual improvement appears to be related to molecular size which in turn is inversely proportional to rate of molecular diffusion. This indicates that there is still diffusion despite the lack of a significant thermodynamic driving force from the the close match of elastomer polarity. We therefore postulate that there is a secondary diffusion step which is caused during vulcanization by accelerator depletion of the faster-curing polymer. Gardiner noted this possibility and explained it by Le Chatlier's principle (5). Table IV shows that the cure rate of natural rubber is far greater than that of EPDM. According to our picture of secondary diffusion, this imbalance of cure rates would cause the accelerator to diffuse out of the EPDM into the natural rubber. Therefore, we have expanded the concept of curative diffusion to include two driving forces: the first is a thermodynamic driving force based on elastomer polarity; the second is attributable to the difference in the cure rates of the respective phases.

Blends of EPDM with Polyisoprene Rubber. In order to verify experimentally the role of cure rate-induced diffusion in elastomer blends, we repeated the previous experiments with synthetic polyisoprene replacing the natural rubber. In this case the cure rates match fairly closely (*see* Table V). Figure 13 clearly shows that the ethylhexylated accelerator gives linear tensile strengths for these blends. This combination of linear tensile strengths with similar cure rates verifies that cure rate-induced diffusion is possible, and that it can be overcome by matching the cure rate of each polymer.

Blends of EPDM and Natural Rubber with Matched Cure Rates. The final experiment in this series was designed to match the cure rates in EPDM and natural rubber using equally alkylated versions of zinc dialkyldithiocarbamate and tetraalkylthiuram disulfides. It is well known that the carbamates are fast accelerators while the thiuram disulfides act much more slowly. Therefore, we made up masterbatches of each polymer using carbamates in the slower-curing EPDM and thiuram disulfides in the faster-curing natural rubber. The cure rates were closely matched for all but the methylated accelerators (Table VI). The masterbatches were blended in a 50/50 ratio; then they were vulcanized and tensile strength was measured. The results are

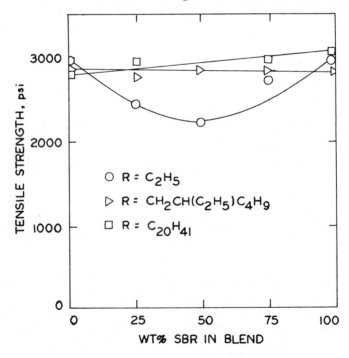

Figure 8.   Tensile strength of EPDM/SBR rubber blends

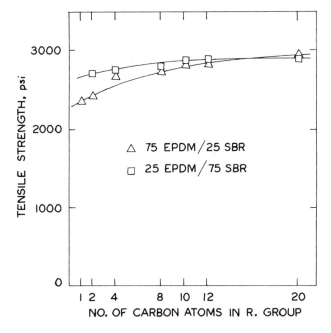

Figure 9.   Tensile strength of 75/25 and 25/75 EPDM/SBR
rubber blends

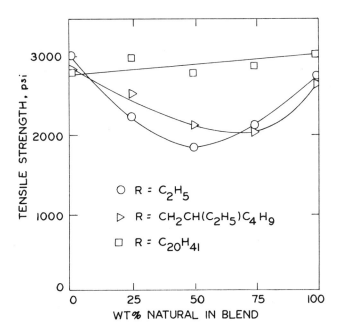

Figure 10.   Tensile strength of EPDM/natural rubber blends

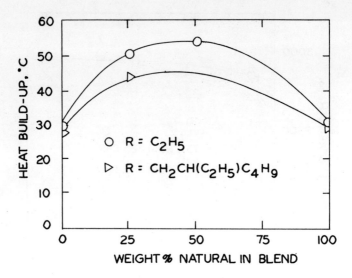

Figure 11.   Flexometer heat buildup in EPDM/natural rubber blends

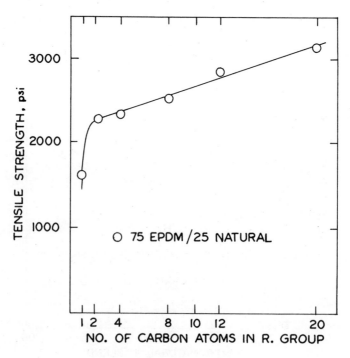

Figure 12.   Tensile strength of 75/25 EPDM/natural rubber blends

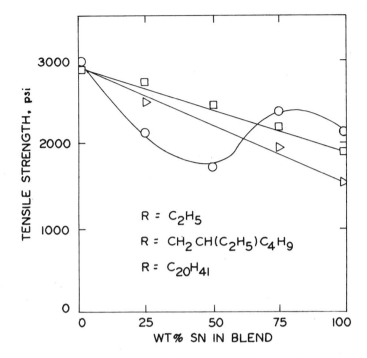

Figure 13. Tensile strength of EPDM/SN rubber blends

Table IV. Cure Rates (in. lb/min) for EPDM and Natural Rubber Containing Zinc Dialkyldithiocarbamates

	Accelerator		
Elastomer	C_2	C_8	C_{20}
EPDM	15	8	10
Natural rubber	20	22	18

presented in Table VII together with data on control compounds in which both phases contained the faster zinc dialkyldithiocarbamates.

Despite the large difference in cure rates, the methylated accelerators improved tensile strength by 500 psi. However, the values are still less than those for covulcanized blends. Improvement was greatest (700 psi) with the matched ethylhexylated accelerators. The tensile strength of 2700 psi is within 10% of that of the covulcanized blends. This difference is within the experi-

Table V. Cure Rates (in. lb/min) for EPDM and Polyisoprene Elastomers Containing Zinc Dialkyldithiocarbamates

	Accelerator		
Elastomer	C_2	C_8	C_{20}
EPDM	15	8	10
Polyisoprene	10	6	11

Table VI.   Cure Rates (in. lb/min) for Epcar-Alkylated Dithiocarbamates and Natural Rubber–Tetraalkylthiuram Disulfides

	Degree of Alkylation		
Elastomer	1	8	20
Epcar–dithiocarbamate	18	10	10.5
Natural rubber–thiuram disulfides	34	10	12

mental error of the test, and it demonstrates clearly that matching of cure rates substantially improves blend properties. The arachidic versions of the accelerator caused no improvement, which, we hypothesize, is because the molecules are too bulky to diffuse at an appreciable rate. Therefore, they would be least affected by the cure rate-induced diffusion.

Table VII.   Tensile Strength (psi) of 50/50 EPDM/NR Blends

Degree of Alkylation	Epcar[a]	NR[b]	50/50 Blend		
			Calcd.	Actual	Control[c]
1	3200	3400	3300	2300	1800
8	3050	2900	2975	2700	2000
20	3000	3400	3200	2750	2750

[a] Contains zinc dialkyldithiocarbamate accelerator.
[b] Contains tetraalkylthiuram disulfide.
[c] Contains zinc dialkyldithiocarbamate in both phases.

Summary and Conclusions

By studying the vulcanization of a series of blends of different elastomers, we demonstrated that covulcanization is possible when proper curatives are used. We also showed that failure to obtain covulcanization is primarily the result of improper partitioning of the accelerators between the phases which is caused by accelerator diffusion. The diffusion arises from two driving forces: thermodynamic, i.e., preference of the curatives for one of the elastomers, and kinetic, i.e., depletion of the curatives in the phase with the greater cure rate and subsequent secondary diffusion of more curatives from the other phase. The design of curatives to overcome both driving forces was demonstrated.

Literature Cited

1. Corish, P. R., Rubber Chem. Tech. (1967) **40**, 324.
2. Rehner, J., Wei, P. E., Rubber Chem. Tech. (1969) **42**, 985.
3. Dougherty, D. J., Forsyth, T. H., Roberts, R. W., Rubber World (1972) **166** (1), 50.
4. Gardiner, J. B., Rubber Chem. Tech. (1968) **41**, 1312.
5. Ibid., (1970) **43**, 370.
6. Zapp, R. L., ADV. CHEM. SER. (1971) **99**, 68.
7. Zapp, R. L., Rubber Chem. Tech. (1973) **46**, 251.
8. Lee, L. H., J. Polym. Sci. Part A-2 (1967) **5**, 1103.
9. Small, P. A., J. Appl. Chem. (1953) **3**, 71.
10. Hertz, D. L., unpublished data.
11. Mastromatteo, R. P., Mitchell, J. P., Brett, T. J., Rubber Chem. Tech. (1971) **44**, 1065.
12. Boonstra, B. B., Appl. Polym. Symp. (1971) **15**, 263.

RECEIVED April 17, 1974.

Dynamic Mechanical and Dielectric Relaxations in Blends of Homopolymers and Block Copolymers

U. MEHRA,[1] L. TOY,[2] K. BILIYAR, and M. SHEN

Department of Chemical Engineering, University of California, Berkeley, Calif. 94720

The relaxation behavior of two series of blends of homopolymers with block copolymers was investigated. One series consisted of blends of poly(ethylene-co-vinyl acetate) with poly(styrene-b-ethylene-co-butylene-b-styrene). The other series consisted of poly(ethylene oxide) blended with poly(ethylene oxide-b-propylene oxide-b-ethylene oxide). Samples were cast from appropriate solvents. Dynamic mechanical and dielectric properties were measured between −150°C and +100°C. In both series, the observed loss peaks were attributable to molecular motions of the homopolymer segments. No intermediate transitions, such as those noted previously in polyblends of polybutadiene with poly-(styrene-b-butadiene-b-styrene), were observed in these two systems.

Most polyblends known today are blends of two or more homopolymers or random copolymers, such as high impact polystyrene and acrylonitrile–butadiene–styrene resins. However, there is now increasing interest in polyblends made by blending homopolymers with block or graft copolymers (*1, 2, 3, 4, 5, 6*). Because of their unique macromolecular structure, block and graft copolymers tend to exert an emulsifying effect in stabilizing the multiphase structure of the resulting polyblends (*7, 8*). In a recent publication (*2*), we reported on the viscoelastic properties of poly(styrene-*b*-butadiene-*b*-styrene) (SBS) with polybutadiene (PB). In addition to the two loss maxima expected for the primary glass transitions of polystyrene (PS) and PB blocks, there was an additional loss peak located at an intermediate temperature. In order to determine if such intermediate transitions are characteristic of blends of homopolymers and block copolymers, two more such blends were made. Their properties were characterized by dynamic mechanical and dielectric measurements.

---

[1] Present address: Indosil, Ltd., Bombay, India.
[2] Present address: Raychem Corp., Menlo Park, Calif. 94025.

Experimental

A sample of the block copolymer Kraton G was supplied by Shell Chemical Co. Kraton G is poly(styrene-*b*-ethylene-*co*-butylene-*b*-styrene (SEBS), the center block being a random copolymer of ethylene and butylene (PEB). Unfortunately, because the manufacturer has not revealed the composition of the center block, it was not possible to determine the styrene content of the polymer by, for instance, IR analysis. Its molecular weight was 10^5 as determined by Fetters by gel permeation chromatography (*9*). Unlike other Kraton samples (Kratons 1101, 1102, and 1107) (*10*), Kraton G contained no diblock copolymers (*9*).

Kraton G was dissolved in toluene with Elvax 350 (EVA) which is manufactured by DuPont Co. EVA is a random copolymer of ethylene and vinyl acetate containing 24.3–25.7% vinyl acetate, with a melt index of 17.3–20.9. Films ($\sim$0.1 mm thick) containing 20 and 40 wt % EVA were cast from solution on a spin caster at 80°C; they were subsequently dried *in vacuo* for one week at room temperature to remove traces of solvents.

Poly(ethylene oxide) (PEO) was received from Union Carbide Corp. as Polyox WSR-N-3000; it has a molecular weight of 400,000. Poly(ethylene oxide-*b*-propylene oxide-*b*-ethylene oxide) (sym-EPE) was supplied by Wyandotte Chemicals Corp. as Pluoronic F108; it has a molecular weight of 15,000 and contains 80 mole % ethylene oxide. Pure PEO and blends containing 25, 50, and 75 wt % sym-EPE were made by dissolving in benzene. Films were spin cast at 100°C and then dried.

Dynamic mechanical properties were measured on a Rheovibron viscoelastometer DDV IIB with a frequency range of 0.01–110 Hz. Dielectric properties were measured over the frequency range of 10^2–10^5 Hz. A General Radio model 716-C capacitance bridge and a 1232A null detector were used with a Hewlett-Packard 1310B oscillator. Appropriate temperature chambers were used for measurements between $-150°$ and $+100°$C.

Results and Discussion

In a previous paper (*2*), we speculated on two potential causes of an additional loss peak in SBS and PB blends. One is the possibility of changes in entanglement coupling of the terminal chains either in the homopolymers or in the extraneous diblock copolymers. The other is that a new mixed phase different from the existing microphases of PS and PB may be present. In this work, we chose the sym-EPE/PEO system because PEO blocks and PEO homopolymer are both crystallizable; thus no entanglement coupling can be observed below melting temperature. Any new loss peak would most likely be caused by formation of a new mixed phase. We selected the SEBS/EVA system because it is possible to form a macroscopically compatible blend from these two polymers, as reported by the manufacturer. On the other hand, the midblock of SEBS, poly(ethylene-*co*-butylene), is perhaps sufficiently different in structure from EVA so that they are not expected to be microscopically compatible, as is the case with PB block in poly(styrene-*b*-butadiene-*b*-styrene) with blended PB. Unfortunately, because of the lack of unsaturated groups, SEBS/EVA cannot be successfully stained for electron microscopic examination to confirm this hypothesis. The EVA used was of sufficiently high molecular weight so that the onset of entanglement coupling relaxation occurred at a high temperature, thus ruling out its contribution to a new peak intermediate between those of the block copolymer. In both systems, one would suspect that the existence of a separate mixed phase is responsible for any new loss peaks.

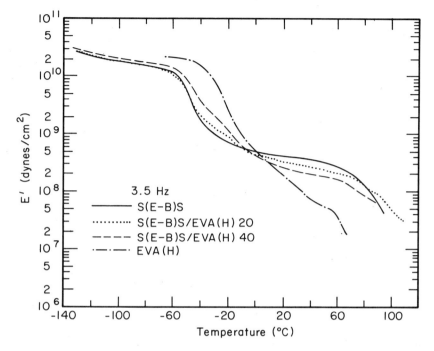

Figure 1. Dynamic storage moduli as a function of temperature for poly(styrene-b-ethylene-co-butylene-b-styrene), poly(vinyl acetate-co-ethylene), and their blends (frequency, 3.5 Hz)

The dynamic storage moduli at 3.5 Hz of the four samples are shown in Figure 1. Pure SEBS and samples blended with EVA all had more extended plateaus at about −20°C whereas EVA declined steadily. The glassy moduli of these systems, however, were quite similar. The transition region slopes of the two homopolymers were similar, but the blended samples showed signs of broadening caused by the presence of two transitions.

Loss tangent curves of the samples are presented in Figure 2. With pure SEBS, the loss peak at −125°C is attributable to the γ relaxation of methylene sequences in the PEB. It was demonstrated (*11, 12*) that melt-crystallized polyethylene has two such relaxations, γ_1 and γ_2, whereas solution-crystallized polyethylene has only one, γ_2. The γ_2 relaxation has been interpreted as a local mode relaxation of the surface molecular segment in a frozen state (*12*). The γ_1 relaxation is attributable to molecular motion in the amorphous phase, such as the crankshaft mechanism proposed by Schatzki (*13, 14*). Since our data did not extend to sufficiently low temperatures, it is difficult to judge whether two superposed γ relaxations are present.

The −45°C loss peak is most likely the β relaxation process of PEB in SEBS. This relaxation was observed in polymers containing the group —CH_2CHRCH_2— where R is a side group such as methyl, butyl, or acetate (*15*). It was reported by a number of workers (*16, 17, 18*) that T_β's of random copolymers of ethylene generally decrease as the ethylene content increases (up to about 50–60 wt %). The minimum T_β of ethylene with propylene,

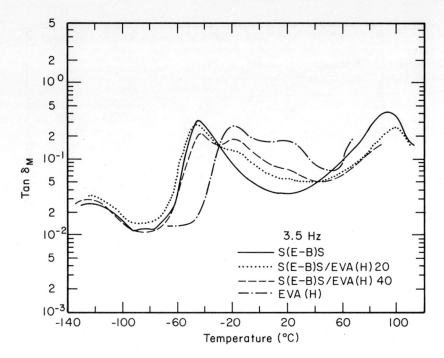

Figure 2.   Mechanical loss tangent as a function of temperature for poly(styrene-b-ethylene-co-butylene-b-styrene), poly(vinyl acetate-co-ethylene), and their blends (frequency, 3.5 Hz)

ethyl acrylate, butyl acrylate, etc., is about $-50°$ to $-60°C$. At greater ethylene content, crystallization intervenes and the observed transition temperature rises. From loss tangent measurements at another frequency (0.1 Hz), the activation energy of this relaxation was estimated at about 23 kcal/mole, a value that is within the range of reported values (16–38 kcal/mole) for the β relaxation of polyethylene (15).

The 90°C relaxation peak results from the glass transition of PS blocks in SEBS. This peak was observed in other styrene-containing block copolymers such as poly(styrene-b-isoprene-b-styrene) (19, 20) and SBS (20).

The most prominent loss peak in pure EVA was located at $-20°C$ (Figure 2). This is also the temperature at which the dynamic storage modulus decreased precipitously (Figure 1), and it can be considered the β relaxation temperature of EVA (15). This observation agrees with that reported by Nielsen (21). Since this mode of molecular motion involves the polar groups of vinyl acetate, the relaxation should be detectable by dielectric measurements. Figure 3 depicts the dielectric loss data of pure EVA. A single relaxation peak was observed at each frequency. The activation energy of the dielectric T_β was 45 kcal/mole. By Vibron experiments at 0.1 and 110 Hz (data not shown), we estimated the activation energy of the mechanical T_β of this polymer at 36 kcal/mole.

It is of interest that the dielectric loss curve of EVA contained only a single peak, whereas the mechanical loss curve (Figure 2) had another peak at

20°C. We attribute the 20°C peak to α relaxation (15) of polyethylene segments in EVA. According to Takayanagi (22), it results from the reorientational motion of methylene units in the crystalline region of polyethylene. Nielsen (21) found that EVA copolymers containing 19 and 27% vinyl acetate had 30 and 20% degrees of crystallinity, respectively. Since our EVA sample contains 24.3–25.7% vinyl acetate, its degree of crystallinity should be in the order of 20%, which would be sufficient to cause the observed α relaxation. Since these motions involve only methylene units, which are non-polar, they are dielectrically inactive and are not observable in our dielectric loss measurements. The α relaxation was absent from the mechanical loss curve of SEBS, which indicates the amorphous nature of the PEB block. The onset of another rise in the mechanical loss curve above 60°C results from crystalline melting of EVA (21).

The loss curves of two polyblends, SEBS/EVA20 and SEBS/EVA40, appeared to be composites of the relaxations in the respective homopolymers. The γ relaxation in the polyblends was comparable in strength to that in SEBS, probably because of the contribution by the methylene segments in EVA. PEB β relaxations were only slightly weaker in the polyblends. On the other hand, the vinyl acetate β and α relaxations were considerably weaker in the polyblends because of the relatively small amount of EVA in the blends. The fact that all the relaxations from the homopolymers were visible in the blends indicates that EVA is not solubilized into SEBS but most likely exists as separate microphase domains.

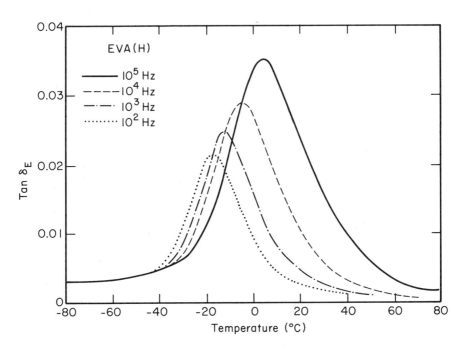

Figure 3.   Dielectric loss tangent as a function of temperature for poly(vinyl acetate-co-ethylene) at various frequencies

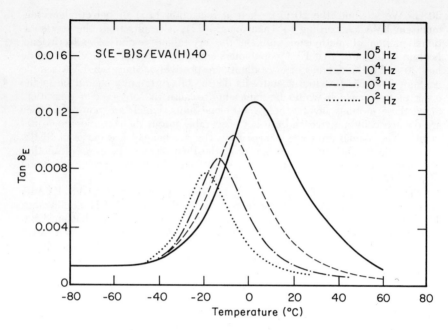

Figure 4. Dielectric loss tangent as a function of temperature for poly(styrene-b-ethylene-co-butylene-b-styrene) blended with 40 wt % poly(vinyl acetate-co-ethylene) at various frequencies

The dielectric loss curves at various frequencies for SEBS/EVA40 blend (Figure 4) were very similar to those for pure EVA (Figure 3). There was no evidence, however, in either mechanical or dielectric data of any extra relaxation peaks not attributable to the homopolymers. This is in contrast to previous reports on SBS blends with PB homopolymer (2).

Regarding the relaxation behavior of another polyblend system, *i.e.*, blends of PEO and sym-EPE, Figure 5 depicts the dynamic storage and loss moduli of PEO and PEO/sym-EPE blends at 110 Hz as a function of temperature. Storage moduli were highest for pure PEO, and they became progressively lower as sym-EPE content increased. The drop in modulus near −60°C is characteristic of primary glass transitions. In fact, the dilatometric T_g of PEO is known (23) to be −67°C. The second drop in moduli at 60°–70°C is associated with the melting of crystallites. The melting points of PEO and sym-EPE are 65° and 57°C, respectively, according to their manufacturers. Since the poly(propylene oxide) (PPO) block is non-crystallizable, melting is attributable only to the PEO crystallites. Booth and Dodgeson (24) demonstrated that T_m's of sym-EPE were always lower than those of PEO because of a rise in interfacial energy near the crystal end interface. In addition, O'Malley *et al.* (25) demonstrated that T_m's of poly(ethylene oxide–styrene–ethylene oxide) decreased as styrene content increased.

The mechanical loss tangent data at 110 Hz are depicted in Figure 6. The rise in tan δ at the high temperature end was also caused by the melting of PEO crystallites. The loss peak of PEO at −50°C was observed by other workers (26, 27, 28); it was designated as β relaxation. Takayanagi *et al.* (27) demon-

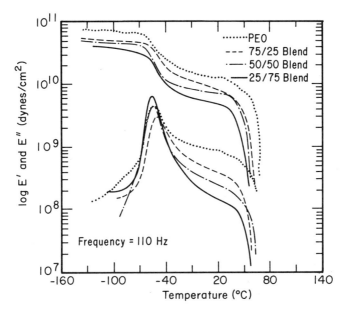

Figure 5.   Dynamic storage (E') and loss (E'') moduli as a function of temperature for poly(ethylene oxide) and its blends with poly(ethylene oxide-b-propylene oxide-b-ethylene oxide) (frequency, 110 Hz)

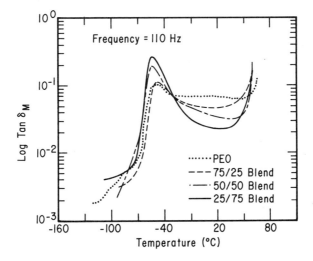

Figure 6.   Mechanical loss tangent as a function of temperature for poly(ethylene oxide) and its blends with poly(ethylene oxide-b-propylene oxide-b-ethylene oxide) (frequency, 110 Hz)

strated that the β peak was present only in melt-crystallized PEO, but not in a single crystal mat. Porter and Boyd (28) found in their dielectric measurements that the β peak persisted at gigahertz frequencies even in the molten state. The evidence supports the conclusion that β relaxation originates from the amorphous phase as a consequence of micro-Brownian motions of the PEO chains.

The β peaks increased in strength and shifted to lower temperatures as sym-EPE content increased (see Figure 6). This is probably attributable to the primary glass transition processes of the PPO blocks. The dilatometric T_g of PPO is known (29) to be $-75°C$, which is $8°C$ lower than that of PEO. In addition, since PPO blocks are non-crystallizable, strengths of this peak can be expected to increase as a result of larger amorphous fractions undergoing glass transition. It is of interest that Read (23) found that the strength of the β peak was greater for completely amorphous PPO than for a semi-crystalline sample.

Dielectric loss data for PEO and PEO/sym-EPE blends determined at 10^4 Hz are depicted in Figure 7. The β peak shifted to lower temperatures and increased in strength as sym-EPE content increased. The rise in dielectric loss above $-20°C$ resulted from dc conductivity effects (8). In order to determine the activation energy of the β process, dynamic mechanical properties of a 50/50 blend were measured at 0.1–110 Hz and dielectric properties at $100–10^5$ Hz. The data were plotted against the reciprocal temperature positions at maximum loss (Figure 8). The relation was linear only at high frequencies, and the curve bent downward at lower frequencies. The shape of this curve was very similar to those for pure PEO and PPO presented by McCrum, Read, and Williams (15). It is well known that the glass transition cannot be represented by an activation energy plot, but by a Williams–Landel–Ferry plot (30). Figure 3 thus provides further support for the hypothesis

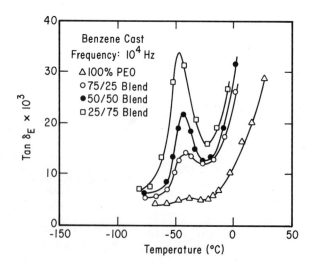

Figure 7.   Dielectric loss tangent as a function of temperature for poly(ethylene oxide) and its blends with poly(ethylene oxide-b-propylene oxide-b-ethylene oxide) (frequency, 10^4 Hz)

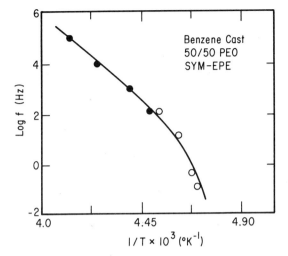

Figure 8. Frequency vs. reciprocal temperature plot for poly(ethylene oxide) blended with 50 wt % poly-(ethylene oxide-b-propylene oxide-b-ethylene oxide)
●, *dielectric data; and* ○, *dynamic mechanical data*

that the observed β peak is a composite of primary glass transition processes of PEO and PPO.

In Figure 6 there was a high loss plateau between the β relaxation and the melting peak of PEO. The height of the plateau, however, decreased as sym-EPE content increased. The presence of a loss process, designated α relaxation, was noted in the dynamic mechanical data of Read (23). Conner et al. (26) attributed this relaxation to a crystal disordering mechanism. No α peak was observable in dielectric loss data because of the ionic conduction at these temperatures (14). The dynamic mechanical data on PPO of Saba et al. (31) demonstrated no loss process in the $-40°$ to $+40°C$ temperature range. Thus one would expect that, as more sym-EPE is blended into PEO, α peak strength must decrease as a result of the reduction in PEO content.

In conclusion, we can state that the relaxation behavior of both blends can be readily explained on the basis of the behavior of their component polymeric species. In contrast to the system of SBS/PB blends (2), no new relaxation peaks were observed in these polyblends. The implication is that the existence of a new loss peak in a homopolymer/block copolymer blend may be attributable to the entanglement coupling mechanism. Further experiments are being conducted, and the results will be reported in a future publication.

Acknowledgment

The authors wish to thank L. J. Fetters for providing the gel permeation chromatography data on Kraton G.

Literature Cited

1. Childers, C. W., Kraus, G., Gruver, J. T., Clark, E., in "Colloidal and Morphological Behavior of Block and Graft Copolymers," G. E. Molau, Ed., p. 193, Plenum, New York, 1971.

2. Choi, G., Kaya, A., Shen, M., *Polym. Eng. Sci.* (1973) **13**, 231.
3. Locke, C. E., Paul, D. R., *J. Appl. Polym. Sci.* (1973) **17**, 2791.
4. Barenstsen, W. M., Heikens, D., Piet, P., *Polymer* (1974) **15**, 119.
5. Inoue, T., Soen, T., Hashimoto, T., Kawai, H., *Macromolecules* (1970) **3**, 87.
6. McIntyre, D., Campos-Lopez, E., in "Block Polymers," S. L. Aggarwal, Ed., p. 19, Plenum, New York, 1970.
7. Riess, G., Kohler, J., Tournut, C., Banderet, A., *Makromol. Chem.* (1967) **101**, 58.
8. Molau, G. E., Wittbrodt, W. M., *Macromolecules* (1968) **1**, 260.
9. Fetters, L. J., private communication.
10. Fetters, L. J., Meyer, B. H., McIntyre, D., *J. Appl. Polym. Sci.* (1972) **16**, 2079.
11. Illers, K. H., *Kolloid Z. Z. Polym.* (1969) **231**, 622.
12. Kakizaki, M., Hideshima, T., *J. Macromol. Sci. Phys.* (1973) **8**, 367.
13. Schatzki, T. F., *J. Polym. Sci.* (1962) **57**, 496.
14. Schatzki, T. F., *Bull. Amer. Phys. Soc.* (1971) **16**, 364.
15. McCrum, N. G., Read, B. E., Williams, G., "Anelastic and Dielectric Effects in Polymeric Solids," Wiley, New York, 1967.
16. Reding, F. P., Faucher, J. A., Whitman, R. D., *J. Polym. Sci.* (1962) **57**, 483.
17. Tuijnman, C. A. E., *J. Polym. Sci. Part C* (1967) **16**, 2379.
18. Schmieder, K., Wolf, K., *Kolloid Z. Z. Polym.* (1953) **134**, 149.
19. Jamieson, R. T., Kaniskin, V. A., Ouano, A. C., Shen, M., in "Advances in Polymer Science and Engineering," K. D. Pae, D. R. Morrow, and Y. Chen, Eds., p. 163, Plenum, New York, 1972.
20. Shen, M., Kaniskin, V. A., Biliyar, K., Boyd, R. H., *J. Polym. Sci. Part A-2* (1973) **11**, 2261.
21. Nielsen, L. E., *J. Polym. Sci.* (1960) **42**, 357.
22. Takayanagi, M., *Mem. Fac. Eng. Kyushu Univ.* (1963) **23**, 1.
23. Read, B. E., *Polymer* (1962) **3**, 529.
24. Booth, C., Dodgeson, D. V., *J. Polym. Sci. Part A-2* (1973) **11**, 265.
25. O'Malley, J. J., Crystal, R. G., Erhardt, P. F., in "Block Polymers," S. L. Aggarwal, Ed., p. 163, Plenum, New York, 1969.
26. Conner, T. M., Read, B. E., Williams, G., *J. Appl. Chem.* (1964) **14**, 74.
27. Ishida, Y., Matsuo, M., Takayanagi, M., *J. Polym. Sci. Part B* (1965) **3**, 321.
28. Porter, C. H., Boyd, R. H., *Macromolecules* (1971) **4**, 589.
29. Williams, G., *Trans. Faraday Soc.* (1965) **61**, 1564.
30. Williams, M. L., Landel, R. F., Ferry, J. D., *J. Amer. Chem. Soc.* (1955) **77**, 3701.
31. Saba, R. G., Sauer, J. A., Woodward, A. E., *J. Polym. Sci. Part A* (1963) **1**, 1483.

RECEIVED July 1, 1974. Work was supported by the Petroleum Research Fund administered by the American Chemical Society and by the Advance Research Projects Agency monitored by the Office of Naval Research (#N00014-69-0200-1053).

The Role of Filler Modulus and Filler–Rubber Adhesion in Rubber Reinforcement

MAURICE MORTON, R. J. MURPHY,[1] and T. C. CHENG[2]

Institute of Polymer Science, University of Akron, Akron, Ohio 44325

Studies are made of styrene–butadiene rubber vulcanizates reinforced with different polymeric fillers that were introduced by latex blending. Experiments with polystyrene, poly(methyl methacrylate), and polyacenaphthylene, which have different moduli of rigidity, demonstrate that a higher filler modulus results in greater strength reinforcement, both tensile and tear. The changes in density of these composites under uniaxial strain reveal that dewetting of the filler–rubber interface is a linear function of the extent of strain and is inversely proportional to both particle size and filler–rubber interfacial energy. The effect of filler–rubber adhesion is further reflected in the reinforcement of the rubber since better adhesion leads to higher strength but greater hysteresis.

Recent studies in these laboratories dealt with the use of model, spherical polymeric fillers in the reinforcement of styrene–butadiene rubber (SBR) and polybutadiene elastomers (*1, 2, 3, 4*), with both fillers and elastomers prepared by emulsion polymerization. With latex blending techniques, true dispersions of the filler in the rubber were prepared, and definitive conclusions could be reached about the effect of the filler on vulcanizate strength. More recently, we investigated the possible effects of filler rigidity and filler–rubber adhesion on reinforcement of the elastomer.

Table I.   Fillers for Modulus Study

Variable	PS	PMMA	PA
Particle diameter, A	570	510	410
T_g, °C	108	110	240
Flex. modulus (E) at 25°C, kg/cm² $\times$ 10⁻⁴	3.2±0.1	2.4±0.1	8.0±0.5

[1] Present address: Education Development Center, Newton, Mass. 02160.
[2] Present address: Air Products and Chemicals, Inc., Middlesex, N. J. 08846.

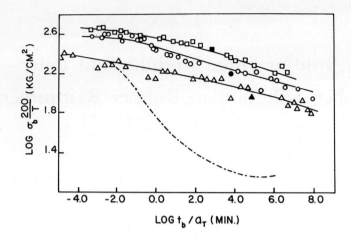

Figure 1. Tensile strength of filled SBR
Filler, 25 vol %; – – –, SBR gum; △, PMMA; ○, PS; and □, PA

Experimental

Three emulsion polymers with different flexural moduli were prepared, *i.e.* polystyrene (PS), poly(methyl methacrylate) (PMMA), and polyacenaphthylene (PA); all had very small particle size (~500 A) suitable for rubber reinforcement. Characteristics of these fillers are listed in Table I.

In addition, a series of PS fillers with particle size of 500–3000 A and a series of PMMA fillers with particle size of 800–2500 A were prepared. Adhesion was studied by determining the decrease in density of the filled vulcanizates under uniaxial strain, on the assumption that this was caused by vacuole formation resulting from dewetting of the filler. For these vulcanizates, benzoyl peroxide (2.5 phr, 5 hrs at 80°C) was used as crosslinking agent in order to avoid using particulate zinc oxide as is customary with sulfur-based compounds (*1, 2, 3*).

Normal techniques for preparing the peroxide vulcanizates used in the adhesion studies involved blending correct proportions of filler and SBR latexes to yield a final composite containing 25 vol % filler on a dry basis. The latex blend was then coagulated in 2-propanol and dried at 40°C *in vacuo* to yield the dry composite. The peroxide (and antioxidant) were then incorporated by milling 2 min on a cold mill. Vulcanization was at 80°C for 12 hrs. It was assumed that this coagulation procedure removed the adsorbed soap from the latex particles. Hence, an alternative procedure of simple evaporation was used with PS latex of 440-A particle size when it was desired that the adsorbed layer of soap, which covered 66% of the surface, remain on the surface of the filler particles and thereby alter the nature of the filler–rubber interface. This filler was designated PS440 (66% S). This filler was further modified by adding soap solution to the PS latex until the surface was saturated with soap; the product was PS440 (100% S).

Effect of Filler Modulus

The effect of the three fillers with different moduli, *i.e.* PS, PMMA, and PA, on the tensile strength of the SBR vulcanizates is depicted in Figure 1. As is customary, tensile strength is expressed as a function of time and temperature on master curves obtained by using an empirical shift factor (*1*) which is generally larger for filled vulcanizates than that predicted by the Williams–Landel–Ferry (WLF) equation. Fillers with higher modulus gave vulcanizates with greater strength. These findings confirmed the preliminary findings reported for another series of polymeric fillers (*4*) and indicate very strongly that filler rigidity does indeed affect reinforcement of the elastomer.

Filler–Rubber Adhesion and Reinforcement

Volume dilation of these vulcanizates appeared to be a linear function of the uniaxial strain; hence a parameter $\Delta V / V\lambda$ could be derived for each filler, where $\Delta V / V$ is the fractional volume change at any strain λ. In Figure 2, this dilation parameter ($\Delta V / V\lambda$) is plotted as a function of filler particle diameter (or of $1/A$, the reciprocal of the surface area) for the various fillers used. It was not surprising that dewetting was a linear function of $1/A$ since it was expected to depend directly on the work required to create a new surface. However, all the data for the soap-free PS fillers were on one line whereas those for the soap-modified PS fillers were on another line together with those for the PMMA fillers. The coincidence of the latter two fillers may be fortuitous, but it indicates an apparent greater adhesion between these fillers and the rubber which might be expected from the known higher surface energy (*5*) of PMMA (39 dynes/cm^2) than of unmodified PS filler (33 dynes/cm^2).

The effect of these adhesion differences on reinforcement of the vulcanizates was determined by measuring uniaxial stress–strain hysteresis as well as tear strength. These two characteristics are presumably directly related (*6, 7*) so their relation to interfacial adhesion in filled vulcanizates was considered

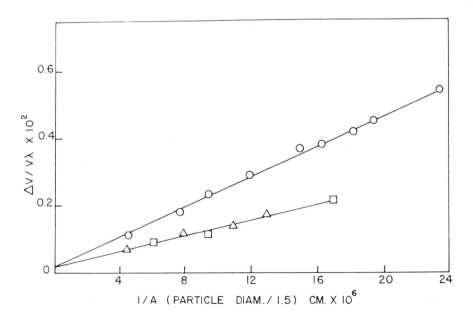

Figure 2. Dilation as a function of particle size and surface
Filler, 25 vol %; O, PS; △, PS (66% S); and □, PMMA

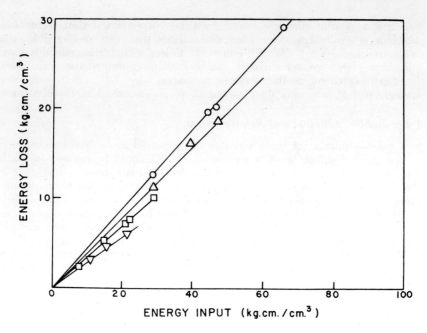

Figure 3.   Third cycle hysteresis of polystyrene-filled SBR
Filler, 25 vol %; O, 440 A; △, 1750 A; □, 3000 A; and ▽, SBR gum

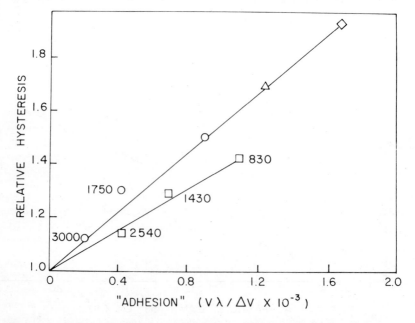

Figure 4.   Relative hysteresis of filled SBR as a function of interface adhesion
Filler, 25 vol %; O, PS 440; △, PS 440 (66% S); ◇, PS 440 (100% S); □, PMMA; and numbers adjacent to points: particle diameters

interesting. The hysteresis of each vulcanizate was calculated as usual from the ratio of the area within the stress–strain hysteresis loop to the total area, using the third stress–strain cycle in order to avoid any stress-softening complications. The effect of PS fillers of different particle size on the third cycle hysteresis is depicted in Figure 3 with various strain ratios used to obtain the different energy inputs. The greater effect of the smaller particle size is obvious.

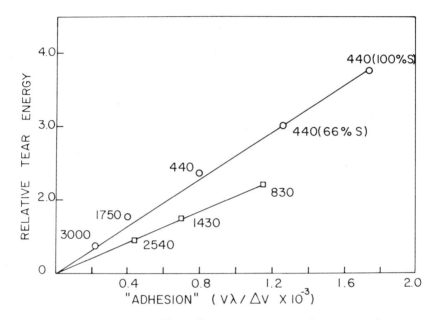

Figure 5.   Effect of filler–rubber adhesion on tear strength of SBR vulcanizates
Filler, 25 vol %; O, PS; □, PMMA; and numbers adjacent to points: particle diameters

Figure 4 illustrates the effect of filler–rubber adhesion (*i.e.,* the reciprocal of the dilation parameter $\Delta V/V\lambda$) on the relative hysteresis (*i.e.,* the ratio of the hysteresis of the filled vulcanizate to that of the gum vulcanizate). The relation is apparently linear, *i.e.* better adhesion gives higher hysteresis. In other words, the restrictions imposed by the filler on the elastic response of the elastomer depend on the degree of adhesion between the two—a reasonable conclusion. The consistently lower hysteresis values of the PMMA filler must, therefore, be attributable to an innate property of this filler, presumably its lower modulus of rigidity, which would impose somewhat less restriction on the elastomer.

The same type of relation is apparent in Figure 5, which demonstrates the effect of degree of adhesion on the tear strength of these composites. Although the strength appeared to be a linear function of the degree of adhesion, dependency was different for the two types of polymeric fillers with the softer PMMA filler contributing less reinforcement. This is an excellent illustration of the effect of both of these filler characteristics, *i.e.* modulus and interfacial adhesion, on the extent of elastomer reinforcement.

It should be noted, in conclusion, that the interfacial adhesion referred to in this study is not complicated or affected by any chemical bonding between the two phases since solvent extraction of the composites revealed no noticeable chemical bonding of the filler to the rubber vulcanizate.

Literature Cited

1. Morton, M., Healy, J. C., Denecour, R. L., "Proceedings of the International Rubber Conference 1967," p. 175, MacLaren & Sons, London, 1968.
2. Morton, M., Healy, J. C., Amer. Chem. Soc., Div. Polym. Chem., Prepr. **8**, 1569 (Chicago, September, 1967).
3. Morton, M., Healy, J. C., Appl. Polym. Symp. (1968) **7**, 155.
4. Morton, M., ADVAN. CHEM. SER. (1971) **99**, 490.
5. Ellison, E. H., Zisman, W. A., J. Phys. Chem. (1954) **58**, 260.
6. Mullins, L., Trans. Inst. Rubber Ind. (1959) **35**, 213.
7. Grosch, K. A., Harwood, J. A. C., Payne, A. R., J. Appl. Polym. Sci. (1968) **12**, 889.

RECEIVED August 7, 1974. Work supported by grants from the B. F. Goodrich Co. and the Phillips Petroleum Co.

Injection Molding of Glass-Filled Polypropylene: Mold Filling Studies

LAWRENCE R. SCHMIDT

General Electric, Corporate Research and Development,
Schenectady, N. Y. 12301

A specially designed mold was used with an Instron capillary rheometer to investigate the shear and extensional flows produced in a mold cavity during injection molding. The materials studied were an impact-grade polypropylene, glass-bead-filled polypropylene, and glass-fiber-filled polypropylene. Flow patterns in the mold were illustrated by a visual tracer technique, and the complete deformation history of the tracers was recorded on movie film. The effect of mold temperature on flow patterns and cavity pressure and temperature was investigated at constant flow rate (2.21 in.³/min). For both composites and the unfilled resin, the mold temperature significantly affects the development of the shear and extensional flow fields. Mold geometry, however, has the greatest effect on the flow.

The injection molding process creates complex flow fields at several points in the process. In the mold cavity, for instance, the flow causes variations in the morphology of a molding, thereby affecting its physical and mechanical properties. A rigorous mathematical analysis of these flows, therefore, would be very informative, but such an analysis is impractical because of the non-isothermal, unsteady conditions which are characteristic of injection molding and the viscoelastic nature of the fluids.

Recently Kamal and Kenig (1) reported on a computer study which simulated the injection molding of a power-law fluid to predict fill times, pressure gradients, and temperature profiles in a semicircular mold cavity. Other studies (2, 3, 4) used a similar approach and solved the same transport equations but with different assumptions and different mold geometries. These investigations were not concerned with the physical and mechanical properties of a molding.

Kantz, Newman, and Stigale (5) were interested in the morphology of injection-molded tensile bars as a function of melt temperature and injection pressure. They used optical microscopy and x-ray diffraction to study molecular orientation in bars molded from polypropylene homopolymer and an ethyl-

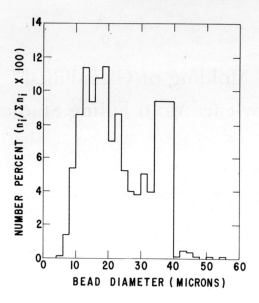

Figure 1.   Distribution of glass bead diameters

ene–propylene copolymer. Three different crystalline zones were identified in the cross section of a molding. The mechanical properties of these test specimens were correlated to the size of the skin or surface zone.

The addition of inert fillers to polymeric (viscoelastic) materials produces composites whose viscous and elastic properties are significantly different from the original material (6, 7, 8, 9 10, 11). When these composites are injection

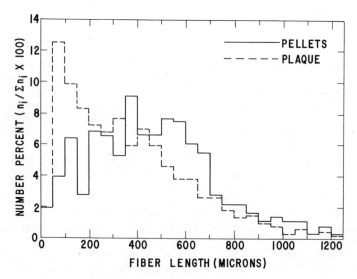

Figure 2.   Distribution of glass fiber lengths before and after molding

molded, the filler alters the flow in the mold, and this ultimately influences the properties of the molding.

Lockett (*12*) and Johnson (*13*) have approached the analysis of melt flow with fillers and reinforcing agents by developing a continuum theory. They applied this theory to different simple flows under isothermal conditions and solved the equation of motion for a fiber-filled melt. They then calculated the velocity profile and fiber orientations. Although this continuum theory has many limitations, it does provide insight into the stability of fiber movement in the melt. To extend this approach to the injection molding of composites would be a formidable task.

Fuccella (*14*) investigated different glass-fiber-filled polymer samples which were injection molded into $4 \times 12 \times 1/8$-in. plaques. He used five rather simple experimental techniques to determine fiber orientation, and he found that it varied considerably throughout the cross section of the plaque and that the mechanical properties of test specimens cut from these plaques had a strong dependency on specimen orientation relative to the bulk flow direction.

This paper focuses on the flow and deformation in a mold and, in particular, how this flow and deformation are altered by the presence of inert fillers. A visual tracer technique was developed to illustrate the flow patterns in the cavity and to provide specific information about the extensional and shear flows which affect the overall properties of a molding. The entire deformation history of the tracers was recorded on movie film.

Experimental

Two different composites were prepared in a Werner-Pfleiderer twin-screw extruder from the same impact-grade polypropylene (Amoco #20-9605 ethylene–propylene copolymer, about 2.5 wt % ethylene). This resin has a narrow molecular weight distribution and a melt flow of 5 (ASTM D 1238). The first composite contained 19 wt % glass beads (Ballotini) plus 2 wt % TiO_2; the second composite contained 16 wt % short glass fibers (Owens-Corning Fiberglas, *K* filament, *E* glass) but no TiO_2. Neither beads nor fibers had been treated with special coatings, but a sizing had been applied to the fibers after spinning. Processing stabilizer ($\sim$ 0.5 wt %) was added to each composite as it was being blended. The distribution of bead diameters is depicted in Figure 1. The fiber length distribution (Figure 2) is skewed to the left after molding because the fibers break up as they are injected into a cavity.

The injection molding machine consists of an Instron capillary rheometer, model MCR, and a specially designed mold attached to the reservoir of the rheometer. The mold has six basic parts—a cavity plate sandwiched between two face plates, a top and bottom yoke, and a capillary—all machined from 347 stainless steel. Figure 3 shows the front view of the mold with solid steel face plates. This mold was not equipped with heater bands for establishing the mold temperature; instead the mold, which is massive (7.5 lbs), was placed in a natural convection oven and allowed to reach the set temperature before being attached to the rheometer reservoir. The plugs in the center of the face plates were replaced with a flush-mounted pressure transducer (Dynisco, model PT-467 with model PT-420A tip) and a flush-mounted fiber optic which was part of an IR thermometer (Vanzetti, model 1017). A second set of face plates with 1-in. thick glass inserts was used for observing flow in the mold cavity.

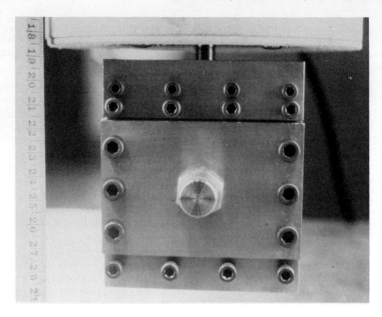

Figure 3. Injection mold with steel face plates attached to capillary extending from rheometer reservoir

Two different cavity plates (1/8- and 1/16-in. thick) were fabricated by machining a hole 2.5-in. square in the center of each flat plate. A straight gate was cut through one end of the cavity plate; it was located midway across the end and was aligned so that the boundary was defined by 90° angles at the entrance to the cavity. The gate in the 1/16-in.-thick plate measured 1/16 × 1/8 × 1/2 in., and that in the 1/8-in. thick plate was 1/8 × 1/8 × 1/2 in.

A tracer technique was developed and refined in order to visualize the flow patterns in the mold. This technique is a modification of the one described by Bagley and Birks (15) who were interested in studying the flow in the entry region of a capillary (converging-channel flow). The polymer or composite was compression molded in the heated rheometer reservoir. After being cooled to room temperature, the rod was removed from the reservoir, and five small holes (diameter, 0.10 in.) were drilled perpendicular to the axis of symmetry to a depth equal to 2/3 the reservoir diameter. Tracer elements consisting of the polypropylene or one of the composites plus less than 1% inorganic pigment were molded into small rods 0.10 in. in diameter. Five different pigments (red, yellow, green, blue, and black) were used. Each rod was cut into several pieces 0.12 in. long. One piece (the tracer) was then inserted into each hole and positioned in the center of the reservoir rod, and the remainder of the drilled hole was plugged with the original material. This is an important modification of the tracer technique of Bagley and Birks, who placed many tracers in a rod, because the length of their tracers was equal to the reservoir diameter. The modification results in tracer elements which are only slightly deformed as they enter the mold cavity since they were positioned in the low shear region of the reservoir. Consequently, by this technique deformation in the reservoir can be distinguished from deformation in the mold cavity. Deformation of the tracer caused by the extensional flow above the capillary and the high shear in the capillary and gate is recovered to a large degree as the tracer decelerates in the diverging channel just beyond the gate. This deformation is similar to the recoverable strain in extrudates which is observed in capillary rheometry studies (*16*), but it occurs to a much greater extent in this case.

Figure 4 shows the location of the tracer elements (each with a different color pigment) in a polymer rod. Just before a run, the rod with the tracers in place was dropped into the reservoir and allowed to reach the set temperature (a small steel plug placed in the capillary during heating prevented material from flowing under the force of gravity). The mold was then attached, and the melt was injected into the cavity at constant flow rate. The crosshead was stopped manually when packing force reached 350 lb$_f$.

Since the tracers are centered about the axis of symmetry of the reservoir, it was not necessary to align the mold with the reservoir. Furthermore, this modified tracer technique can be used in radial flow experiments where the melt is injected normal to the plane of the cavity.

Results

The two composites as well as the unfilled resin were molded into plaques at constant flow rate (2.21 in.^{3}/min) but three different mold temperatures. Figure 5 shows nine 1/8-in.-thick plaques; the melt temperature for each was 240°C. It is apparent that the flow patterns remain distinct, even when the mold temperature equals the melt temperature. For certain runs, the entire history of the tracers in the cavity as well as the contour of the flow front was recorded on 16-mm movie film with a Bolex camera, model REX-16, run at 24 to 48 frames/sec.

A duplicate set of experiments was run in the mold fitted with the face plates containing a pressure transducer and a fiber optic which were positioned 1 3/8 in. from the gate midway across the channel. Pigmented tracers were not used in these experiments because pigments change the emissivity of the

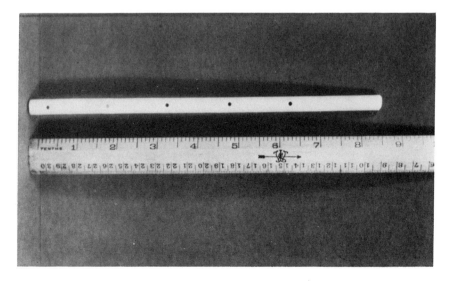

Figure 4.   Premolded reservoir rod with five tracer elements in place (left to right: red, yellow, green, blue, and black)

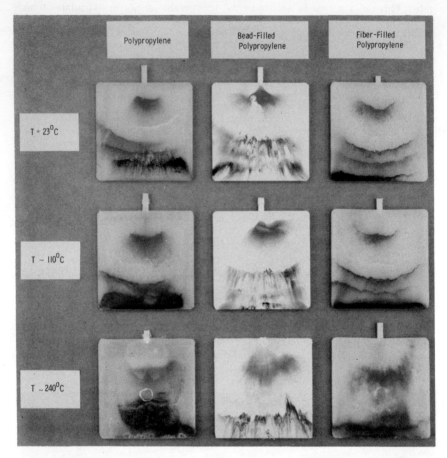

Figure 5.   Injection-molded plaques showing the effects of filler and mold temperature on flow patterns

The red tracer entered first and is found closest to the gate. The yellow tracer (not clearly shown in the black and white photograph) was second, and it is downstream from the red. The green, blue, and black tracers follow.

material and affect the output of the IR thermometer. Pressure and temperature traces are presented in Figure 6. The plunger pressure is the plunger force (from load cell on crosshead) divided by the cross-sectional area of the plunger. Similar traces were obtained for each material at the three mold temperatures. The plunger-force trace (Figure 7) has two constant slope portions before packing. The irregular curve at the end of the run (after packing) is the result of free play in the crosshead drive after it was stopped. The point at which the slope changes (which corresponds to the time at which the flow front has extended to the side walls) is referred to as the crossover time. Figure 8 shows the effect of mold temperature on the two slopes and the crossover time for the fiber-filled polypropylene.

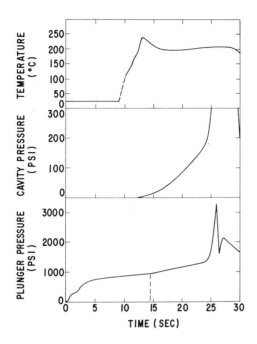

Figure 6.   Pressure and temperature traces for fiber-filled polypropylene (T_{mold} = $23°C$)

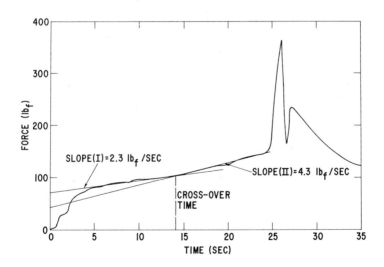

Figure 7.   Plunger force vs. *processing time for fiber-filled polypropylene* (T_{mold} = $23°C$)

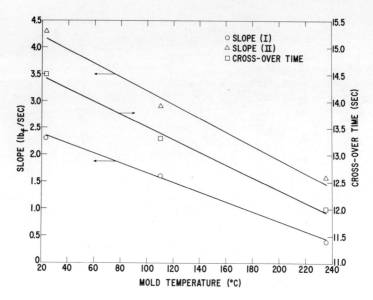

Figure 8. Force trace slopes and crossover time vs. mold temperature for fiber-filled polypropylene

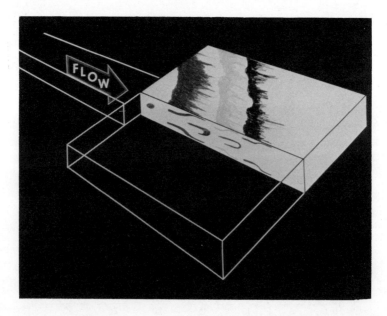

Figure 9.   Schematic of the flow in the mold cavity for fiber-filled material

The fourth tracer (blue) has entered the cavity as a sphere, and it moves through the core to the flow front. The previous three tracers (red, yellow, and green) are part of the skin near the mold walls.

Discussion

The plaques (Figure 5) demonstrate that a small amount of filler (approximately 5 vol % for a 240°C melt temperature) does not significantly alter the flow patterns in the mold. The mold temperature, however, determines the skin thickness next to the cavity walls, and it greatly affects the flow when the temperature is near the melt temperature. Cavity geometry generally has the greatest effect on the flow.

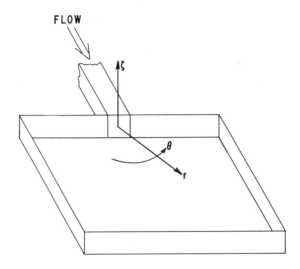

Figure 10.   Cylindrical coordinate system with its origin lying on the cavity midplane positioned at entrance to cavity

Figure 9 depicts the flow in the mold for the fiber-filled sample. At this point, the red (first), yellow (second), and green (third) tracers have entered the cavity, and they exist as a skin (either solid or liquid depending on mold temperature). The blue (fourth) tracer is entering as a sphere which is almost ellipsoidal because of shear forces in the capillary and gate. Near the gate there is a diverging flow channel which decelerates the fluid since the cross sectional area of flow increases with distance from the gate, but the flow rate remains constant. The tracer begins to deform into a circular arc with a U-shaped cross section. This type of flow and deformation continue until the tracer reaches the flow front. Here the material flowing on the midplane at the highest velocity splits the tracer and forces the leading edges to the mold surfaces. The S-shaped tracer marks become permanent in the skin leaving the core open for flow.

Flow patterns result directly from different shear and extensional flows which will be discussed in terms of the components of velocity and the rate-of-deformation tensor. Figure 10 depicts a cylindrical coordinate system (r, θ, ζ) with its origin at the entrance to the cavity and lying on the midplane. The ζ direction is thickness, channel width is W, and height is H. The components of the rate-of-deformation tensor, d_{ij}, are given in cylindrical coordinates (17).

$$d_{rr} = 2 \frac{\partial v_r}{\partial r} \qquad\qquad d_{r\theta} = d_{\theta r} = r \frac{\partial}{\partial r} \left(\frac{v_\theta}{r}\right) + \frac{1}{r} \frac{\partial v_r}{\partial \theta}$$

$$d_{\theta\theta} = 2 \left(\frac{1}{r} \frac{\partial v_\theta}{\partial \theta} + \frac{v_r}{r}\right) \qquad d_{\theta\zeta} = d_{\zeta\theta} = \frac{\partial v_\theta}{\partial \zeta} + \frac{1}{r} \frac{\partial v_\zeta}{\partial \theta} \qquad\qquad (1)$$

$$d_{\zeta\zeta} = 2 \left(\frac{\partial v_\zeta}{\partial \zeta}\right) \qquad\qquad d_{\zeta r} = d_{r\zeta} = \frac{\partial v_\zeta}{\partial r} + \frac{\partial v_r}{\partial \zeta}$$

The components on the left side of Equation 1 lie on the diagonal and are the extensional flow components. Those on the right side of Equation 1 are the off-diagonal, shear components. The diagonal components are not completely independent of each other since they must satisfy the continuity equation:

$$\frac{1}{r} \frac{\partial}{\partial r} (r \, v_r) + \frac{1}{r} \frac{\partial v_\theta}{\partial \theta} + \frac{\partial v_\zeta}{\partial \zeta} = 0 \qquad\qquad (2)$$

Equation 2 is written for an incompressible fluid. Table I defines three extensional flows which follow from Equations 1 and 2 where a is an unspecified constant. Associated with each of these flows is a different resistance to flow, i.e., viscosity. Dealy (18) discussed the relation between steady extensional viscosities and the steady shear viscosity for Newtonian and non-Newtonian fluids, and he reviewed their importance in polymer processing. Injection molding is not a steady process dominated by either shear flow or extensional flow, but it is clearly a dynamic process with some combination of both types of flow (represented by a rate-of-deformation tensor with both non-zero diagonal and off-diagonal components). The precise mathematical definitions presented in Table I permit a detailed analysis of the flow patterns in the mold although this was not done quantitatively in this paper.

Table I. Extensional Flows

Planar Extension (pure shear)	Simple Extension	Biaxial Extension
$D = \begin{pmatrix} a & 0 & 0 \\ 0 & -a & 0 \\ 0 & 0 & 0 \end{pmatrix}$	$D = \begin{pmatrix} a & 0 & 0 \\ 0 & -\frac{1}{2}a & 0 \\ 0 & 0 & -\frac{1}{2}a \end{pmatrix}$	$D = \begin{pmatrix} a & 0 & 0 \\ 0 & a & 0 \\ 0 & 0 & -2a \end{pmatrix}$

In the vicinity of the gate, the cross-sectional area $(\pi r H)$ for flow increases with the distance, r, from the gate until the flow front extends to the side walls (see Figure 9). Hence, $v_r = v_r (r, \zeta)$ and v_θ and v_ζ are zero. This yields $d_{rr} = -d_{\theta\theta}$ and $d_{\zeta r} = d_{r\zeta}$ as the only non-zero components. Thus, before the flow front reaches the side walls, there is a planar extensional flow and a simple shear flow. The shear component deforms the tracer into a U shape in cross section, and the U is continually drawn out until it reaches the flow front. At this point, the drag force on the trailing portion of the tracer is sufficiently great so that the material flowing on the midplane splits the flow front, forcing the leading portions of the tracer to the cavity walls. With fiber-filled material, this results in S-shaped marks or curves in cross section. The splitting operation creates a velocity component in the ζ direction which is a function of ζ and which changes the type of extensional flow to a combination of simple and biaxial extension. Adjacent to the mold walls, the material is drawn out in simple extension in the r direction.

Beyond the gate region, the side walls affect the flow and make the shear components more important (except at the flow front). The cross-sectional area for flow (WH) remains constant thereby eliminating the planar extensional flow. However, the extensional flows continue to be important at the flow front.

Similar features were observed with unfilled and bead-filled polypropylene (Figure 5) although in cross section the in-stream tails of the S-shaped curves are swept forward, and the tracers resemble V-marks. The V-marks form because the leading edge of the tracer is drawn out on the mold surfaces and is immobilized before the trailing edges of tracer become stationary. Fibers apparently increase the resistance to flow and deformation, and they prevent further change of the S-shaped curves.

The different shear and extensional flows have drastic effects on fiber orientation. In the gate region where there is planar extensional flow, there are primarily three different orientations. Within a few mils of the surface, the fibers are oriented along radii from the gate. Beneath the surface, corresponding to the in-stream tail of the tracer mark shown in Figure 9 (about 10 mils for the plaques molded at a 23°C mold temperature), the fibers lie tangential to the circular arcs. In the core, orientation is less ordered than in the other two layers, but it is definitely not random. These orientations were verified by a burn-out technique similar to that described by Fuccella (*14*) and by x-ray studies (*19*).

Beyond the gate region, the surface fibers are oriented parallel to the side walls of the mold while those in the layer below the surface are oriented orthogonal to the surface layer fibers. Here, too, the core is less ordered.

These different fiber orientations prevent a molded part from shrinking uniformly upon removal from a mold. Hence, there is a high level of residual stress in the plaques. If the core section is not thick enough for the surface area, the part will distort or warp to relieve some of the residual stress. In this study, 1/8-in.-thick plaques did not warp whereas 1/16-in.-thick plaques, which had been molded under identical conditions, did warp even though the flow was symmetric about the cavity midplane. The flow patterns were nearly identical in both. Annealing experiments at various temperatures did not reduce the warpage. Neither unfilled nor bead-filled polypropylene plaques warped when cooled.

The effect of mold temperature on flow patterns is apparent in Figure 5. Below 110°C, there is little if any difference in flow patterns. When the mold temperature is approximately 240°C (melt temperature), the immobile skin is very thin, and the first three tracers are drawn out on the surface while the last two tracers are found in the core. This was characteristic of all three materials.

All runs were made at a constant crosshead speed of 20 in./min; this is equivalent to a flow rate of 2.21 in.³/min which is considerably less than typical commercial flow rates. However, the flow patterns reported in this paper are entirely consistent with the findings for high flow rates reported by Gilmore and Spencer (*20*) and by Kamal and Kenig (*21*) for example. Figure 6 shows the pressure on the plunger as a function of processing time. In the gate region, force increases at a constant rate as the flow path increases in length, despite the increasing cross-sectional area of the flow. As the front extends to the side walls, the flow cross-sectional area becomes constant, and the plunger pressure increases with a higher slope. The crossover point indicates that the flow front has extended to the side walls of the mold cavity. The cavity pressure transducer, which was located beyond the gate region, indicated that cavity pressure starts to increase before the crossover time is reached. This occurs because the flow front is not semicircular, but rather somewhat extended in the r direction. The pressure transducer, therefore, provides information about the flow in the gate region and in the rectangular channel beyond the gate region, and about the transition between the two regions. Consequently, this trace is made up of three linear portions before packing. Plunger pressure is too insensitive to isolate the transition portion.

Temperature of the molten flow front was detected by an IR thermometer. As the hot flow front approaches the IR probe (fiber optic), the output voltage begins to increase. Since this instrument correlates intensity of radiation and temperature, meaningful results are not obtained until the entire field of view is filled with the target. In addition, the instrument is unreliable below 100°C. The IR thermometer was carefully calibrated for each material by fixing a sample with an embedded thermocouple on a hot plate and then placing the fiber optic in contact with the sample near the thermocouple; output voltage from the thermometer was then correlated to the thermocouple temperature.

From the temperature trace in Figure 6, it is evident that the cold mold wall rapidly cools the immobile skin, but that some heat is transferred to the skin from the molten core thereby raising its average temperature slightly. As mold temperature increases, the quenching effect in the temperature trace diminishes. Similar traces were obtained with each material.

Figure 7 reveals three interesting features of the plunger force trace. The initial constant slope portion, slope I, refers to the gate region while beyond the gate there is a second constant slope portion, slope II. Associated with the transition from one region to the other is a crossover time. Figure 8 shows that these two slopes and the crossover time decrease with increasing mold temperature for the fiber-filled material, a pattern that was expected and that was also observed for the unfilled and the bead-filled polypropylene. The findings demonstrate that as the mold temperature increases, it takes less energy to inject a composite into the cavity and the flow front extends to the side walls earlier in the process. This occurs because the skin (solid or liquid) next to the cavity walls is thinner at higher mold temperatures, and consequently a larger core channel is created. However, the seemingly linear relations were unexpected. In Table II slope and crossover time data for all three materials

Table II. Effect of Mold Temperature on Plunger Force Slopes and Crossover Time

Material	Mold Temperature, °C	Slope I lb_f/sec	Slope II lb_f/sec	Crossover Time, sec
Polypropylene	23	2.0	2.1	14.4
	114	0.7	1.1	13.1
	240	0.2	1.2	12.3
Bead-filled polypropylene	23	3.5	5.0	13.8
	110	1.5	3.3	13.3
	240	0.6	0.8	13.2
Fiber-filled polypropylene	23	2.3	4.3	14.5
	110	1.6	2.9	13.3
	235	0.4	1.6	12.0

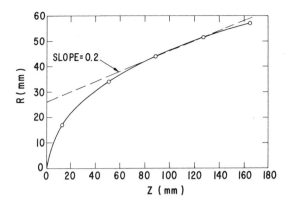

Figure 11. Location of tracers in molded plaques as a function of initial location in rheometer reservoir (glass-fiber-filled polypropylene, $T_{mold} = 23°C$)

are summarized. Although certain sets of these data are closely fitted by a straight line, additional data at intermediate temperatures are needed to ascertain the type of relations.

The final location of the tracer in a plaque depends on its initial location in the rheometer reservoir. The initial position of the tracers in the reservoir is measured by Z which coincides with the reservoir axis. $Z = 0$ is at the origin of the coordinate system shown in Figure 10. The radial measure, R, locates the tracers in the plaques and is measured in the r direction. When R is plotted as a function of Z (Figure 11), the relation is nonlinear in the gate region, then linear. Figure 12 shows that R^2 is a linear function of Z in the gate region. These findings were observed for each material at each of the three mold temperatures. Table III summarizes the data for all nine samples.

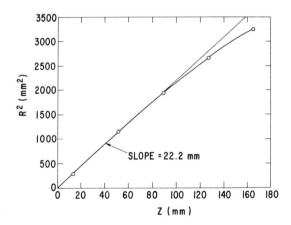

Figure 12. R^2 vs. Z (glass-fiber-filled polypropylene, $T_{mold} = 23°C$)

Table III.   Summary of Tracer Data

		Tracer				
		Red	Yellow	Green	Blue	Black
Initial Location Z, mm:		12.7	50.8	88.9	127.0	165.1
Material	Mold Temperature, °C	Final Location R, mm				
Polypropylene	23	7	28	39	46	52
	114	16	34	46	54	61
	240	18	43	56	—	—
Bead-filled polypropylene	23	9	29	35	43	51
	110	10	31	40	52	59
	240	18	41	51	60	—
Fiber-filled polypropylene	23	17	34	44	52	57
	110	16	35	44	52	60
	235	23	45	56	61	—

A simple model can be used to explain the trends in the R and R^2 data. Consider a skin of thickness H_s and width dR formed in the gate region at each of the two mold surfaces (side wall effects are assumed negligible for thin-walled moldings). The volume of each skin is $\pi R H_s dR$. Assume that a cylindrical slug of material was injected from a reservoir with diameter D_R to produce this skin. For an incompressible fluid these two volumes must be equal; hence

$$\pi R(2H_s)dR = (\pi D^2{}_R/4)\ dZ \qquad (3)$$

Integration of Equation 3 yields

$$R^2 = \frac{D^2{}_R}{4\ H_s}\ Z + C_1 \qquad (4)$$

For boundary condition #1 at $R = 0$, $Z = 0 \rightarrow C_1 = 0$,

$$R^2 = \frac{D^2{}_R}{4\ H_s}\ Z \qquad (5)$$

The skin thickness was approximated from the tracer marks in the cross section of a plaque; the reservoir diameter was measured. When D_R and H_s are substituted into Equation 5, the predicted line does not agree with the data. It is necessary to assume a value for D_R which is considerably less than the actual reservoir diameter for agreement with the data. For the fiber-filled polypropylene plaque, the assumed reservoir diameter is one-half the actual diameter at a mold temperature of 23°C.

Beyond the gate region, the cross-sectional area of flow is constant. The r axis is used to locate tracers in this region by directing it so that it divides the midplane into two equal parts. In this region Equations 1 and 2 become

$$W\ (2H_s)\ dR = (\pi D_R{}^2/4)\ dZ \qquad (6)$$

$$R = \frac{\pi D^2{}_R}{8W\ H_s}\ Z + C_2 \qquad (7)$$

For boundary condition #2 at $R = W/2$, $Z = H_s\ (W/D_R)^2$. Equation 7 becomes

$$R = \frac{\pi D^2{}_R}{8W\,H_s}\,Z + \left(\frac{4-\pi}{8}\right)W \tag{8}$$

The second boundary condition assumes that the shape of the flow front is semicircular in the gate region. From the movie it is clear that the flow front takes on different shapes for different materials and at different mold temperatures. However, the flow front shapes are all close to semicircular, and thus this assumption does not seriously affect the results. Equation 8 is in good agreement with the data for the fiber-filled plaque (mold temperature, 23°C) when D_R is assumed equal to one-fourth the actual reservoir diameter. This indicates that beyond the gate region there is increased resistance to flow that is caused primarily by the longer path length and the constant cross-sectional area for flow. Once again, similar results were obtained with the other samples. The fictitious D_R values varied by $\pm 5\%$ for the different samples. These findings indicate that the effect of the velocity gradient in the reservoir is significant, and that it allows the material near the center to enter the cavity before the bulk of the polymer near the walls of the reservoir (somewhat like a telescope being extended).

Summary

Unfilled polypropylene, glass-bead-filled polypropylene, and glass-fiber-filled polypropylene were injection molded at constant flow rate at three different mold temperatures ranging from room temperature to the melt temperature. The injection molding machine consisted of an Instron capillary rheometer and a specially designed mold which was attached to a capillary extending from the reservoir.

The two fillers (approximately 5 vol % in each composite) have little effect on the flow patterns (as was demonstrated by a tracer technique) and on the temperature gradient in a molding. Larger injection forces are required, though, for the filled materials than for the unfilled resin. Mold temperature determines the thickness of the immobile skin next to the cavity walls and consequently the dimensions of the core channel. Geometry of the mold cavity has the greatest effect on the flow. With the fiber-filled composite, the fiber orientations exhibit a strong dependency on the shear and extensional flows which, in turn, depend on the cavity geometry.

The location of the five tracer elements in the molded plaques is related to their initial location in the rheometer reservoir. A simple, selective volume addition scheme agrees well with the experimental data; this indicates that the velocity gradient in the reservoir is significant in the injection molding process.

Acknowledgment

The assistance of J. L. Henkes in filming the flow in the mold is gratefully acknowledged.

Literature Cited

1. Kamal, M. R., Kenig, S., *Polym. Eng. Sci.* (1972) **12** (4), 294.
2. Berger, J. L., Gogos, C. G., *Polym. Eng. Sci.* (1973) **13** (2), 102.
3. Harry, D. H., Parrott, R. G., *Polym. Eng. Sci.* (1970) **10** (4), 209.
4. Pearson, J. R. A., "Mechanical Principles of Polymer Melt Processing," p. 128, Pergamon, Oxford, 1966.

5. Kantz, M. R., Newman, Jr., H. D., Stigale, F. H., *J. Appl. Polym. Sci.* (1972) **16,** 1249.
6. Newman, S., Trementozzi, Q. A., *J. Appl. Polym. Sci.* (1965) **9,** 3071.
7. Schmidt, L. R., M.S. Thesis, Washington University, St. Louis, 1969.
8. Rosen, S. L., Rodriguez, F., *J. Appl. Polym. Sci.* (1965) **9,** 1615.
9. Kerner, E. H., *Proc. Phys. Soc. London* (1956) **69,** 808.
10. Mooney, M., *J. Colloid Sci.* (1951) **6,** 162.
11. Nielsen, L. E., *J. Compos. Mater.* (1967) **1,** 100.
12. Lockett, F. J., Nat. Phys. Lab., Teddington, Middlesex, Eng. (1972) **NPL Rep. Mat. App.–25.**
13. Johnson, A. F., Nat. Phys. Lab., Teddington, Middlesex, Eng. (1973) **NPL Rep. Mat. App.–26.**
14. Fuccella, D. C., *SPI Ann. Tech. Conf., 27th,* Washington, D.C., February, 1972.
15. Bagley, E. B., Birks, A. M., *J. Appl. Phys.* (1960) **31,** 556.
16. Nakajima, N., Shida, M., *Trans. Soc. Rheol.* (1966) **10,** 299.
17. Middleman, S., "The Flow of High Polymers," Interscience, New York, 1968.
18. Dealy, J. M., *Polym. Eng. Sci.* (1971) **11** (6), 433.
19. LeGrand, D. G., private communication.
20. Gilmore, G. D., Spencer, R. S., *Mod. Plast.* (1951) **28,** 117.
21. Kamal, M. R., Kenig, S., *Polym. Eng. Sci.* (1972) **12** (4), 302.

RECEIVED June 7, 1974.

Concrete–Polymer Composite Materials and Their Potential for Construction, Urban Waste Utilization, and Nuclear Waste Storage

MEYER STEINBERG

Department of Applied Science, Brookhaven National Laboratory, Upton, N. Y. 11973

A range of concrete–polymer composite materials is under investigation. The old technology of hydraulic cement concrete is combined with the new technology of polymers. Polymer-impregnated precast concrete (PIC) is the most highly developed composite, and it exhibits the highest degree of strength and durability. Polymer concrete (PC), an aggregate bound with polymer, is potentially a most promising material for cast-in-place applications. PC with solid waste aggregate offers interesting possibilities for converting urban waste into commercially valuable construction materials. PIC and PC also show potential for immobilizing radioactive waste from the nuclear power industry for long-term engineered storage.

The concrete–polymer composite materials program at Brookhaven National Laboratory is directed at developing both improved and new concrete materials by combining the ancient technology of hydraulic cement concrete formation with the more modern technology of polymer chemistry. A range of concrete–polymer composites is being investigated.

Polymer-Impregnated Concrete Materials Development

Polymer-impregnated concrete (PIC) is a precast and cured hydrated cement concrete which has been impregnated with a low viscosity monomer and polymerized *in situ*. This material is the most highly developed composite. The greatest improvements in structural and durability properties have been attained with PIC. With conventional concrete (28-day water-cured), compressive strengths can be increased from 5000 psi ($352 \ \mathrm{kg/cm^2}$) to 20,000 psi ($1410 \ \mathrm{kg/cm^2}$). Water absorption is reduced by 99%, and freeze–thaw resistance is enormously improved. With high silica cement, strong basaltic aggregate, and high temperature steam curing, strength can be increased from 12,000 psi ($845 \ \mathrm{kg/cm^2}$) to more than 38,000 psi ($2630 \ \mathrm{kg/cm^2}$). The tensile strength of PIC is approximately one-tenth the compressive strength similar in relationship to conventional concrete. A maximum tensile strength of 3500 psi (238 kg/cm^2) has been obtained with the steam-cured concrete. In steam-cured concrete, polymer loadings [poly(methyl methacrylate), (PMMA)] are about 8 wt % of dried concrete. PIC and conventional concrete were tested for freeze-thaw effects (Figure 1) and for resistance to chemical attack by acids (Figure 2).

CONCRETE - POLYMER
3,650 - CYCLES

CONTROL
690 - CYCLES

Figure 1.   Freeze–thaw test
Weight loss: PIC (6 wt % PMMA), 0.5%; control (conventional concrete), 26.5%

In contrast to conventional concrete, PIC exhibits essentially zero creep properties (*see* Figure 3). Furthermore, polymer impregnation transforms conventional concrete from a plastic material to essentially an elastic material with at least a doubling in the modulus of elasticity. This is indicated by the linearity of the stress–strain plot for PIC in Figure 4. The ability to vary the shape of the stress–strain curve offers some interesting possibilities for tailoring desired properties of concrete for particular structural application. This may be achieved by adding plasticizers to the monomer systems or by varying the type and shape of aggregate (*e.g.*, steel fiber aggregate).

PIC is formed by drying cured conventional concrete by the most convenient and economical processing technique (hot air, oven, steam, dielectric heating, etc.), displacing the air from the open cell void volume (vacuum or monomer displacement and pressure), diffusing a low viscosity monomer

Figure 2.   Resistance to chemical attack (15% HCl)
Weight loss: PIC, 7% after 497 days; control, 25% after 105 days

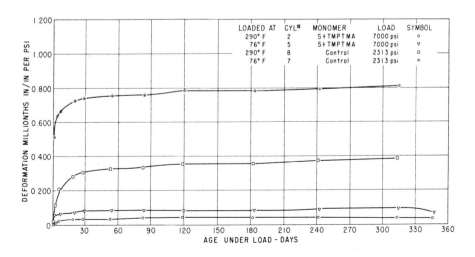

Figure 3.   Creep strain characteristics of PIC

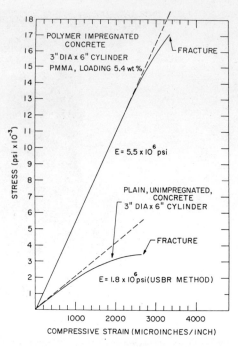

Figure 4.   Compressive stress–strain curve for PMMA-impregnated concrete
Impregnated: elastic behavior; unimpregnated: plastic behavior

(<10 cps) through the open cell structure, saturating the concrete with the monomer, and then polymerizing the monomer to a polymer *in situ* by the most convenient and economical means (irradiation, thermal, or chemical initiation). A schematic flow sheet of the simplest process is presented in

Table I.   Classification of

Material	Polymer Loading, wt % PMMA	Density	
		lbs/ft³	g/cm³
Conventional concrete	0.0	150	2.40
Surface coating (paint or overlay)	0.0	150	2.40
Coating in depth	1.0	150	2.40
Polymer cement concrete			
monomer premix	35.0	130	2.08
polymer premix	1.0	150	2.40
Polymer-impregnated concrete			
standard aggregate			
undried-dipped	2.0	153	2.45
dried-evac.-filled	6.0	159	2.55
hi-silica steam cured	8.0	159	2.55
lightweight aggregate			
structural	15.0	130	2.08
insulating	65.0	60	0.96
Polymer concrete (cementless)	6.0	150	2.40

PRODUCTION OF POLYMER IMPREGNATED CONCRETE

PIC

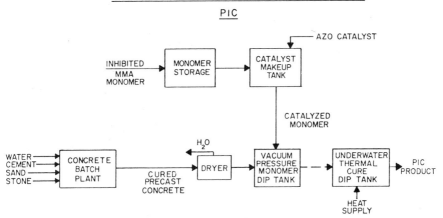

Figure 5. Schematic of PIC process

Figure 5 where underwater thermal–catalytic polymerization (curing) is mentioned.

The free radical, vinyl type monomers (*e.g.*, methyl methacrylate styrene, acrylonitrile, and tert-butyl styrene) and other thermoplastic monomers are used primarily. For increased thermal stability, crosslinking agents and thermosetting monomers such as styrene–trimethylol propane trimethacrylate (TMPTMA) and polyester–styrene are used. The more important criteria are that the monomer should be relatively cheap and readily available and should have a low viscosity. Much information on the formation, structure, and durability of PIC has been reported during the past five years in the United States (*1, 2, 3, 4*). In Table I are a brief summary and classification of PIC

Concrete–Polymer Materials

Compressive Strength,		Strength: Weight Ratio	Durability	Benefit: Cost Index
lbs/in.²	*kg/cm²*			
5,000	353	33	poor	1.0
5,000	353	33	limited	1.1
6,000	423	40	good	1.3
7,500	528	58	fair	0.4
10,000	705	49	better	1.5
10,000	705	49	fair	1.4
20,000	1410	126	very good	2.0
38,000	2680	240	very good	3.0
25,000	1760	193	very good	2.5
5,000	353	84	very good	2.5
20,000	1410	133	excellent	4.0

materials and properties. A U.S. patent has also been issued on the production of PIC (5).

Polymer Cement Concrete

Polymer cement concrete (PCC) is a premixture of hydrated cement paste and aggregate to which a monomer is added prior to setting and curing. The introduction of various organic materials to a concrete mix was tried numerous times in the past by others (6) as well as by the Brookhaven National Laboratory. Improvements in strength and durability were either disappointing or relatively modest. In many cases, materials poorer than concrete were obtained. Under the best conditions, compressive strength improved by *ca.* 50% with relatively high polymer concentrations in the order of *ca.* 30 wt %. Polyester–styrene, furans, and vinylidene chloride have been used in PCC with limited success, probably because most organic materials are incompatible with aqueous systems, and in many cases polymerization either is inhibited by the alkalinity of the cement phase or interferes with the cement hydration process. In addition, porosity develops as a result of shrinkage during the curing process. With PIC, the polymer fills the voids, and the strength, which is a function of the unfilled pore fraction, is greatly increased.

There is incentive to attain an improved premix concrete material because it can be cast in place for field applications whereas PIC requires a precast structure.

A variation of PCC is the addition of a modest amount (less than 4 wt %) of polymer latex such as styrene–butadiene or poly(vinylidene chloride) emulsion to the fresh concrete mix. No polymerization takes place, however, and the polymer particles coalesce during curing and coat the pore structure of the concrete. Maximum strength increases only by a factor of two, but durability is significantly increased (6, 7). The answer for cast-in-place concrete–polymer materials lies in PC development.

Polymer Concrete Materials Development

Polymer concrete (PC) is an aggregate bound with a polymer binder. This material can be cast and formed in the field. It is called a concrete because, by general definition, concrete consists of any aggregate bound with a binder. The cheapest binder, portland cement, costs about 1.25¢/lb in the United States. A polymer can also serve as binder; however, it is more expensive than hydrated cement with most commercial polymers costing 5–30¢/lb.

Polymer filled with aggregate (*e.g.*, powdered walnut shells in plastics for table tops and furniture products) has been known for a long time. PC refers to an aggregate filled with a polymer. The technique in producing PC is to minimize void volume in the aggregate mass in order to reduce the quantity of relatively expensive polymer needed for binding the aggregate. This is accomplished mainly by grading and mixing the aggregates to minimize void volume. For example, to obtain less than 20 vol % voids, a stone aggregate mix of 3/8–1/2 in. stone (60.7 wt %), 20–30 mesh sand (23.0 wt %), 40–60 sand (10.2 wt %), and 170–270 sand (6.1 wt %) are mixed and vibrated together in a form. Monomer may then be allowed to diffuse up through the mixed aggregate, and polymerization is initiated by either irradiation or chemical means. Conventional concrete mixing equipment may also be used for making up the mix. Safety precautions are necessary when flammable monomer is handled.

There is also another reason for anticipating the development of this new class of PC materials. A problem with conventional concrete is the alkaline portland hydraulic cement which forms voids and cracks during hydration and curing. Water can intrude and crack the concrete; furthermore the alkaline cement is attacked by acidic media with resulting severe deterioration. With a polymer as binder, most of these difficulties are overcome. The polymer can be made compact with a minimum of open voids, and most polymers are hydrophobic and resist chemical attack. PC compressive strengths can be as high as those with PIC (*ca.* 20,000 psi, 1410 kg/cm^2) and with similar monomer loadings in the order of 6% (*see* Table I). A silane coupling agent is added to the monomer to improve the bond strength between polymer and aggregate. The main problems arise from the viscoelastic properties of the polymer. Polymers usually have a low modulus of elasticity, that is they are flexible and exhibit creep properties. Consequently plastics are not used alone in structural members. Using a polymer as binder with aggregates overcomes some of these difficulties, and there is much hope for developing an important new class of materials with high benefit-to-cost ratios. PC development is more advanced in the USSR than in the United States (8), whereas PIC is much more advanced here. More investigation and experience with PC are required to make it a reliable, acceptable construction material.

Aggregate Compositions

Consideration can be given to the aggregate in binder–aggregate concretes or composites. An aggregate is any readily available bulk material. The cheapest and most abundantly available aggregates are natural stone and sand which are widely used throughout the world. There are a number of types and grades of stone aggregates which basically relate to igneous (granite), metamorphic (slate), and sedimentary (sandstone) stone. Aggregates are also

Table II. Solid Waste and Sewage-Containing Polymer Concrete

Type	Water	Portland Cement	Aggregate	Polymer Loading[a]	Compressive Strength, psi	Tensile Strength, psi
Composition, wt %						
Concretes						
standard	6	14	80[b]	0	4,500	450
sewage–cement	38	46	16 solids[c]	0	2,200	—
sewage–cement–polymer	28	60	12 solids[c]	24	11,300	—
refuse–cement–polymer	17	33	50 refuse	10	4,000	—
sewage–refuse–cement–polymer	18	28	54 refuse[c]	10	3,700	—
Composites						
glass–polymer[d]	0	0	100 glass bottles	7	16,000	1200
paper–polymer (paper plywood)	0	0	100 newsprint	23	7,300	7500

[a] Poly (methyl methacrylate), wt % of unloaded dried material.
[b] 33% sand, 67% stone.
[c] Content of sewage sludge: 70% water, 30% solids.
[d] Acid resistance (5-wk, 5% H_2SO_4), 0.2% weight gain; 5-wk water absorption, no gain.

classified as standard weight, structural lightweight, and insulating lightweight. Among the various lightweight polymer-impregnated concrete materials, perlite or expanded shale (which has essentially no structural properties) has been used for producing a material which is lighter than wood and compressively as strong as concrete. Thus, there are possibilities for producing lightweight, structurally strong mortars and concretes which are buoyant with a high strength-to-weight ratio.

Table I summarizes the properties of the various concrete–polymer materials produced to date including surface-coated (SC) and partially-penetrated PIC (coated-in-depth concrete, CID). Strength, durability, and manufacturing cost are combined to give a benefits cost rating for each material.

Urban Waste Utilization

Aggregates can also include urban solid waste such as garbage, refuse, sewage, paper, glass, and metal. Sewage and solid waste refuse–polymer concrete (SRPIC) has been produced using garbage as aggregate and sewage as hydrating media for the cement; setting and curing the concrete are followed by drying, impregnating, and *in situ* polymerization of the monomer in the precast concrete mix by irradiation. Compressive strengths as high as that of conventional concrete can be obtained (*see* Table II).

Flat paper newsprint has been soaked in monomer and polymerized *in situ* under pressure to produce a material called paper polymer composite (PPC) or paper plywood because it has very good tensile strength along the plane of the paper (7500 psi, 510 kg/cm²).

Figure 6. Glass–polymer composite (GPC) showing various particle sizes in mixture and specimen of GPC in lower right corner (composition: 90% crushed waste glass, 10% styrene–polyester)

Figure 7.   Glass–polymer composite "eco-pipe" for sewer lines

Non-returnable glass bottles have been crushed and graded, and the mixed particulate glass was filled with monomer in a manner similar to that for PC. Polymer-bound broken glass is almost as strong as PIC, and it is highly resistant to attack by corrosive media. The various glass particle sizes and a sample of the composite material are shown in Figure 6. One application of this material is for sewer pipe, especially for acid wastes or under aerobic conditions when the hydrogen sulfide gas coming from sewage is oxidized to sulfuric acid. We call this material glass–polymer composite (GPC) and the sewer pipe "ecopipe" (Figure 7). Conventional concrete is unsuitable for these purposes, and so asbestos cement, vitreous clay, or cast iron pipe is usually specified. Decorative bricks and facings for buildings can also be made from GPC. Incinerator ash has also been used as aggregate to produce a structurally sound material.

Table II summarizes some of the solid waste refuse polymer–concrete material properties. Hopefully development work on this material will lead to the conversion of an ecological detriment into a positive asset in the form of valuable construction materials. This is of great importance in the recycling and reuse of man's solid waste.

Applications Development

A number of applications of concrete–polymers are being investigated. For the U.S. Federal Highway Administration, advanced prestressed–post-tensioned bridge deck designs are being investigated; the use of high strength PIC leads to some highly interesting streamlined designs. The U.S. Office of Saline Water has been investigating corrosion-resistant PIC for constructing economical and durable distillation vessels used in producing fresh water from the sea. The U.S. Bureau of Mines is investigating the chemical stabilization of coal mine roof supports by impregnation of stone (9) with monomers followed by *in situ* polymerization, and also in developing a pumpable roof bolt. PIC tunnel linings are interesting to the U.S. Bureau of Reclamation. The U.S. Navy has been interested in PIC for underwater buoys and piling. PIC is also being investigated for use in beams and building blocks for durable housing. The American Concrete Pipe Association in cooperation with the U.S. Atomic Energy Commission (USAEC) and the Bureau of Reclamation is investigating the use of PIC for sewer and pressure pipe. There is also interest in PIC railroad crossties. A number of industry associations in the United States are investigating concrete–polymer materials for their own particular applications.

In Japan an industrial construction company together with a chemical company is reported to have constructed a 15 ton/day pilot plant for experimental production of PIC using thermal–catalytic initiation. Beams, panels, and pipe have been produced from this material (Powercrete), and a number of structural and durability tests have been performed.

A firm in the Union of South Africa is reported to have constructed a pilot plant for the thermal–catalytic production of PIC using a special type of concrete with smooth surfaces (mirror concrete). In addition to pipe and building concrete, a market for domestic sanitaryware (*i.e.*, wash basins, bathtubs, sinks, etc.) is being explored.

A cement association in Italy and the University of Rome are collaborating in developing a PIC using a high silica cement for high strength concrete. High pressure steam cured concrete is being used with underwater polymer curing. Possible uses include ship plate, sewer pipe, nuts and bolts, and such items as screw-ended concrete sewer pipe.

Investigators in Norway, Denmark, Belgium, France, Spain, England, and Israel are known to be exploring PIC applications including heated road panels, curbing, base plates for pumps, and window sills.

Extensive work has been conducted in the USSR on developing and applying polmer–concrete (PC). There are indications (8) of significant advances in a number of applications including tunnel supports. PC is also being investigated in the United States, but to a much lesser extent than PIC. Interestingly enough, the Russian literature reports little work on PIC in the USSR. A number of universities around the world have initiated studies on concrete–polymer materials.

In addition to conventional design procedures, sophisticated three-dimensional finite element computer code structural analysis has been developed at Brookhaven National Laboratory (BNL) (10), and it is being applied to aid in the design and analysis of structures using these new concrete–polymer materials.

Preliminary cost estimates indicate that PIC materials could be competitive with conventional construction materials. The feeling is that concrete–polymer, which mates an ancient materials technology (concrete) with a new materials technology (polymer) is a growing and exciting material. With additional investigation, demonstration, and experience, concrete–polymers can become an important class of construction materials.

Storage of Nuclear Waste Materials

Another potentially important application for hydraulic cement concrete in combination with polymers in PIC and PC is the storage of long-lived radioactive waste from the nuclear industry. A major unsolved problem facing the exponentially growing nuclear power industry is the safe disposal of fission product wastes. A technically and economically reasonable approach taken by the AEC is immobilizing fissionable and fission products in long-term durable materials in an engineered storage system. Concrete appears to be an attractive material for this purpose. The wastes must be stored for 1000 years before they can be considered biologically safe. In some environments concrete has lasted for much longer periods of time. Concrete ingredients are inexpensive and readily available, and the new dimension of PIC and PC can ensure additional durability and strength. The radioactive waste materials requiring storage are in the form of soluble salts, aqueous solution (nitrates), oxides, glasses, and contaminated process equipment. At BNL, the USAEC has established a program for determining the radiation stability, leachability, thermal stability, and structural integrity of promising formulations of conventional concrete, PIC, and PC for incorporating radioactive waste materials. Much more will have to be learned before concrete can be confidently recommended as a material that will last 1000 years or more. Some promising formulations of aqueous nitrates with calcium aluminates in high early strength mortars and concretes have been produced and impregnated with styrene–divinylbenzene; they have shown radiation stability to 10^{10} rads which is the total integrated dose expected during a 1000-year exposure. Crosslinked polystyrene is especially radiation resistant. The compressive strength of these materials is about 13,000 psi. Oxide materials incorporated in styrene–divinylbenzene for PC composites also show promise.

Literature Cited

1. Steinberg, M., *et al.,* "Concrete Polymer Materials, First Topical Report," Brookhaven National Laboratory, Upton, N. Y. (1968) **BNL 50134.**
2. Steinberg, M., *et al.,* "Concrete Polymer Materials, Second Topical Report," Brookhaven National Laboratory, Upton, N. Y. (1969) **BNL 50218.**
3. Dikeou, J., *et al.,* "Concrete Polymer Materials, Third Topical Report," Federal Center, Denver, Colo. (1971) **BNL 50275.**
4. Kukacka, L. E., *et al.,* "Concrete–Polymer Materials, Fourth Topical Report," Federal Center, Denver, Colo. (1972) **BNL 50328.**
5. Steinberg, M., Colombo, P., Kukacka, L. E., USAEC, "Method of Producing Plastic Impregnated Concrete," U.S. Patent **3,567,496** (1971).
6. Wagner, H. B., *Chem. Tech.* (1973) 105-118.
7. Emig, G. L., "Latex Polymer Cement Concrete—Structural Properties and Applications, ACI Seminar on Concrete with Polymers, Denver, April, 1973.
8. Moshchanskii, N. A., Paturoev, Y. V., "Structural Chemically Stable Polymer Concretes," Moscow, 1970, NSF transl. **TT 71-50007.**
9. Steinberg, M., Colombo, P., "Preliminary Survey of Polymer Impregnated Stone," Brookhaven National Laboratory, Upton, N. Y. (1970) **BNL 50255.**
10. Reich, M., Koplick, B., Hendrie, J. M., "Finite Element Approach to Polymer Concrete Bridge Deck Designs and Analysis," Brookhaven National Laboratory, Upton, N. Y. (1972) **BNL 16890.**

RECEIVED June 7, 1974.

38

High Temperature Laminating Resins Based on Melt Fusible Polyimides

HUGH H. GIBBS and C. V. BREDER[1]

Plastics Department, E. I. duPont de Nemours & Co., Wilmington, Del. 19898

Incorporation of the perfluoroisopropylidene group as the flexibilizing linkage in the aromatic dianhydride 2,2-bis(3',4'-dicarboxyphenyl)hexafluoropropane dianhydride (6F) has resulted in a new class of melt fusible polyimide binders. Polyimides based on 6F and aromatic diamines are ideally suited for preparing laminates having very low void levels. The low molecular weight prepolymer undergoes thermally induced condensation polymerization to produce a high molecular weight polyimide which is both linear and amorphous. Voids formed as a result of the elimination of volatiles can then be readily removed by applying pressure above the glass transition temperature (T_g). Varying the structure of the aromatic diamine varies the T_g over the range 229°–385°C. Such polymers are tough and have unusually good thermal-oxidative stability.

For some time now, there has been a need for a polyimide binder which could be used to make low void (less than 3%) laminates, that would have a high degree of property retention and thermal-oxidative stability in the 260°–371°C temperature range, and that would not be crosslinked. The last feature would not only serve to promote melt fusibility but would also contribute to toughness. In addition, the binder should be based on a high solids, low viscosity solution of the kind normally used in prepregging operations. It should also be capable of being used in rapid, high pressure, matched-die molding as well as in lower pressure (100–200 psi) vacuum bag–autoclave molding. No known polyimide binder met these criteria completely.

Conventional aromatic polyimide binders polymerize by thermally induced condensation reactions. In the molding of a typical prepreg, as much as 20% weight loss can occur. This loss is attributable to a combination of residual solvent such as N-methylpyrrolidone (NMP) or dimethylformamide (DMF) and the water and alcohol evolved in the polymerization and imidization reactions (*see* Reaction 1).

[1] Present address: Food and Drug Administration, Department of Health, Education, and Welfare, Washington, D.C.

$$
\begin{array}{c}
\text{R'O—C} \quad \text{C—OR'} \\
\text{O} \qquad \text{O} \\
\diagdown \quad \diagup \\
\text{R} \\
\diagup \quad \diagdown \\
\text{HO—C} \quad \text{C—OH} \\
\text{O} \qquad \text{O}
\end{array}
\quad + \quad H_2N\text{—R''—}NH_2 \quad + \quad \text{solvent}
$$

Diester Diacid

Binder Solution

Δ —solvent
—2R'OH
—2H$_2$O

(1)

$$
\left[\begin{array}{c}
\quad \text{O} \qquad \text{O} \\
\quad \| \qquad \| \\
\quad \text{C} \qquad \text{C} \\
\diagup \quad \diagdown \quad \diagup \quad \diagdown \\
\text{—N} \qquad \text{R} \qquad \text{N—R''—} \\
\diagdown \quad \diagup \quad \diagdown \quad \diagup \\
\quad \text{C} \qquad \text{C} \\
\quad \| \qquad \| \\
\quad \text{O} \qquad \text{O}
\end{array}\right]_n
$$

Polyimide

If the dianhydride is allowed to prereact with alcohol to form the diester diacid derivative, no polymerization occurs when the diamine is added. If this ester-forming reaction were not allowed, the polyamide acid formation would occur immediately. In this case the molecular weight would depend only on the molar imbalance of the monomers (Reaction 2).

It is generally not desirable to have a high molecular weight polymer solution as binder because at high solids content the solution is too viscous to handle, and at low solids concentration there is usually insufficient pickup in one pass in the prepregging operation.

In the early stages of cure, which normally take place at 150°–200°C, the molecular weight rapidly advances to the point at which the highly fluid binder solidifies (gels). Since this occurs before all of the volatiles have been eliminated, the net result is a laminate that generally has a high void content (10–20%). When 3,3',4,4'-benzophenone tetracarboxylic dianhydride (BTDA) is used as the dianhydride, it is very difficult to eliminate these voids, even at elevated temperatures and pressures. Side reactions apparently occur during polymerization which could lead to long chain branching and crosslinking, possibly *via* the interaction of the central carbonyl group in BTDA and an amine group leading to an azomethine ($>C=N$—) linkage. Such a reaction would greatly reduce the melt fusbility of the cured polyimide. The adverse effects of the voids so produced in such laminates have been discussed previously (*1, 2, 3, 4, 5*).

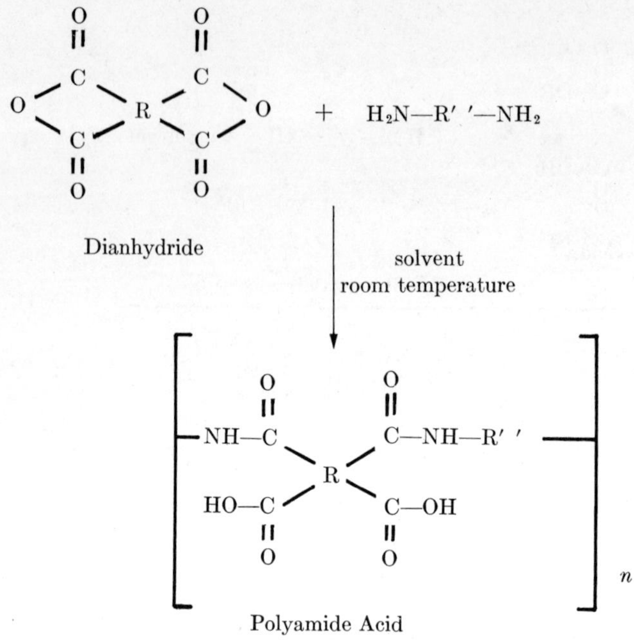

(2)

Dianhydride

solvent
room temperature

Polyamide Acid

A novel approach to eliminating voids has been to develop binders based on imide oligomers containing unsaturated aliphatic end groups (6, 7). At elevated temperatures these oligomers melt and undergo a free radical polymerization reaction which does not involve the loss of any volatile byproducts. For this reason very low void laminates can be produced. However, this class of polyimides is inherently very brittle because of extensive crosslinking which occurs during polymerization. Also, such materials cannot be used for long much above 225°C because of the oxidative instability resulting from the presence of aliphatic hydrogen atoms in the polymer chains.

The objective of this study was to develop a polyimide binder system which would not undergo side reactions during polymerization but would result in a linear high molecular weight amorphous polymer. Such a cured polyimide would have to have sufficient melt fusibility to undergo coalescence readily in order to eliminate trapped voids when heated under pressure above the glass transition temperature (T_g). The approach has been to prepare a new family of polyimides, all based on 2,2-bis(3',4'-dicarboxyphenyl)hexafluoropropane dianhydride (6F) and various aromatic diamines.

Some work involving polyimides prepared from monomers containing perfluorocarbon groups has already been done. Rogers (8) prepared a series

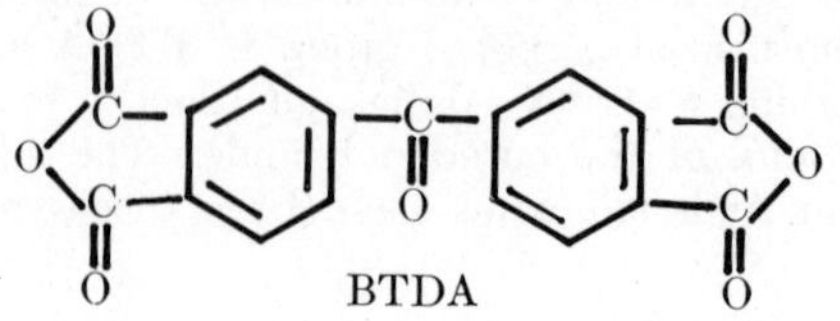

BTDA

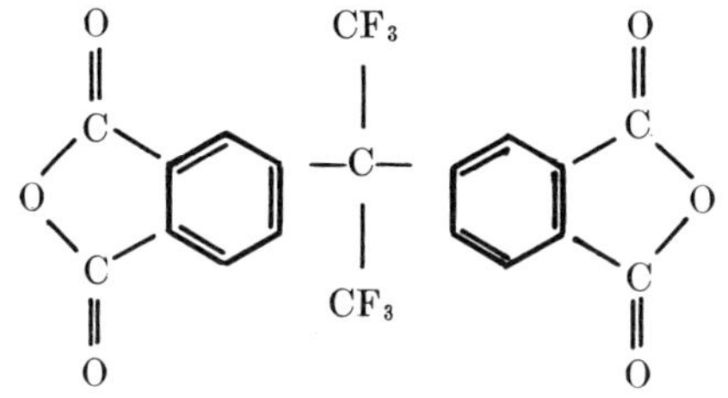

$$\text{6F structure}$$

6F

of polyimides in which the common denominator was the hexafluoroisopropyli-dene bridged 6FDA. Although the 6F/6FDA polyimide was prepared, there was no report of polyimides prepared without the fluorocarbon bridged diamine.

Critchley *et al.* (9, 10) have described polyimides containing linear per-fluoroalkylene bridges in either one or both of the monomers. This paper, therefore, describes the preparation and properties of polyimides based on 6F and aromatic diamines which have unexpected utility in preparing low-void laminates that are tough, retain their properties at elevated temperatures, and possess an unusual degree of thermal-oxidative stability.

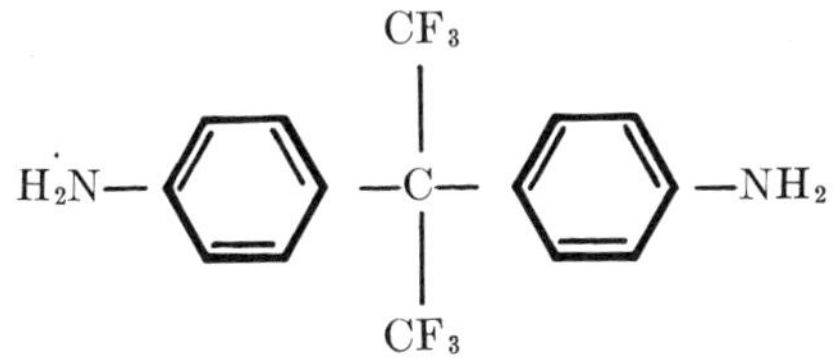

$$\text{6FDA structure}$$

6FDA

Experimental

Preparation of Unreinforced Polyimides. The polyimides characterized in this study were prepared in two ways. When laminating operations were not going to be carried out, the polymers were most readily prepared *via* the amide acid route in pyridine. When the laminate-forming characteristics were to be studied ultimately, the neat resins were prepared by simple thermal polymeri-zation of binder solutions.

Amide Acid Route. To a 3-liter, 3-necked round-bottomed flask equipped with a thermometer, mechanical stirrer, condenser, and nitrogen purge was added 100 g (0.500 mole, 6% molar excess) 4,4'-oxydianiline (ODA). The diamine was washed in with 700 ml distilled pyridine which had been dried over molceular sieves (5A) just prior to use. The mixture was stirred and heated to 50°–60°C to effect solution. Then 6F (213.6 g, 0.48 mole) was added as a dry solid in small portions during about 5 min with stirring. Last traces were washed in with 100 ml pyridine. The slightly exothermic reaction was easily maintained at 50°–60°C by intermittent cooling with an ice bath. At the end of 45 min, all the dianhydride was in solution and the polyamide acid containing amine end groups was produced. Phthalic anhydride capping agent (5.69 g, 0.0385 mole) was added all at once and washed in with 100 ml pyridine, stirring was continued for an additional 45 min, and then acetic anhydride (330 ml) was added. Imidization occurred almost instantly. The temperature rose from about 30° to 45°C. The 6F/ODA polyimide solution was stirred for 30 min, then cooled to room temperature and poured slowly, with vigorous stirring, into a large excess of methanol in a Waring blender. This part of the operation was usually carried out in three stages using a 1-gallon blender. The light cream-colored granular polymer which came out of solution was separated by filtration and washed on the filter with three volumes of methanol. It was dried overnight in a vacuum oven at 150°C and then heated to 260°C for 2 hrs to remove last traces of volatiles. The inherent viscosity was 0.42. The molecular weight was normally controlled by using a 4–6% molar excess of diamine. Note that the use of phthalic anhydride as a capping agent was not essential. Without it, acetamide end groups would have been formed. The more inert phenylene end groups were preferred in order to minimize molecular weight changes during molding of the polymer.

Preparation of Binder Solutions. In a typical run, 6F (444 g, 1.00 mole) was charged into a 2-liter 3-necked round bottomed flask fitted with a thermometer, stirrer, and condenser. Then, 92.0 g (2.00 mole) absolute ethanol and 357 ml DMF were added (NMP was also used as a solvent in this study). The slurry was heated with stirring to 110°–120°C at which point all of the dianhydride was in solution, and the diester was formed. The dark amber solution was stirred at this temperature for 30 min, then cooled to 80°C; ODA (200 g, 1.00 mole) was added slowly with vigorous stirring over a 5-min period. The ODA was washed in with an additional 100 ml DMF. The temperature was maintained at 75°–85°C for 30 min. At the end of this time, all of the diamine was in solution, and the binder was ready for use. Solution viscosity was 4–6 poises at 25°C as determined with the Brookfield viscometer. The calculated concentration was 52 wt % on a cured resin basis. In all cases, stoichiometric concentrations of the monomers were used.

Polymerization of Binder Solutions. To assess the properties of the unreinforced resins derived from binder solution, the solutions were placed in an aluminum pan and then cured in a vacuum oven. The foamed mass of high molecular weight polymer (inherent viscosity greater than 0.5) was then mechanically cut into a finely divided granular product. The cure cycle for polyimides with a T_g less than 300°C was 2 hrs at 200°C followed by 3 hrs at 300°C. For those with a T_g of 300°–350°C, an additional 3 hrs at 360°C was used. For members of the family with the highest softening temperatures (350°–385°C), the cure was extended even further to include a 3-hr period at 371°C. All polymerizations of this type were run under vacuum.

Compression Molding of Unreinforced Resins. The finely divided polymer was charged into a positive pressure mold between two sheets of graphite-coated aluminum foil, heated to the molding temperature at 4000 psi, and held there for 10 min. After being cooled under pressure to room temperature, the compression-molded specimen was removed. Test bars could then be cut from the plaque which was normally about ⅛-inch thick. The molding temperature was usually about 50°C higher than the T_g of the particular polyimide being used.

Table I.   Typical Final Molding Conditions for Laminates

Matrix Polymer	Rapid Matched-Die Cycle, min/°C/psi	Vacuum Bag– Autoclave Cycle, hrs/°C/psi
6F/ODA	10/325/2500	4/316/200
6F/PPD	10/371/2500	3/371/200
6F/1, 5-ND	30/440/2500	1/400/200

Characterization of Unreinforced Resins. The inherent viscosities were determined in concentrated (95–97%) sulfuric acid at 25°C at a polymer concentration of 0.5%. For thermogravimetric analyses (TGA), the Dupont model 950 thermogravimetric analyzer was operated at a 15°C/min heating rate. The T_g values of unreinforced resins were determined with the Dupont model 900 differential scanning calorimeter cell (DSC) operated at a heating rate of 10°C/min.

Preparation of Laminates. The stacked plies of prepreg were laid up in a vacuum bag according to the procedure previously outlined (*11, 12, 13*). Slight vacuum (1–2 psi) was applied, and the temperature was brought to 200°C at the rate of about 5°–8°C/min. These conditions were maintained for one hr, and then the vacuum was increased to 12–15 psi for another 45 min. When 6F/ODA binder was used, the partly cured laminate was discharged from the vacuum bag and cut into bars ½-in. wide and 5–6 in. long. Laminates made with polymers having higher T_g's were usually heated for an additional 2 hrs at 250°C before being discharged from the vacuum bag. At this point, the T_g's were generally low (160°–180°C) and several wt % of volatiles were not yet driven off. Normally, if an autoclave were to be used, the laminating cycle would be continued in the original vacuum bag. In this study, however, all laminations were run in a press. To keep laminates (especially those which were unidirectional) from moving out sideways with the application of pressure, it was necessary to place the machined bars in a positive pressure mold, rebag the mold, and then continue the cycle. In this way the sides of the mold constrained the laminate. The remainder of the molding cycle varied according to the particular binder and the pressure used (*see* Table I for a summary).

Preliminary Characterization of Laminates. All laminates were routinely checked for void content by comparing the actual density with that calculated from the concentrations of the components and their densities. The resin content could be determined by Kjeldahl nitrogen analysis. The T_g was most conveniently determined with the Dupont model 941 thermomechanical analyzer operated at 5°C/min in the expansion mode (TMA-expansion).

Discussion

Properties of Polyimides. EFFECT OF STRUCTURE ON T_g. A summary of the polyimides prepared in this study by the amide acid route is given in Table II. It is apparent that the T_g increases with decreasing flexibility of the diamine. 6F/RBA, the most flexible combination, had the lowest T_g (229°C). At the other extreme, 6F/1,5-ND, which contains the bulky, rigid naphthalene ring, had the highest T_g (365°C). It is of interest to note that the diamines that contain the similarly sized —O—, —S—, and —CH$_2$— linkages all gave polyimides having similar T_g values (280°–291°C). Change in the position of the amino group in phenylenediamine from the meta to the para position caused the T_g to increase significantly (297°–326°C).

Table II.   Glass Transition Temperatures of 6F Polyimides Made *via* Amide Acid Route

Diamine	Code	Inherent Viscosity	T_g, °C
H₂N–⟨⟩–O–⟨⟩–O–⟨⟩–NH₂	RBA	0.35	229
H₂N–⟨⟩–O–⟨⟩–NH₂	ODA	0.46	285
H₂N–⟨⟩–S–⟨⟩–NH₂	TDA	0.35	283
H₂N–⟨⟩–CH₂–⟨⟩–NH₂	MDA	0.38	291
H₂N–⟨⟩–NH₂ (meta)	MPD	0.41	297
H₂N–⟨⟩–NH₂	PPD	0.35	326
H₂N–⟨⟩–SO₂–⟨⟩–NH₂	SDA	0.31	336
H₂N–⟨⟩–⟨⟩–NH₂	BDA	0.40	337
naphthalene-1,5-diamine	1,5-ND	0.64	365

All polyimides prepared using 6F as the dianhydride were found to be amorphous. All were readily soluble in solvents such as pyridine, DMF, and NMP, as well as in concentrated sulfuric acid. Although DSC curves sometimes showed weak melting endotherms on freshly precipitated polymer, the endotherm was permanently destroyed by melting and could not be detected in a

Table III. Room Temperature Mechanical Properties of 6F Copolyimides Made *via* Amide Acid Route

Diamines[a]	Molar Ratio	Inherent Viscosity	T_g, °C	Tensile Strength,[b] psi	Elongation,[b] %
ODA/RBA	90/10	0.39	277	15,000	11
	80/20	0.40	269	14,700	13
	60/40	0.39	254	14,600	12
ODA/1, 5-ND	65/35	0.35	311	15,700	6
ODA/PPD	62/38	0.38	298	14,500	5
	51/49	0.34	304	—	—
	41/59	0.36	307	—	—
	25/75	0.36	315	—	—

[a] See Table II.
[b] ASTM D-1708.

rerun of the sample. Furthermore, two of the polyimides (6F/ODA and 6F/1,5-ND) were tested for crystallinity by x-ray diffraction; none was detected.

The ability to vary the T_g over a broad range is of considerable practical significance. Laminates based on polyimides having low T_g values would be relatively easy to prepare but would suffer in terms of high temperature mechanical property retention. This is because such properties generally fall off quite rapidly in the vicinity of the T_g. On the other hand, if a polyimide with a high T_g were used, properties could be maintained through much higher temperatures, but then the fabrication temperature would have to be correspondingly increased. This is because, for optimum results involving laminates, the final molding temperature is usually higher than the T_g. In any particular application it is necessary to balance off ease of processing *vs.* end-use performance. From the data cited in Table II, it is apparent that a wide range of T_g values was attainable in the 6F family of polyimides simply by choosing the appropriate diamine.

Yet another way of varying the T_g was through the preparation of copolyimides using mixed diamines (Table III). The data indicate that there is an

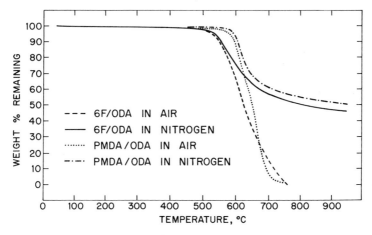

Figure 1. Programmed temperature thermogravimetric analysis of polyimides

approximately linear relationship between the T_g values of the individual polyimides and the concentration of the diamines in the copolyimide.

THERMAL-OXIDATIVE STABILITY BY TGA. Typical programmed temperature TGA curves run in air and nitrogen for 6F/ODA are presented in Figure
1. For comparison, curves for PMDA/ODA are also given. It is apparent that

Pyromellitic dianhydride (PMDA)

while both polymers were quite stable, the latter polyimide performed somewhat better in this kind of test. In isothermal weight loss studies carried out
at 390°C, both polymers degraded at the rate of 0.2–0.3%/hr in nitrogen and
0.4–0.6%/hr in air.

MECHANICAL PROPERTIES OF UNREINFORCED POLYIMIDES. Typical room
temperature properties of the homopolyimides and the copolyimides are presented in Table IV and Table III respectively. It can be seen that as the
diamine moiety becomes bulkier and less flexible, there is a progressive increase
in the T_g values, and the polymers also become stiffer and somewhat more
brittle. At 250°C, the 6F/ODA polyimide still had a tensile strength of 4800
psi, a tensile modulus of 148,000 psi, and 78% elongation.

LONG TERM AIR AGING OF UNREINFORCED POLYIMIDES. A summary of
the data on air oven aging of 6F/ODA, 6F/PPD, and 6F/1,5-ND run at
260°, 316°, and 371°C respectively is given in Table V. Note that 6F/ODA
retained 77% of its original tensile strength after 10,000 hrs at 260°C.

Table IV. Room Temperature Mechanical Properties of Polyimides
Made by Thermally Curing DMF-based Binder Solutions

Property	Polyimide			
	6F/RBA	6F/ODA	6F/PPD	6F/1,5-ND
T_g, °C	229	285[a]	326	365[b]
Tensile strength,[c] psi	13,400	17,600	17,500	16,600
Tensile modulus,[c] psi	—	496,000	570,000	624,000
Flexural strength,[d] psi	—	22,000	—	27,000
Flexural modulus,[d] psi	—	500,000	—	581,000
Compressive strength,[e] psi	—	62,400	—	—
Elongation,[c] %	11	15	8	6
Notched Izod impact,[f] ft-lb/in.	1.0	1.0	0.8	0.8
Density, g/cm³	—	1.42	1.42	1.40

[a] T_g values as high as 300°C were obtained when the solvent was NMP.
[b] T_g values as high as 385°C were obtained when the solvent was NMP.
[c] ASTM D-1708.
[d] ASTM D-790.
[e] ASTM D-695.
[f] ASTM D-256.

Table V.   Air Oven Aging of Unreinforced 6F Based Polyimides Made *via* Amide Acid Route

Polymer	Aging Conditions, hrs	°C	Tensile Strength,[a] psi	Elongation, %	Weight Loss, %
6F/ODA	0	—	15,300	7	—
	10,000	260	11,800	4	2.8
6F/PPD	0	—	17,400	8	—
	1,000	316	12,000	3	4.9
6F/1, 5-ND	0	—	16,600	3	—
	100	371	6,300	2	2.3

[a] Measured at room temperature. ASTM D-1708.

6F/PPD still had 69% of this property after 1000 hrs at 316°C. After 100 hrs at 371°C, 6F/1,5-ND still had a tensile strength of just over 6000 psi.

HYDROLYTIC STABILITY. 6F/ODA bars (⅛-inch thick) boiled in water for one week had a 2% weight gain. Before being retested, the bars were dried in an oven at 200°C for 16 hrs. Both tensile strength (15,000 psi) and elongation (7%) remained unchanged.

ELECTRICAL PROPERTIES. 6F/ODA polyimide has been shown to have very low dielectric constant and loss tangent values at room temperature. Furthermore, these values were maintained out to elevated temperatures (*see* Table VI for properties measured out to 218°C and 9.375 GH$_z$). The volume resistivity was 2×10^6 ohm-cm. The arc resistance determination afforded a wide spread; values ranging from 17 to 123 sec were obtained.

MISCELLANEOUS PROPERTIES OF UNREINFORCED POLYIMIDES. Additional tests run on 6F/ODA indicate that the equilibrium water content at 50% RH was 0.9%. The coefficient of linear thermal expansion was 4.7×10^{-5} in./in. °C, and the Rockwell hardness was 67 (E-scale). The oxygen index was 60, and it had an Underwriters Laboratory SE rating of zero.

Properties of Laminates. UNIDIRECTIONAL GRAPHITE FIBER LAMINATES. Typical room temperature properties are presented in Table VII. Low void levels could be achieved in all samples except the one based on PMDA/ODA in which the intractable nature of the matrix polymer prevented coalescence from occurring. It can also be seen that a low void level was essential for the attainment of high short beam shear strength values. The differences in flexural properties, and to a lesser degree the short beam shear strengths, were in large part attributable to the different types of graphite fibers used in the study.

Note particularly the high level of Sonntag flexural fatigue endurance (60.5 $\times 10^6$ cycles to failure at a stress of 50,000 psi) for high modulus graphite–6F/ODA laminates. This type of test, run at a frequency of 1800 cycles/min, illustrates the considerable toughness of the matrix.

Table VI.   Electrical Properties of 6F/ODA[a] Measured at 9.375 GH$_z$[b]

Temperature, °C	Dielectric Constant	Loss Tangent
22	2.914	0.0016
107	2.921	0.0025
190	2.909	0.0042
218	2.878	0.0066

[a] Polymer derived from thermally curing DMF-based binder solution.
[b] Data supplied by Hitco, Gardena, Calif.

Table VII.   Room Temperature Mechanical Properties of Unidirectional Laminates

Property	Unidirectional Laminate			
	6F/ODA	6F/PPD	6F/1,5-ND	PMDA/ODA
Fibers, vol %	60	55	52	45
Voids, vol %	<1	<1	<1	14
Flexural strength,[a] psi	132,000	206,000	139,000	21,000
Flexural modulus,[a] psi $\times 10^6$	29	19	16	15
Short beam, shear strength,[b] psi	10,000	12,000	11,000	2,900
Izod impact,[c] ft-lb/in.				
notched	14	—	—	—
unnotched	30	—	—	—
Flexural fatigue life, cycles to failure at 50,000 psi stress	60.5×10^6	—	—	—
Density, g/cm³	1.70	1.60	1.52	1.42
Graphite fiber type	Grafil HM	Modmor type III	Modmor type II	Grafil HM
Final molding conditions, hrs/°C/psi	0.1/375/2500	3/371/2500	0.5/440/2500	0.1/450/2600
Binder solvent	DMF	DMF	DMF	DMF

[a] ASTM D-790.
[b] 4/1 span/depth ratio.
[c] ASTM D-256.

As the temperature was increased, there was a gradual decrease in the level of mechanical properties until the T_g was reached. At this point, the properties rapidly deteriorated. This is illustrated in Figure 2 in which the stiffness of a graphite fiber–6F/ODA laminate was tested at temperatures as high as 350°C. Out to the T_g (275°C), there was a 90% retention of flexural modulus. The effect of temperature on the short beam shear strengths of 6F/PPD and 6F/1,5-ND graphite fiber laminates is evidenced in Table VIII. The ability to maintain a high level of short beam shear strength out to elevated

Table VIII.   Effect of Temperature on Short Beam Shear Strength of Graphite Fiber Laminates

Variable	Laminate	
	6F/PPD	6F/1,5-ND
Short beam shear strength,[a] psi		
Temperature, °C		
23	12,000	10,600
260	10,200	—
288	8,400	10,500
316	6,500	10,300
342	—	9,500
371	—	3,800
Characterization of Laminates		
T_g, °C	335	357
Binder solvent	DMF	NMP
Final molding conditions, hr/°C/psi	0.5/440/500	0.5/440/2500
Post-cure	yes (to 371°C)	no
Fibers, vol %	49	53
Voids, vol %	<3	<1

[a] 4/1 span/depth ratio.

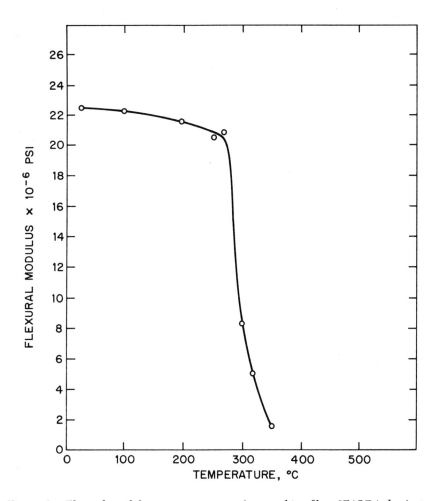

Figure 2. Flexural modulus vs. *temperature for graphite fiber–6F/ODA laminate*

temperatures (*e.g.*, 90% retention for 6F/1,5-ND at 342°C) means that such laminates could be used for structural applications at these temperatures.

LAMINATE STABILITY. Graphite fiber–6F/1,5-ND laminates (⅛-inch thick) were aged for a total of 1200 hrs at 316°C in air. After this period of time, there was a 1.9% weight loss and the short beam shear strength, as measured at 316°C, decreased from 10,000 to 8500 psi. The laminate density actually increased from 1.55 to 1.57 g/ml during this aging period. This indicates that, as the polyimide slowly decomposes, there is gradual coalescence to fill the voids left by degradation. The net result is a laminate with a higher fiber content and, therefore, a higher density. The aging test was conducted on test bars molded at 440°C and 2500 psi for a period of 30 min. The fiber (Modmor type II) concentration was 50–55 vol %, and the void level was less than 2%. The solvent used was DMF.

Table IX.   Mechanical Properties of Quartz Fabric–6F/ODA Laminates[a]

Property	Test Temperature 25°C	204°C	250°C
Tensile strength,[b] psi	80,400	—	—
Tensile modulus,[b] psi $\times$ 10^{-6}	3.20	—	—
Flexural strength,[c] psi	133,000	76,900	63,900
Flexural modulus,[c] psi $\times$ 10^{-6}	4.11	4.55	3.99
Compressive strength,[d] psi	82,200	—	—
Short beam shear strength,[e] psi	10,500	9,400	7,200
Izod impact strength,[f] ft-lb/in.			
notched	26	—	—
unnotched	39	—	—
T_g, °C	277	—	—
Density, g/cm³	1.89	—	—

[a] Used Stevens 581 Style cloth, A-1100 finish and DMF based binder solution. Final molding conditions: 3 hrs/316°C/200 psi. Laminate contained 66 vol % fibers, 3 vol % voids.
[b] ASTM D-1708.
[c] ASTM D-790.
[d] Celanese method.
[e] 4/1 span/depth ratio.
[f] ASTM D-256.

QUARTZ FABRIC LAMINATES. Properties of quartz fabric–6F/ODA laminates (DMF solvent) were measured at room temperature, 204°, and 250°C (see Table IX). In general, all values were at a particularly high level.

The effects of long term air aging at 260°C were examined; the laminates were tested before and after the exposure at room temperature and at 250°C. The results are summarized in Table X. It should be observed that after 5000 hrs of exposure, there was an increase in both flexural strength (114%) and short beam shear strength (124%) when these properties were measured at 250°C. There were some reductions in these values, however, in the samples tested at room temperature. It is quite likely that aliphatic A-1100 silane coupling agent on the quartz constitutes a weak link in this system.

Table X.   Air Aging of Quartz Fabric–6F/ODA Laminates[a] at 260°C

Property	Tested at 23°C Initial	1100 hrs	5000 hrs	Tested at 250°C Initial	1100 hrs	5000 hrs
Flex strength,[b] psi	89,800	77,200	78,700	53,100	53,600	60,300
Flex modulus,[b] psi $\times$ 10^{-6}	3.5	3.2	3.2	3.2	2.5	3.1
Compressive strength,[c] psi	84,800	71,400	65,500	—	—	—
Shear strength,[d] psi	13,600	11,400	9,700	4,200	5,100	5,200
Tensile strength,[e] psi	74,600	—	70,100	—	—	—
Tensile modulus,[e] psi $\times$ 10^{-6}	2.7	—	3.7	—	—	—
Izod impact,[f] ft-lb/in.						
notched	24	21	—	—	—	—
unnotched	>38	29	—	—	—	—
T_g, °C	270	270	270	—	—	—
Weight loss, %	—	0.34	1.06	—	—	—
Density, g/cm³	1.78	1.81	1.81	—	—	—

[a] Used Stevens 581 style cloth, A-1100 finish. Final molding conditions: 3 hrs/316°C/200 psi. Laminate contained 47 vol % fibers, 1.5 vol % voids.
[b] ASTM D-790.
[c] Celanese method.
[d] 4/1 span/depth ratio.
[e] ASTM D-1708.
[f] ASTM D-256.

The electrical properties of quartz fabric–6F/ODA laminates were measured at 9.375 GH$_z$ and are summarized in Table XI. Both the dielectric constant and loss tangent values were low initially, and they remained remarkably constant all the way out to 246°C.

Table XI.   Electrical Properties[a] of Quartz Fabric–6F/ODA Laminates[b] Measured at 9.375 GH$_z$

Temperature, °C	Dielectric Constant	Loss Tangent
22	3.400	0.0048
107	3.318	0.0039
190	3.384	0.0029
246	3.353	0.0036

[a] Data supplied by Hitco, Gardena, Calif.
[b] Used Stevens 581 style cloth, A-1100 finish. Final molding conditions: 3 hrs/316°C/200 psi. Laminate contained 50 vol % fibers.

EFFECT OF POST-CURE. It has long been recognized that the properties of polyimide laminates can be upgraded by post-cure at elevated temperatures. This normally involves a slow programmed temperature heating of the laminate out to 316°–371°C in an air oven. With 6F-containing polyimides, post-cure was found to have a significant effect on increasing the T_g when the proper value was not obtained under the original molding conditions. This is illustrated by the data in Table XII—the T_g can be increased as much as 62°C by post-cure. During this procedure, it is important to control the heat-up rate so that no permanent swelling will occur. While the cycles described in Table XII worked quite well, no attempt was made to minimize the time involved. Under these conditions, permanent swelling was normally kept to less than 1%. The weight loss usually amounted to 1–2%. Post-cure most likely promotes the last stages of condensation polymerization. Because of the very high melt viscosity of the matrix polymer, temperatures close to the T_g must be reached before the polymer chains have sufficient mobility to interact with one another. Also, at very high temperatures it is possible that there could be a small amount of oxidative crosslinking when the operation is carried out in air. One indication of this might be the significant reduction in the coefficient of linear thermal expansion on the post-T_g part of the TMA-expansion curve of graphite fiber–6F/1,5-ND unidirectional laminates. The coefficient decreased from 8.2×10^{-4} to 1.5×10^{-4} in./in. °C as the result of post-cure. The actual increase in T_g in this case was not large (357° to 376°C) which indicates that the initial laminate had already achieved relatively high molecular weight in the original molding operation.

HYDROLYTIC STABILITY. Unidirectional laminates containing about 60 vol % HMG-50 graphite fibers and 6F/ODA were exposed to boiling water for 1000 hrs, and then put through 10 freeze (dry ice)–thaw (boiling water) cycles (10–40 min/cycle). Properties were tested with the laminate in the wet condition. The flexural modulus (22.7×10^6 psi), flexural strength (96,800 psi), and short beam shear strength (5700 psi) values were essentially the same as those of the unexposed controls. In addition to demonstrating good hydrolytic stability, the test illustrates the toughness of the matrix polymer. In spite of the extensive temperature cycling, the stresses which inevitably developed at the matrix–graphite fiber interface did not result in any matrix cracking; if present, this would have been reflected in a significant reduction in the short beam shear strength.

Table XII.   Effect of Post-Cure on Glass Transition Temperature

Variable	6F/ODA	6F/PPD	6F/1, 5-ND
T_g, °C			
before	218	278	357
after	280	335	376
Post-cure cycle, hrs/at °C	4/230	2/225	same as for
	20/260	2/250	6F/PPD
		4/275	
		16/288	
		8/300	
		16/316	
		8/325	
		16/335	
		8/350	
		16/371	
Characterization of Laminates			
Final molding condition, hrs/°C/psi	2/316/150	0.5/440/500	0.5/440/2500
Fibers, vol %	52	49	50
Voids, vol %	<2	<3	<2
Fiber type	Magnamite A	Modmor II	Magnamite A
Binder solvent	NMP	DMF	NMP

Utility

Several of the 6F/all aromatic diamine combinations were selected for more intensive investigation of their utility. These experimental materials are currently available (from the Plastics Department, E. I. duPont de Nemours & Co., Inc., Wilmington, Del. 19898) under the code name NR-150, and they include polyimides with T_g values ranging from 285° to 385°C. In view of their unique and useful combination of properties, NR-150 polyimide binders are expected to find broad applications in vacuum bag–autoclave molding and matched-die molding of high performance composites reinforced with graphite, boron, quartz, and glass fibers wherever high levels of strength, stability, and toughness are required. Specific areas would include structural parts on supersonic aircraft, jet engine blades and vanes, as well as other components near the hot sections of the engine. Another area would be tough, low void ablative heat shields. The combination of outstanding electrical and mechanical properties with low void levels would indicate that NR-150 binders should be ideally suited for making electromagnetic windows (e.g., radomes). In addition to laminates, NR-150 binders are expected to be useful as high temperature adhesives. For instance, Novak et al. (14) have successfully used NR-150 to bond sheets of boron-reinforced aluminum. Other areas of potential application include coatings, chopped fiber moldings, filament windings, brake linings, pultrusion, and thermoforming.

Summary

6F-based polyimide binders offer a new approach to high performance composites. Whereas other polyimide binders rapidly become relatively intractable during cure, such is not the case with those containing 6F. The low molecular weight prepolymer undergoes thermally induced condensation polymerization to produce linear, high molecular weight amorphous polyimide. Voids formed as a result of the elimination of volatile byproducts can be readily eliminated by applying pressure above the T_g. Varying the structure of the aromatic diamine can vary the T_g value over a broad temperature range (229°–385°C). Such composites can either be vacuum bag–autoclave or matched-die molded.

Some of the important features which can be achieved in laminates are: outstanding long term thermal-oxidative stability; very low void content (< 1%); high shear strength levels with such fibers as glass, quartz, and graphite; excellent property retention out to temperatures as high as 342°C; and excellent toughness and fatigue resistance.

Several of the 6F/diamine combinations have been selected for more intensive investigation. These experimental materials are currently available under the code name NR-150 and include polyimides with T_g values ranging from 285° to 385°C.

Literature Cited

1. Petker, I., Sakakura, R. T., Segimoto, M., "Low Void Content Polyimide Composites," Ann. Tech. Conf., 23rd, SPI Reinf. Plast. Compos. Inst. (Washington, February, 1968) Sect. 17-B.
2. Copeland, R. L., Beeler, D. R., Chase, V. A., "Polyimides—Reinforced and Unreinforced," Ann. Tech. Conf., 23rd, SPI Reinf. Plast. Compos. Inst. (Washington, February, 1968) Sect. 3-C.
3. Pike, R. A., DeCrescente, M. A., "Elevated Temperature Characteristics of Boron and Graphite Fiber/Polyimide Composites," Ann. Tech. Conf., 23rd, SPI Reinf. Plast. Compos. Inst. (Washington, February, 1971) Sect. 13-D.
4. Hirsch, S. S., Darmory, F. P., "Condensation Polymerization," Amer. Chem. Soc., Div. Polym. Chem., Prepr. 4 (2), p. 9 (Washington, September, 1971).
5. Stuckey, J. M., "Development of Graphite Fiber Polyimide Composites," National SAMPE Tech. Conf. Space Shuttle Mater. (Huntsville, October, 1971) p. 747.
6. Lubowitz, H. R., U.S. Patent **3,528,950** (1970).
7. Bargain, M., Grosjean, P., U.S. Patent **3,562,223** (1971).
8. Rogers, F. E., U.S. Patent **3,356,648** (1967).
9. Critchley, J. P., Grattan, P. A., White, M. A., Pippett, J. S., J. Polym. Sci., Part A-1 (1972) **10**, 1789.
10. Critchley, J. P., White, M. A., J. Polym. Sci., Part A-1, **10**, 1809 (1972).
11. Gibbs, H. H., "Low Void Composites Based on NR-150 Polyimide Binders," Ann. Conf., 28th, SPI Reinf. Plast. Compos. Inst. (Washington, February, 1973) Sect. 2-D, p. 1.
12. Gibbs, H. H., "Processing and Properties of Composites Based on NR-150 Polyimide Binders," 18th Nat. SAMPE Symp. (Los Angeles, April, 1973) p. 307.
13. Gibbs, H. H., "NR-150 Polyimide Binders—A New Approach to High Performance Composites," 17th Nat. SAMPE Symp. (Los Angeles, April, 1972) p. 111-B-6.
14. Novak, R. C., Pike, R. A., Prewo, K. M., Kreider, K. G., "Resin-Bonded Boron–Aluminum Composites," 17th Nat. SAMPE Symp. (Los Angeles, April, 1972) p. V-A-5.

RECEIVED February 20, 1974.

39

Phthalonitrile Resins

JAMES R. GRIFFITH, JACQUES G. O'REAR, and
THEODORE R. WALTON

Naval Research Laboratory, Washington, D. C. 20375

*A series of aliphatic diamides with chain lengths of 4–18 carbon
atoms terminated with ortho-phthalonitrile units on the amide
nitrogen atoms was synthesized. Melting points of the pure compounds were determined. At temperatures above the melting
points, these compounds are resinous liquids of moderate viscosity which are similar in appearance and reaction behavior to
liquid epoxy resins. At about 170°C, the liquids react to form
intensely green, polymeric products which are believed to be
poly(phthalocyanines). These polymerized products have useful
mechanical properties similar to those of cured epoxy resins, and
they are thermally stable well above 200°C.*

Ever since Linstead and co-workers investigated phthalocyanines in the 1930's
(1), there has been some interest in the reaction of aromatic ortho-dinitriles
for polymerization. Marvel and Rassweiler (2) and Marvel and Martin (3)
studied the possibilities of producing high molecular weight phthalocyanines
from tetranitriles such as 3,3′,4,4′-tetracyanodiphenyl ether. The phthalocyanine formation apparently progressed to a limited extent, and a metal-free
derivative was found which decomposed in air at 350°C.

We felt that steric advantages favoring extensive polymerization would
ensue if the phthalonitrile units were separated farther and placed at the ends
of a long, flexible molecular chain (*see* Structure 1). Freedom of movement

(1)

of the reacting ends should allow proper orientation for the inclusion of an
end into the phthalocyanine structure. A thermal ceiling in air of only 300°C
would not preclude interest in such polymers since most resins with convenient
use properties, such as epoxies, have a practical use ceiling of about 200°C.

458

Syntheses

Several synthesis pathways should lead to structures similar to Structure 1. Any number of tetraacids (Structure 2) can be synthesized, and by following

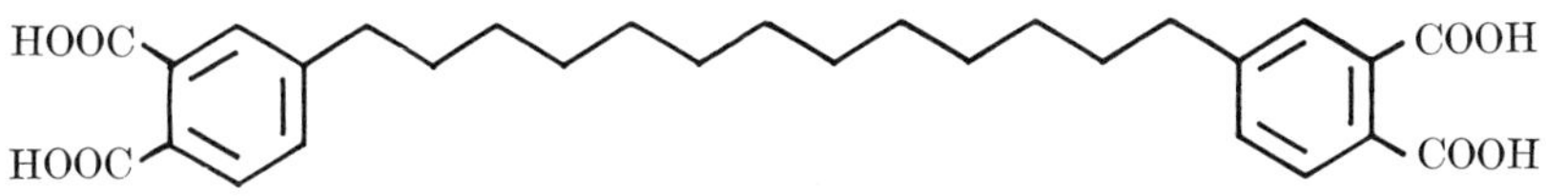

(2)

the route through the tetraamides, tetranitriles can be synthesized in principle. This plan did not appear attractive because of purification difficulties with the intermediates and the necessity that all structures in the central portion be stable under the reaction conditions.

Another possibility is the direct placement of a functional group, X, on phthalonitrile (Reaction 3). Several attempts to nitrate or halogenate phthalo-

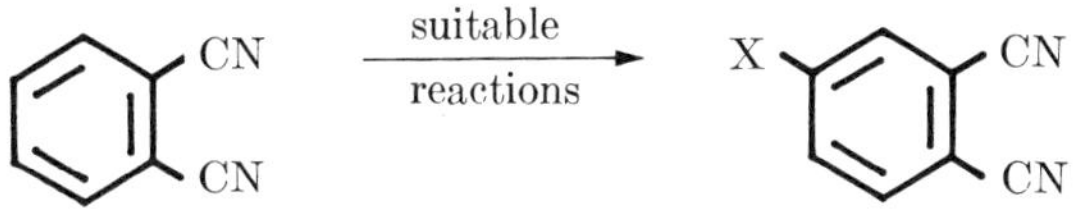

(3)

nitrile were unsuccessful, and, although the method is attractive, suitable reactants and conditions for the placement of a useful functional group were not found. The nitro group may be acquired in the preferred 4 position by diamide formation and dehydration from 4-nitrophthalimide (Reaction 4). The

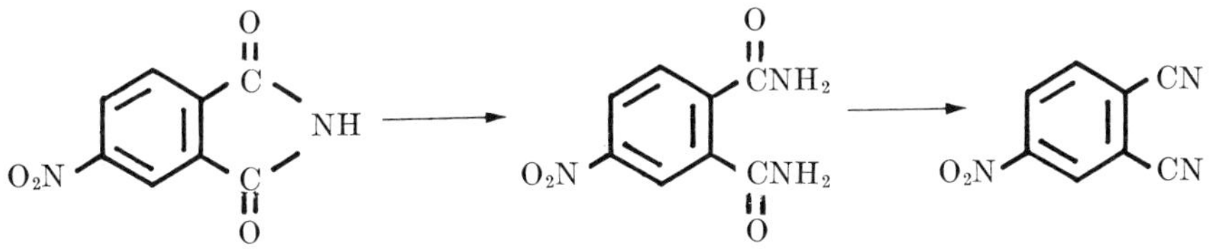

(4)

nitro group is not favorable for subsequent attachment of structure, but it can be differentially reduced in the presence of the two reactive nitrile groups to yield 4-aminophthalonitrile (Reaction 5) (4). The amino group offers a

(5)

point of reaction for a number of possible courses, the most direct of which is diamide formation (Reaction 6).

Table I.   N,N'-Bis(3,4-Dicyanophenyl)

Compound	n	Yield, %	Melting Point, °C	Recryst. Solvent	Empirical Formula
IV	4	55	300–302	C_6H_6–DMF 70:30	$C_{20}H_{12}N_6O_2$
V	5	68	256–259	DMF	$C_{21}H_{14}N_6O_2$
VI	6	73	320–323	DMF	$C_{22}H_{16}N_6O_2$
VII	6 (3 Me)[b]	61	203–206	MeOH	$C_{23}H_{18}N_6O_2$
VIII	7	63	227–230	MeOH–H_2O 90:10	$C_{23}H_{18}N_6O_2$
IX	8	53	255–258	CH_3CN	$C_{24}H_{22}N_6O_2$
X	9	63	181–183	MeOH	$C_{25}H_{22}N_6O_2$
XI	10	63	192–194	MeOH	$C_{26}H_{24}N_6O_2$
XII	12	73	163–165	CH_3CN	$C_{28}H_{28}N_6O_2$
XIII	13	71	178–181	CH_3CN	$C_{29}H_{30}N_6O_2$
XIV	14	50	163–165	CH_3CN	$C_{30}H_{32}N_6O_2$
XV	18	60	163–166	CH_3CN	$C_{34}H_{40}N_6O_2$

[a] Structure:

[b] The linkage between the two carbonyl groups is —$CH_2CH(CH_3)CH_2CH_2$—.

(6)

This affords an example of the general type of structure desired, and, although the amide linkage may not be the most desirable for high thermal stability, this class of phthalonitrile was considered sufficiently promising that a series was synthesized to determine some properties and reaction behavior.

Resin Properties

The 12 diamides listed in Table I were derived from 4-aminophthalonitrile and aliphatic diacid chlorides. These are white, crystalline solids with melting points above 160°C. The melts are syrups with free-flow characteristics at the melt temperature; glasses may be obtained in some cases by rapid cooling of the melt to room temperature. A plot of the melting points has the

Alkanediamides[a]

Calculated, %			Found, %		
C	H	N	C	H	N
65.21	3.28	22.83	64.96	3.20	22.75
65.96	3.69	21.98	65.86	3.59	21.83
66.66	4.07	21.20	66.41	4.02	21.03
67.31	4.42	20.48	67.47	4.44	20.40
67.31	4.42	20.48	67.57	4.24	20.42
67.91	4.75	19.80	67.71	4.79	19.74
68.48	5.06	19.17	68.60	5.12	19.08
69.01	5.35	18.57	69.05	5.23	18.39
69.98	5.87	17.49	69.92	5.65	17.35
70.42	6.11	16.99	70.28	5.97	16.79
70.84	6.34	16.52	71.04	6.38	16.29
72.31	7.14	14.88	72.22	7.10	14.75

familiar odd–even zig-zag as the aliphatic chain increases in length, with a plateau beyond C_{12} (*see* Figure 1).

Polymerization

When a melt of phthalonitrile containing a suspension of clean copper flake (activated copper bronze pigment) is heated above 170°C, a green tint soon develops uniformly throughout the melt if the copper is evenly distributed. Areas of melt which are copper-poor do not become green as rapidly as those where the metal is located, although the color will develop at higher temperatures (220°C) even in the absence of added metal. The sharp, isolated IR absorption of the nitrile group at about 2240 cm[-1] affords a convenient measure of the degree of reaction, although the total spectrum has not been useful as proof that the reaction is phthalocyanine formation. The green color which becomes progressively deeper with solidification of the melt is a strong indication that phthalocyanine formation is significant. IR spectra of unsubstituted phthalocyanines depend to some extent on crystal modifications as well as on composition and structure (5). In the resin cure, phthalocyanine units should be isolated in the developing network, and the spectral changes which occur as phthalonitriles become phthalocyanine polymer have not so far been definitive except with regard to nitrile diminution. Primarily on the basis of the obvious intense color development and physical changes (liquid → solid transition) it is believed that substituted phthalocyanine forms by Reaction 7. Decrease in nitrile IR absorption and physical strength development both indicate that polymerization of approximately 48 hrs at 220°C is required to develop the polymeric network to the maximum extent. The hypothetical cure process is depicted in Reaction 8.

Cured Polymer Properties

Properties of the cured phthalonitriles are now under investigation, and the most interesting properties, the thermo-mechanical, will require extensive evaluation. After full cure, polymers appear much the same as cured epoxies except that the very deep green color makes the bulk material appear black.

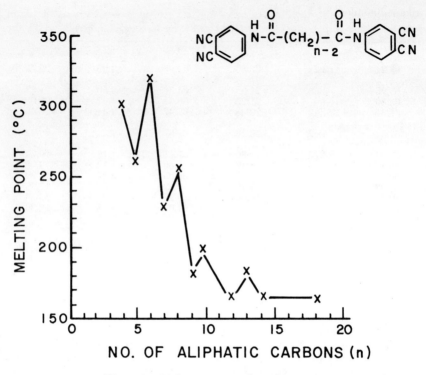

Figure 1. Melting points of diamide resins

Some difficulty in avoiding bubbles at the high cure temperatures is frequently encountered, and all ramifications of this problem are not fully understood. More than a stoichiometric amount of copper flake in the reaction mixture tends to cause foaming, and there is some indication that a component of the atmosphere (probably oxygen) may affect the course of a polymerization. For

Reaction 7. The phthalocyanine reaction

example, it was occasionally observed that a resin melt between cover plates on a hot stage will first begin to turn green near the edges exposed to the atmosphere.

At 300°C in air, there is a gradual weight loss of approximately 2%/day, and filings of a sample will slowly acquire a brownish tint instead of the distinct green of the original polymer. It is now unknown whether this decomposition is occurring at the phthalocyanine nucleus, at the amide linkage, or as a result of unreacted ortho-dinitrile in the polymeric network. In spite of this process, the structural integrity of the polymeric material is retained for a number of days at this temperature.

Reaction 8.   Hypothesis for cure of phthalonitrile resins

□, *phthalocyanine nucleus;* R, *connecting structure, usually hydrocarbon;* X, *a large, indefinite number* (e.g. 2×10^{25})

A suitable means to estimate the molecular weight of the cured resins is yet to be devised. Solvents such as acetonitrile will slowly leach some soluble material from the polymer. A bulk sample does not initially appear soluble in concentrated sulfuric acid, but considerable dissolving will occur over several days. This probably results from degradation at the amide linkage.

Although mechanical property studies are not completed, the polymer strength levels are sufficiently high to be useful in practical applications, particularly that of matrix materials for glass and carbon fiber reinforced composites. The adhesive properties do not appear as generally useful as those of epoxies (except above the decomposition temperatures of epoxies), but coupling agents (similar to the amino silanes used in epoxy applications) which are designed for phthalonitrile resins should improve their suitability for use as metal-to-metal adhesives.

Experimental

4-Nitrophthalamide (I). This compound was prepared by a modified Bogert and Boroshek (6) method. Commercial 4-nitrophthalimide (500 g;

2.60 moles; mp, 196°–200°C) is added quickly to a stirred solution of 15.6N
NH₄OH (3500 ml). With continued stirring, external heating (water bath,
45°C) is applied until frothing ceases (1 hr). The resulting precipitate is
collected, washed with ice water (2 × 400 ml), and dried to yield 440 g of I as
pale yellow crystals. Yield, 81%; mp, 200°–202°C; lit mp, 200°C (6).

4-Nitrophthalonitrile (II). 4-Nitrophthalamide was converted into 4-nitro-
phthalonitrile by a modification of the method of Scalera and Bouillard (7).
4-Nitrophthalamide (220 g; 1.05 moles) is suspended in pyridine (840 ml).
Into this vigorously stirred suspension, phosphorus oxychloride (220 ml; 2.40
moles) is added at such a rate that the reaction maintains a temperature range
of 65°–70°C during the 30-min addition time. The same temperature is maintained an additional hour by external heating. The resulting mixture is poured
onto crushed ice (3000 g), and the ice mixture is neutralized with 12N HCl
(150 ml). The purple solid is collected by filtration and dried. The residue
is extracted with ethyl acetate (3 × 700 ml), and the purple extract is de-
colorized with Nuchar C-190 and washed, first with 0.4N NaOH saturated
with NaCl (3 × 70 ml) and then with water (4 × 70 ml). Drying and evaporation of the extract leaves 91 g of II as pale yellow crystals. Yield, 50%;
mp, 142°–144°C; lit mp, 138°–139°C (7) and 142°C (4).

4-Aminophthalonitrile (III). The procedure provides analytically pure III.
4-Nitrophthalonitrile (100 g; 0.576 mole) is dissolved in pyridine (400 ml).
The pyridine solution is added dropwise during 1 hr to a stirred solution of
90% sodium dithionite (457 g; 2.36 moles) and water (2700 ml). The resulting mixture is stirred for an additional hour. Then water (900 ml) is
added, and the mixture is cooled to 4°C. The white precipitate is collected
and washed with cold water (2 × 400 ml). Recrystallization from water (2700
ml) leads to 27.3 g of III as light cream colored crystals. Yield, 33.1%; mp,
181°–183°C; C₈H₅N₃ analysis: calculated: 66.93% C, 3.47% H, and 29.58%
N, and found: 67.03% C, 3.58% H, and 29.47% N.

N,N'-Bis(3,4-Dicyanophenyl) Alkanediamides (IV–XV). Data on the 12
alkanediamides analyzed in the present study are presented in Table I. Included are structure, melting point, recrystallization solvent, yield, and elemental
assays. All are new compounds, and all were prepared by appropriate modification of the following procedure.

N,N'-Bis(3,4-Dicyanophenyl) Tridecanediamide (XIII). A solution of
brassylyl chloride (9.81 g; 0.035 mole) in CCl₄ (25 ml) is added dropwise
to a stirred solution of 4-aminophthalonitrile (10.0 g; 0.070 mole) in pyridine
(250 ml) during 25 min. CCl₄ (225 ml) is added, and the final mixture is
cooled in an ice bath. The resulting white precipitate is collected, washed
with CCl₄ (2 × 15 ml), and dried. The residue is dispersed in water (200
ml), collected, water washed, and dried to yield 15.9 g crude diamide. Three
recrystallizations from acetonitrile (3 × 160 ml) yield 11.5 g of XIII as
analytical white crystals (*see* Table I for properties).

Literature Cited

1. Barrett, P. A., Dent, C. E., Linstead, R. P., *J. Chem. Soc. London* (1937) 1719.
2. Marvel, C. S., Rassweiler, J. H., *J. Amer. Chem. Soc.* (1958) **80,** 1196.
3. Marvel, C. S., Martin, M. M., *J. Amer. Chem. Soc.* (1958) **80,** 6600.
4. Drew, H. D. K., Kelley, D. B., *J. Chem. Soc. London* (1941) 637.
5. Ebert, Jr., A. A., Gottlieb, H. B., *J. Amer. Chem. Soc.* (1952) **74,** 2806.
6. Bogert, M. T., Boroshek, L., *J. Amer. Chem. Soc.* (1901) **23,** 756.
7. Scalera, M., Bouillard, R. E., U.S. Patent **2,525,620** (1950).

Received March 27, 1974.

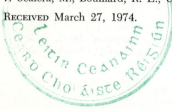

INDEX

INDEX

A

467

The text of this book is set in 9 point Caledonia with one point of leading. The chapter numerals are set in 30 point Garamond; the chapter titles are set in 18 point Garamond Bold.

The book is printed offset on Danforth 550 Machine Blue White text, 50-pound. The cover is Joanna Book Binding blue linen.

Jacket design by Linda McKnight.
Editing and production by Spencer Lockson.

The book was composed by the Mills-Frizell-Evans Co., Baltimore, Md., and printed and bound by The Maple Press Co., York, Pa.